Nuclear Physics in Astrophysics Conference (NPA VII)

Journal of Physics: Conference Series Volume 940

York, United Kingdom
18 – 22 May 2015

ISBN: 978-1-5108-5750-6
ISSN: 1742-6588

Printed from e-media with permission by:

Curran Associates, Inc.
57 Morehouse Lane
Red Hook, NY 12571

Some format issues inherent in the e-media version may also appear in this print version.

Printed by Curran Associates, Inc. (2018)

For permission requests, please contact the Institute of Physics
at the address below.

Institute of Physics
Dirac House, Temple Back
Bristol BS1 6BE UK

Phone: 44 1 17 929 7481
Fax: 44 1 17 920 0979

techtracking@iop.org

Additional copies of this publication are available from:

Curran Associates, Inc.
57 Morehouse Lane
Red Hook, NY 12571 USA
Phone: 845-758-0400
Fax: 845-758-2633
Email: curran@proceedings.com
Web: www.proceedings.com

Table of contents

Volume 940

Nuclear Physics in Astrophysics Conference (NPA VII)
18–22 May 2015, York, UK

Accepted papers received: 14 December 2017
Published online: 31 January 2018

Preface

Nuclear Physics in Astrophysics Conference (NPA VII)

Peer review

Papers

PUSHing core-collapse simulations to explosion

C Fröhlich, A Perego, M Hempe, K Ebinger, M Eichler, J Casanova, M Liebendörfer and F-K Thielemann.....1

Neutrino-nucleus reactions and their role in supernova dynamics and nucleosynthesis

Karlheinz Langanke and Gabriel Martinez-Pinedo.....5

Measurement of ^{19}Ne spectroscopic properties via a new method of inelastic scattering to study novae

F. Boulay, B. Bastin, F. de Oliveira Santos, T. Davinson, A. Lemasson, P. Ujic, J. Mrazek, A. M. Sánchez-Benítez, E. Traykov, D. Ramos, L. Achouri, L. Acosta, C. Borcea, M. Ciemala, P. Delahaye, A. Estrade, G. de France, J. Grinyer, B. Jacquot, O. Kamalou, A. Laird, V. Margerin, G. Marquinez-Duran, F. Negoita, M. Rejmund, C. Rodriguez-Tajes, F. Rotaru, C. Schmitt, N. de Séréville, O. Sorlin, M. Stanoiu, J.C. Thomas and P.J. Woods.....11

Study of excited states of ^{35}Ar through β-decay of ^{35}K for nucleosynthesis in novae and X-ray bursts

A. Saastamoinen, G. J. Lotay, A. Kankainen, B. T. Roeder, R. Chyzh, M. Dag, E. McCleskey, A. Spiridon and R. E. Tribble.....15

Nuclear physics uncertainties of the astrophysical γ-process studied through the $^{64}Zn(p,\alpha)^{61}Cu$ and $^{64}Zn(p,\gamma)^{65}Ga$ reactions

Gy. Gyürky, Zs. Fülöp, Z. Halász, G. G. Kiss and T. Szücs.....19

Online tools for nucleosynthesis studies

K Göbel, J Glorius, A Koloczek, M Pignatari, R Plag, R Reifarth, C Ritter, S Schmidt, K Sonnabend, B Thomas and C Travaglio.....23

Gamma-ray line diagnostics of supernova explosions - SN2014J and Cas A

Thomas Siegert.....27

Using stellar observations to trace the formation processes of Mo, Ru, Pd, and Ag

Camilla J. Hansen, Terese T. Hansen, Almudena Arcones and Anja C. Andersen.....33

Electron capture rates in stars studied with heavy ion charge exchange reactions

C.A. Bertulani.....37

New direct investigation of the $^{19}F(p,\alpha_0)^{16}O$ down to 0.2 MeV

I. Lombardo, D. Dell'Aquila, A. Di Leva, I. Indelicato, M. La Cognata, M. La Commara, A. Ordine, V. Rigato, M. Romoli, E. Rosato, G. Spadaccini, C. Spitaleri, A. Tumino and M. Vigilante.....43

Theoretical cross sections in the $^{12}C(\alpha,\gamma)^{16}O$ reaction

M. Katsuma.....47

Microscopic description of fission properties for r-process nuclei

S A Giuliani, G Martínez-Pinedo and L M Robledo.....51

Beyond BCS pairing in high-density neutron matter

A. Rios, D. Ding, H. Dussan, W. H. Dickhoff, S. J. Witte and A. Polls.....55

α-induced reaction cross sections in the mass range A $\approx 20 - 50$: a critical review

Peter Mohr.....59

Search for resonant states in ^{10}C and ^{11}C and their impact on the primordial ^{7}Li abundance

F Hammache, A Coc, N de Séréville, I Stefan, P Roussel, M Assié, L Audouin, D Beaumel, S Franchoo, B Fernandez-Dominguez, S Fox, C Hamadache, J Kiener, A Laird, B Le Crom, A. Lefebvre-Schuhl, L Lefebvre, I Matea, A Matta, G Mavilla, J Mrazek, P Morfouace, F de Oliveira Santos, A Parikh, L Perrot, A M Sanchez-Benitez, D Suzuki, V Tatischeff, P Ujic and Marine Vandebrouck.....63

Trojan Horse cross section measurements and their impact on primordial nucleosynthesis

R.G. Pizzone, R. Spartá, C. Bertulani, C. Spitaleri, M. La Cognata, L. Lamia, A. Mukhamedzhanov and A. Tumino.....67

υ-driven winds from the remnant of binary neutron star mergers

A Perego.....71

Ion implant-β-decay correlation half-lives in a pulsed beam for isotopes beyond N=126

R. Caballero-Folch, C. Domingo-Pardo and GSI - S410 collaboration.....75

New theory of stellar convection without the mixing-length parameter: new stellar atmosphere model

Stefano Pasetto, Cesare Chiosi, Mark Cropper and Eva K. Grebel.....79

Nucleosynthesis in the first massive stars

Arthur Choplin, Georges Meynet, André Maeder, Raphael Hirschi and Cristina Chiappini.....83

γ-process reaction studies via in-beam γ-ray spectroscopy at HORUS

Philipp Scholz, Jan Mayer, Lars Netterdon and Andreas Zilges.....89

Detection of MeV scale neutrinos and the solar energy paradigm

Aldo Ianni.....93

Nuclear astrophysics at FRANZ

R Reifarth, S Dababneh, S Fiebiger, J Glorius, K Göbel, M Heil, P Hillmann, T Heftrich, C Langer, O Meusel, R Plag, S Schmidt, Z Slavkovská, D Veltum, M Weigand, C Wiesner, C Wolf and A Zadeh.....99

Photodisintegration reactions for nuclear astrophysics studies at ELI-NP

C Matei, D Balabanski, D M Filipescu and O Tesileanu.....105

Latest results from LUNA

Rosanna Depalo for the LUNA collaboration.....109

Direct underground measurement of the $^{17}O(p,\alpha)^{14}N$ reaction at LUNA

C G Bruno for the LUNA Collaboration.....115

The Feasibility of Studying $^{44}Ti(\alpha, p)^{47}V$ Reaction at Astrophysical Energies

Tariq Al-Abdullah, D. Bemmerer, Z. Elekes and D. Schumann.....119

Investigation of interaction processes of light nuclei with ^{14}N

D. Alimov, N. Burtebayev, I. Boztosun, Z. Kerimkulov, J. Burtebayeva, M. Karakoc, Y. Mukhamejanov, D. Janseitov, A. Bahtibayev, A. Pattayev and Sh. Hamada.....125

Nucleosynthesis of Mo and Ru isotopes in neutrino-driven winds

Julia Bliss and Almudena Arcones.....128

Backgrounds in a BGO detector underground

A Boeltzig, A Best, A Di Leva, G Imbriani, M Junker for the LUNA collaboration.....131

Anharmonic oscillator potentials in the prolate γ-rigid regime of the collective geometrical model

R Budaca.....134

Investigation of α-particle scattering from ^{13}C at 29 MeV

Nassrulla Burtebayev, Daniyar Janseitov, Zhambul Kerimkulov, Alla Demyanova, Andrei Danilov, Dilshod Alimov and Yerzhan Mukhamejanov.....137

Investigation of the elastic and inelastic scattering of ^{4}He from ^{11}B in the energy range 29-50.5 MeV

N. Burtebayev, Zh. Kerimkulov, M. Bakhtybayev, Sh. Hamada, Y. Mukhamejanov, M. Nassurlla, D. Alimov, A. Morzabayev, D. Janseitov and W. Trzaska.....140

Nuclear Astrophysics at LNL: the $^{25}Mg(\alpha,n)^{28}Si$ and $^{10}B(p,\alpha)^{7}Be$ cases

Antonio Caciolli.....143

Detection of solar neutrinos with a torsion balance with sapphire crystal

M Cruceru and G Nicolescu.....146

Structure of ^{10}Be and ^{16}C nuclei via break-up reactions studied with the 4π Chimera array

D. Dell'Aquila, L. Acosta, F. Amorini, R. Andolina, L. Auditore, I. Berceanu, G. Cardella, M.B. Chatterjiee, E. De Filippo, L. Francalanza, B. Gnoffo, A. Grzeszczuk, G. Lanzalone, I. Lombardo, N. Martorana, T. Minniti, A. Pagano, E.V. Pagano, M. Papa, S. Pirrone, G. Politi, A. Pop, F. Porto, L. Quattrocchi, F. Rizzo, E. Rosato, P. Russotto, A. Trifirò, M. Trimarchi, G. Verde and M. Vigilante.....149

Impact of rotation and magnetic fields in low mass AGB stars

J W den Hartogh, R Hirschi, C Georgy, P Eggenberger, F Herwig, M Pignatari, U Battino, Jared Keown and C Ritter.....152

Explosion dynamics of parametrized spherically symmetric core-collapse supernova simulations

K Ebinger, A Perego, M Hempel, C Fröhlich, M Eichler, J Casanova, M Liebendörfer and F-K Thielemann.....155

Explosive nucleosynthesis in core-collapse supernovae and the titanium problem

M Eichler, A Perego, M Hempel, C Fröhlich, K Ebinger, J Casanova, M Liebendörfer and F-K Thielemann.....158

Measurement of the strengths of the resonances at 417, 458, 611, 632 and 1222 keV in the ^{22}Ne(p, γ)^{23}Na reaction

Federico Ferraro.....161

Neutron capture cross section of ^{85}Kr

S Fiebiger, Z Slavkovská, U Giesen, T Heftrich, M Krtička, R Reifarth, S Schmidt, B Thomas, M Weigand and A Zadeh.....164

Rotational excitation of the Hoyle state in ^{12}C

R Garg, C Barton, C Aa Diget, S Courtin, G Fruet, H O U Fynbo, A Howard, A Illana, D G Jenkins, I Marroquin, O S Kirsebom, M V Lund, I Moore, A Perea, J Refsgaard, J E Riley, S Rinta-Antila, L Sinclair, O Tengblad and The IGISOL Collaboration.....167

Maxwellian Neutron Spectrum generation and Stellar Cross-Section measurements: measurement of the $^{197}Au(n,\gamma)$ MACS.

P Jiménez-Bonilla, J Praena and J M Quesada.....170

Nucleosynthesis of intermediate mass stars: inferences from the observed abundances in photoionized nebulae of the Local Group

W. J. Maciel, R. D. D. Costa and O. Cavichia.....173

Effectiveness of using a magnetic spectrograph with the Trojan Horse method

S. Manwell, A. Parikh, A. A. Chen, N. de Séréville, P. Adsley, D. Irvine, F. Hammache, I. Stefan, R. F. Longland, J. Tomlinson, P. Morfuace and B. Le Crom.....176

Synthesis of Heavy Elements in the Ejecta of Neutron Star Mergers

Dirk Martin, Albino Perego, Almudena Arcones, Friedrich-Karl Thielemann, Oleg Korobkin and Stephan Rosswog.....179

TACTIC: the TRIUMF Annular Chamber for Tracking and Identification of Charged particles

L Martin, G Ruprecht, U Hager, P-A Amaudruz, L Buchmann, S P Fox, A M Laird and E Martin.....182

Manganese Abundances in the Stars with Metallicities $-1 < [Fe/H] < +0.3$

T Mishenina, T Gorbaneva, M Pignatari, F-K Thielemann and S Korotin.....185

Role of nuclear reactions on stellar evolution of intermediate-mass stars

H Möller, S Jones, T Fischer and G Martínez-Pinedo.....188

Sensitivity to neutron captures and β-decays of the enhanced s-process in rotating massive stars at low metallicities

N Nishimura, R Hirschi and T Rauscher.....191

Towards an activation cross section measurement of the $^{17}O(p,\gamma)^{18}F$ reaction in a wide energy range

A. Ornelas, Gy. Gyürky, Z. Halász, Cs.M. Oláh, Zs. Fülöp, I. Rajta and I. Vajda.....194

Beta-decay half-lives for short neutron rich nuclei involved into the r-process

I Panov, Yu Lutostansky and F-K Thielemann.....197

Neutrino nucleosynthesis in core-collapse Supernova explosions

A. Sieverding, L. Huther, G. Martínez-Pinedo, K. Langanke and A. Heger.....200

Alpha-Induced Production Cross Sections of $^{77,79}Kr$ and ^{77}Br

Zuzana Slavkovská, Stefan Fiebiger, Ulrich Giesen, Tanja Heftrich, René Reifarth, Stefan Schmidt, Benedikt Thomas and Mario Weigand.....204

Test of a SiPM-scintillator-based muon detector at the Gran Sasso National Laboratory

I Stark, V Könye, R Antolini, F Arneodo, A Boeltzig, A Candela, A Di Giovanni, Zs Fülöp and M Junker.....207

Simple potential model for interaction of dark particles with massive bodies

Nurgali Takibayev.....211

Excited nuclei, resonances and reactions in neutron star crusts

N Takibayev, D Nasirova, K Katō and V Kurmangaliyeva.....214

Study of the $^{2}H(p,\gamma)^{3}He$ reaction in the BBN energy range at LUNA

Davide Trezzi for the LUNA collaboration.....217

Development of a detector in order to investigate (n,γ)-cross sections by ToF method with a very short flight path

C Wolf, J Glorius, R Reifarth and M Weigand.....220

IOP Conf. Series: Journal of Physics: Conf. Series **940** (2017) 011001 doi:10.1088/1742-6596/940/1/011001

The seventh edition of the Nuclear Physics in Astrophysics conference took place on 18-22 May in York, United Kingdom. The conference was jointly organized by the Universities of York and Edinburgh and brought together over 100 participants from 18 different countries. The conference provided an opportunity to review the latest advances in many different areas, from Big Bang Nucleosynthesis, to explosive scenarios, stellar evolution, nuclear structure and theory, as well as experimental tools, techniques and facilities.

The scientific programme consisted of plenary sessions and a poster session (with over 60 contributions) and included also an industrial session with presentations on links between industry and academia and opportunities for both parties to engage in closer collaborations.

About 40% of participants were PhD students and young post-docs. Thus, the conference provided an excellent opportunity for them to meet and interact with more senior colleagues and leading figures in the field. Participants enjoyed plenty of occasions to discuss about their research, during coffee breaks and lunches, as these were held in the same rooms as the posters displays.

The best three posters by students received prizes kindly offered by the European Physical Society (top prize, to R Garg, University of York, UK) and the Institute of Physics Nuclear Physics Group (two runner ups, to F Ferraro, University of Genova, Italy; and to J Bliss, Technische Universitaet Darmstadt, Germany).

The social programme included a very enjoyable whisky tasting and an excellent tour to Helmsley Castle and Rievaulx Abbey. The conference dinner took place at the Merchant Adventures' Hall, one of the finest medieval guildhall in the world.

The conference was sponsored by Canberra, EDF, EPL, EPS, Hamamatsu, IoP, JINA, Kromek, Mesytech, Micron, NAVI, RAS, STFC, and SUPA. The organizers wish to thank Amy Light and Dawn Stewart from the Institute of Physics for their excellent support. Thanks are also due to the other members of the Local Organizing Committee: Charles Burton (York), Raphael Hirschi (Keele), Alex Murphy (Edinburgh).

Conference Chairs
Alison Laird and Marialuisa Aliotta

Peer review statement

All papers published in this volume of *Journal of Physics: Conference Series* have been peer reviewed through processes administered by the proceedings Editors. Reviews were conducted by expert referees to the professional and scientific standards expected of a proceedings journal published by IOP Publishing.

IOP Conf. Series: Journal of Physics: Conf. Series **940** (2017) 012001 doi:10.1088/1742-6596/940/1/012001

PUSHing core-collapse simulations to explosion

C Fröhlich[1], A Perego[2], M Hempel[3], K Ebinger[3], M Eichler[3], J Casanova[1], M Liebendörfer[3] and F-K Thielemann[3]

[1] Department of Physics, North Carolina State University, Raleigh NC 27695, USA
[2] Institut für Kernphysik, Technische Universität Darmstadt, D-64289 Darmstart, Germany
[3] Departement für Physik, Univestität Basel, CH-4056 Basel, Switzerland

E-mail: cfrohli@ncsu.edu

Abstract. We report on the PUSH method for artificially triggering core-collapse supernova explosions of massive stars in spherical symmetry. The PUSH method increases the energy deposition in the gain region proportionally to the heavy flavor neutrino fluxes.We summarize the parameter dependence of the method and calibrate PUSH to reproduce SN 1987A observables. We identify a best-fit progenitor and set of parameters that fit the explosion properties of SN 1987A, assuming 0.1 $M_\odot$ of fallback. For the explored progenitor range of 18-21 $M_\odot$, we find correlations between explosion properties and the compactness of the progenitor model.

1. Introduction

At the end of the life of a massive star ($M_{\rm ZAMS} \gtrsim 8 M_\odot$), the iron core collapses. In this process, a proto-neutron star is formed and an outgoing shock wave is triggered. However, this shock wave looses energy to the dissociation of nuclei and neutrino emission, and thus stalls. It is thought that the shock is revived by fluid instabilities such as convection, turbulence, and the standing accretion shock instability and by the resulting more efficient neutrino-energy deposition. Core-collapse supernova simulations in spherical symmetry even with detailed neutrino-transport and general relativity fail to explode self-consistently. There are many ongoing efforts using multi-dimensional simulations. While such models are needed to fully investigate and understand the explosion mechanism, they are currently computationally too expensive for investigations that naturally require large numbers of simulations, such as "What are the explosive nucleosynthesis yields of massive stars?" or "What is the connection between the progenitor and the compact remnant?". For these questions, a large number of simulations is required. Hence, computationally feasible, exploding models in spherical symmetry are of key importance. Such models have evolved from adding energy to a pre-collapse star either via a piston or via thermal energy to physically motivated models [1, 2, 3, 4, 5, 6, 7, 8]. Here, we describe a new approach, PUSH, to artificially trigger explosions in massive stars in spherical symmetry [9]. The PUSH method captures the relevant physics and is computationally efficient to allow for large and systematic parameter variations.

2. Method

For our simulations, we use the general relativistic, spherically symmetric hydrodynamics code AGILE [10]. We employ spectral neutrino transport [11, 12], a microphysics nuclear equation of state (EOS) [13], and a simplified α-network. We use the solar-metallicity, non-rotating stellar

IOP Conf. Series: Journal of Physics: Conf. Series **940** (2017) 012001 doi:10.1088/1742-6596/940/1/012001

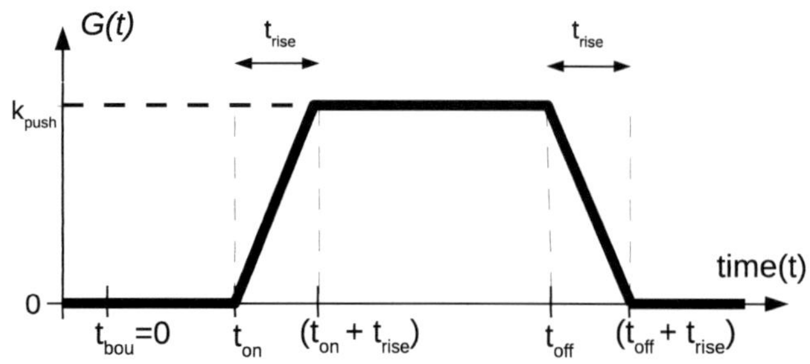

Figure 1. Function $\mathcal{G}(t)$ determines the temporal behavior of the PUSH heating. Two parameters are robustly set from multi-dimensional simulations: $t_{\rm on} = 80$ ms and $t_{\rm off} = 1$ s. The free parameters are $k_{\rm PUSH}$ and $t_{\rm rise}$.

models from [14]. All our models include the progenitor star up to the helium shell. We simulate the collapse, bounce, and onset of the explosion for a total time of 5 s.

To trigger explosions in otherwise non-exploding models, we rely on the neutrino-driven mechanism where energy deposition from neutrinos inside the gain region is thought to revive the stalled shock. The PUSH method mimics this increased net neutrino heating which is expected in multi-dimensional simulations. PUSH provides additional neutrino heating in the gain region by depositing a fraction of the luminosity of heavy flavor neutrinos behind the shock. The rationale for using the heavy flavor neutrinos is that they are one of the largest energy reservoirs available and they do not directly change the electron faction (unlike the electron flavor neutrinos). In addition, all the neutrino luminosities are calculated consistently within the model and do not have to be modified. As such, they include dynamical feedback from the history of the model and change between different progenitor models. The enhanced neutrino-heating is achieved through a local heating term which is proportional to the spectral $\nu_{\mu,\tau}$ flux and which is applied only inside the gain region (i.e. if $\dot{e}_{\nu_e,\bar{\nu}_e} > 0$). The temporal dependence of the additional heating is given by the function $\mathcal{G}(t)$, as shown in Figure 1. The two free parameters are $k_{\rm PUSH}$ and $t_{\rm rise}$. The former directly controls the amount of extra heating provided by PUSH. The latter sets the time scale over which $\mathcal{G}(t)$ increases from zero to $k_{\rm PUSH}$.

3. Results

3.1. Parameter studies

In order to test the PUSH method and to investigate the impact of the free parameters on the explosion properties, we performed a large number of runs for a wide range of parameter values. We include 16 progenitor models in the mass range of 18–21 $M_\odot$, corresponding to the typical progenitor mass of SN 1987A. In the following, we characterize these models by their compactness at $M = 1.75$, defined as

$$\xi_M \equiv \frac{M/M_\odot}{R(M)/1000\mathrm{km}}. \quad (1)$$

As expected intuitively, $k_{\rm PUSH}$ has a strong and direct impact on the explosion: Larger values of $k_{\rm PUSH}$ result in more energetic and faster explosions and as a consequence, a lower remnant mass (assuming all other parameters are fixed). The detailed behavior also depends on the compactness of the progenitor models. We found different behaviors for the low-compactness (LC; $\xi_{1.75} < 0.45$) and the high-compactness (HC) models in our sample ($\xi_{1.75} > 0.45$). For different values of $k_{\rm PUSH}$, the LC models explode slightly weaker and faster, with less variability in the explosion energy and explosion time. The HC models explode stronger and later, with larger variations in the explosion properties. We find explosion energies of $\gtrsim 1$ Bethe only for the HC models in our sample. These trends do not depend on the choice of $t_{\rm rise}$. Figure 2 shows the explosion energy as function of progenitor compactness for $t_{\rm rise} = 150$ ms and different values of $k_{\rm PUSH}$. A fixed explosion energy, e.g. 1 Bethe for the HC models, can be achieve by several parameter combinations in the $k_{\rm PUSH}$-$t_{\rm rise}$ plane. Generally, a longer $t_{\rm rise}$ requires a larger $k_{\rm PUSH}$. This can be understood from the function $\mathcal{G}(t)$: with a longer $t_{\rm rise}$, PUSH takes longer to reach its maximum, at which time the neutrino luminosities have already decreased. To compensate for this, a larger $k_{\rm PUSH}$ is required. A detailed discussion of the contributions to the explosion energy can be found in Ebinger et al (this volume) and in [9].

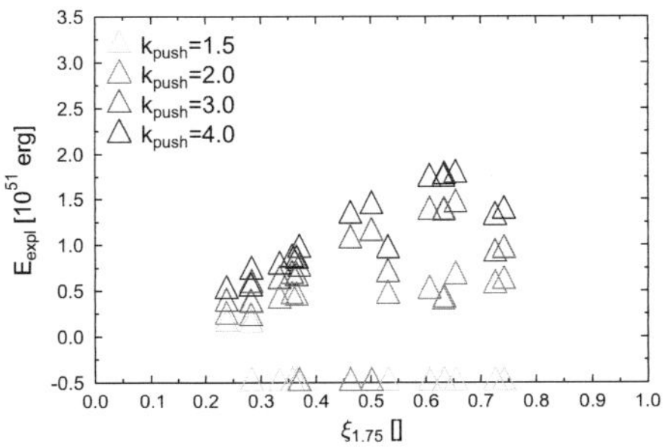

Figure 2. Explosion energies as function of compactness for $k_{\rm PUSH}$ of 1.5, 2.0, 3.0, and 4.0 and $t_{\rm rise}$ fixed at 150 ms.

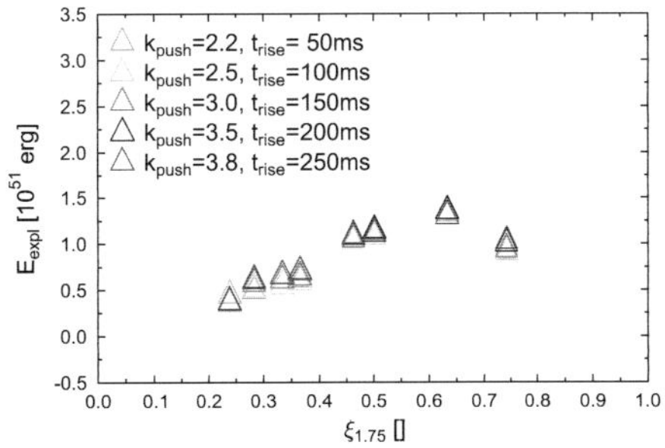

Figure 3. Explosion energies as function of compactness for pairs of $k_{\rm PUSH}$ and $t_{\rm rise}$ for select progenitor models.

3.2. Calibration to SN 1987A

We have calibrated the PUSH method to reproduce the observed properties (explosion energy and Ni ejecta mass) of SN 1987A. The observational properties that we use are summarized in Table 1, including the references. Figure 4 shows the mass of ^{56}Ni as function of explosion energy for different parameter combinations and progenitor models. The error box indicates the observational uncertainties. In order to fit both the explosion energy and the ejecta nickel masses, a fallback of 0.1 $M_\odot$ is required. Our best fit model for SN 1987A is obtained for the 18.0 $M_\odot$ progenitor model, $k_{\rm PUSH} = 3.5$ and $t_{\rm rise} = 150$ ms. The yields of ^{57}Ni and ^{58}Ni strongly depend on the location of the mass cut within the progenitor structure. All our models underproduce ^{44}Ti. For a detailed discussion of the nucleosynthesis yields, see Eichler et al (this volume) and also [9].

Quantity	Value	Ref.
$E_{\rm expl}$	$(1.1 \pm 0.3) \times 10^{51}$ erg	[15]
$m_{\rm prog}$	18-21 $M_\odot$	[16]
$m(^{56}{\rm Ni})$	(0.071 ± 0.003) $M_\odot$	[17]
$m(^{57}{\rm Ni})$	(0.0041 ± 0.0018) $M_\odot$	[17]
$m(^{58}{\rm Ni})$	0.006 $M_\odot$	[18]
$m(^{44}{\rm Ti})$	$(0.55 \pm 0.17) \times 10^{-4}$ $M_\odot$	[17]

Table 1. Observational properties of SN 1987A. For details, see text.

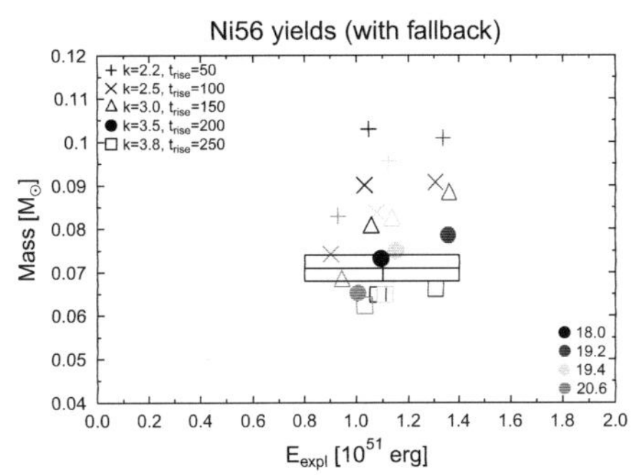

Figure 4. Ejected ^{56}Ni mass and explosion energy for representative HC models.

3.3. Progenitor explorations

From our investigations of the 18-21 $M_{\odot}$ progenitor models we found interesting correlations between different quantities. For these investigations, we use the parameters of our best-fit model ($k_{\mathrm{PUSH}} = 3.5$ and $t_{\mathrm{rise}} = 150$ ms). Even within this relatively narrow mass range of progenitor models we find a large diversity of explosion properties. However, we find indications of at least a weak correlation with the compactness parameters $\xi_{1.75}$. While there are significant deviations, one can identify general trends: For example, the explosion energy increases with larger progenitor compactness, see Figure 5. The explosion times are relatively constant within each of the sub-samples (LC models and HC models). As expected, the remnant mass increases with compactness, see Figure 6.

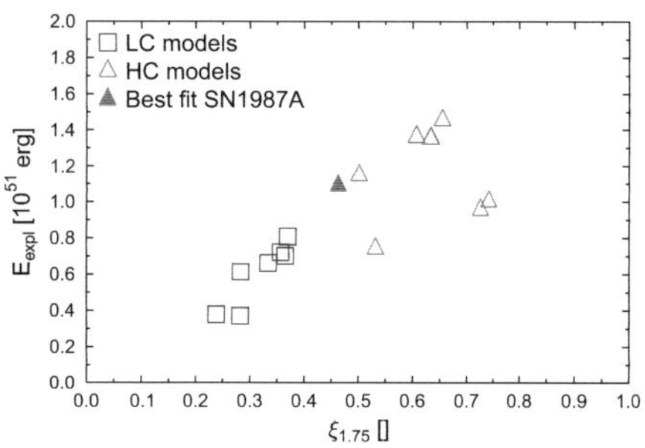

Figure 5. Explosion energies as function of compactness for PUSH parameters of our best-fit model.

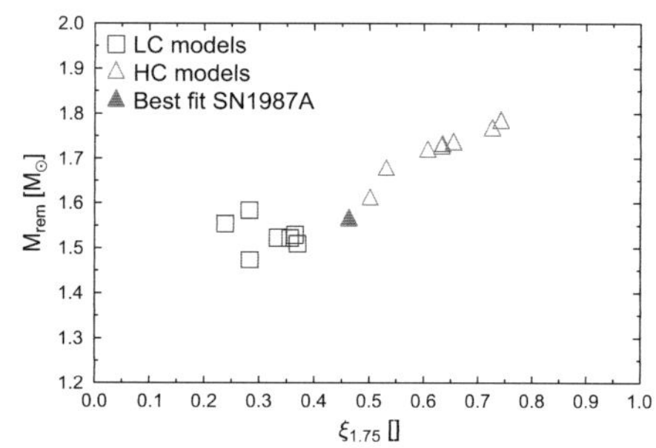

Figure 6. Remnant masses as function of compactness for PUSH parameters of our best-fit model.

Acknowledgments

C.F. acknowledges support from the Department of Energy through an Early Career Award (DOE grant no. SC0010263).

References

[1] Woosley S E and Weaver T A 1995 *ApJS* **101** 181–+
[2] Limongi M and Chieffi A 2006 *ApJ* **647** 483–500
[3] Thielemann F K, Nomoto K and Hashimoto M A 1996 *ApJ* **460** 408–+
[4] Umeda H and Nomoto K 2008 *ApJ* **673** 1014–1022
[5] Fröhlich C, Hauser P, Liebendörfer M and *et al* 2006 *ApJ* **637** 415–426
[6] Fischer T, Whitehouse S C, Mezzacappa A, Thielemann F K and Liebendörfer M 2010 *A. & A.* **517** A80
[7] Ugliano M, Janka H T, Marek A and Arcones A 2012 *ApJ* **757** 69
[8] Ertl T, Janka H T, Woosley S E, Sukhbold T and Ugliano M 2015 *ArXiv e-prints* (*Preprint* 1503.07522)
[9] Perego A, Hempel M, Fröhlich C, Ebinger K, Eichler M, Casanova J, Liebendörfer M and Thielemann F K 2015 *ApJ* **806** 275
[10] Liebendörfer M, Mezzacappa A and Thielemann F K 2001 *PRD* **63** 104003–+
[11] Liebendörfer M, Whitehouse S C and Fischer T 2009 *ApJ* **698** 1174–1190
[12] Perego A, Rosswog S, Cabezón R M, Korobkin O, Käppeli R, Arcones A and Liebendörfer M 2014 *MNRAS* **443** 3134–3156
[13] Hempel M and Schaffner-Bielich J 2010 *Nuclear Physics A* **837** 210–254
[14] Woosley S E, Heger A and Weaver T A 2002 *Reviews of Modern Physics* **74** 1015–1071
[15] Blinnikov S, Lundqvist P, Bartunov O, Nomoto K and Iwamoto K 2000 *ApJ* **532** 1132–1149
[16] Shigeyama T and Nomoto K 1990 *ApJ* **360** 242–256
[17] Seitenzahl I R, Timmes F X and Magkotsios G 2014 *ApJ* **792** 10
[18] Fransson C and Kozma C 2002 *New Astronomy Review* **46** 487–492

Neutrino-nucleus reactions and their role in supernova dynamics and nucleosynthesis

Karlheinz Langanke and Gabriel Martinez-Pinedo
GSI Helmholtzzentrum für Schwerionenforschung, Darmstadt, Germany, and
Institut für Theoretische Physik, Technische Universität Darmstadt, Darmstadt, Germany
E-mail: k.langanke@gsi.de

Abstract. Neutrino reactions on nuclei play important roles for the dynamics of supernovae and their associated nucleosynthesis. This manuscript summarizes the current status in deriving the relevant cross sections for supernova neutrinos and briefly discusses a few recent advances in supernova simulations where these reactions play a role.

1. Introduction

Neutrinos are key-players for the supernova dynamics [1, 2]. During the collapse phase the main neutrino source is electron capture on nuclei [3, 4]. By lowering the electron-to-nucleon ratio Y_e this process reduces the pressure which electrons can stem against the gravitational collapse of the core. Furthermore, at sufficiently low densities the electron neutrinos generated by the capture process can leave the star unhindered keeping the core at relatively low entropies so that heavy nuclei survive the collapse. At densities in excess of about 10^{12} g/cm^3 neutrinos get trapped in the core, mainly by elastic scattering on nuclei. The thermalization of neutrinos with the other core matter occurs by energy exchange via inelastic scattering on electrons and, in a lesser extent, on nuclei. In the final collapse phase at high densities pair production of neutrinos of all flavors becomes relevant. This occurs mainly by nucleon-nucleon bremsstrahlung [5, 6], but also nuclear deexcitation has been identified recently as an important additional source of neutrinos other than electron neutrinos [7]. After core bounce, energy transport by neutrinos from hotter core regions to the matter behind the stalled shock helps to revive the shock [8, 2]. The dominating processes are absorption of electron neutrinos and anti-neutrinos on neutrons and protons. The competition of these two absorption processes also determines the proton-to-neutron ratio for the subsequent explosive nucleosynthesis which might occur either in proton-rich environment (νp process [9, 10, 11]) or in neutron-rich environment. The later scenario has been favored for many years as the possible site for the astrophysical r-process (neutrino-driven wind model, [12]), but recent supernova simulations indicate that the conditions in the neutrino-driven wind are probably only sufficient to support a weak r-process which contributes to the observed r-process abundances up to the barium mass region (second r-process peak, e.g. [13]). There have been several suggestions how neutrino reactions on nuclei might contribute to supernova r-process nucleosynthesis (e.g. [14, 15, 16]), but recent studies point to no significant influence of these processes. Neutrino-induced spallation reactions, however, are crucial, for the production of selected nuclei (ν process [17]). Finally, the observation of supernova neutrinos by earthbound detectors is an emminent tool to verify our understanding of the supernova dynamics and mechanism. One requisite here is the knowledge of the neutrino reaction cross sections for the nuclei comprising the detector material.

There has been a recent extensive review of neutrino-nucleus reactions and their role in supernovae which might be consulted for more details [18].

2. Cross section models

Supernova neutrinos of interest here have relatively low energies (up to a few 10's of MeV). At these energies the cross sections are dominated by allowed transitions. Forbidden transitions become relevant at the higher neutrino energies and in cases where allowed transitions are strongly suppressed [19]. The Fermi contribution to the cross sections are defined by the position of the Isobaric Analog State and the respective sum rule. Charge-exchange experiments have progressed our understanding of Gamow-Teller (GT) distributions significantly in the last two decades [20]. The distributions are strongly fragmented. This is caused by nucleon-nucleon correlations and is well described by nuclear models like the diagonalization shell model which accounts for such correlations [21, 4, 22]. In fact, the combined progress due to experimental GT data from charge-exchange experiments and their detailed description by shell model calculations (except for a constant renormalization factor) have led to a rather reliable description of stellar electron capture [23, 24, 25, 26] which is the dominating weak interaction process during the collapse phase [3, 2]. The absorption of neutrinos on nuclei is the inverse process of electron and positron capture. Its calculation has also benefitted from the advances in describing GT distributions. In supernova simulations it is incorporated via detailed balance with the inverse capture processes. Forbidden transitions become relevant at neutrino energies high enough that reliable cross section calculations only require the reproduction of the energy centroids and total strengths of the respective transitions distributions, but not their detailed description. These requirements are fulfilled by the Random Phase Approximation (RPA). Hence a 'hybrid model' has been proposed in which the allowed contributions to the neutrino-nucleus cross sections are calculated by the shell model and the forbidden contributions within the RPA formalism [27].

Inelastic neutrino-nucleus scattering can be also evaluated within the hybrid model ansatz for temperature $T = 0$. Validation for this procedure can be derived from precision M1 data for spherical nuclei, measured by inelastic electron scattering, which are dominated by the same nuclear transitions [28]. At stellar temperatures, however, transitions mediated from thermally populated excited nuclear states modify the cross sections at low neutrino energies significantly. Two approaches have been proposed to incorporate these modifications: i) by including selected GT transitions involving excited states [28] and ii) within the consistent extension of the RPA to finite temperatures (Thermal Quasiparticle RPA) [29, 30].

Neutrino-nucleus reactions often excite the daughter nucleus to states above particle thresholds which then subsequently decay by particle emissions. The probabilities for decay into different particle channels can be calculate within the statistical model. Nuclear spallation reactions are important for supernova nucleosynthesis and potentially also as detection signal for certain supernova neutrino detectors.

3. Neutrino-nucleus reactions in supernova dynamic and nucleosynthesis

In this section we briefly summarize selected recent examples in which the role of neutrino-nucleus reactions have been investigated for the dynamics of core-collapse supernovae, for the production of selected nuclei in the ν nucleosynthesis process and for the spectrum of supernovae neutrinos and their observation by earthbound detectors.

3.1. Nuclear deexcitation by neutrino pair production

Meyer and Fuller have proposed that the deexcitation of thermally populated states by neutrino-pair production might be an important cooling mechanism of the supernova core during the collapse phase [31]. Recently this process has been incorporated for the first time into a supernova simulation and its role studied [7]. The relevant cross sections have been derived within two different models: i) on the basis of the independent particle model [31] and ii) by parametrization of the allowed and forbidden transition strengths guided by experiment or model calculations [7]. Although these cross section descriptions are both quite simplistic, they are reasonable enough to draw several important conclusions from the supernova simulation. The most important result is that nuclear deexcitation by neutrino-pair production does not influence the supernova dynamics. This is confirmed in Fig. 1 which shows that the luminosities of electron neutrinos, arising mainly from electron capture, are about 4 orders of magnitude larger than those of heavy-flavor neutrinos during the collapse phase. After bounce, nuclei are dissociated into free nucleons and neutrino-pair nuclear deexcitation becomes irrelevant. However, the simulation implies that this process is the main source of μ and τ neutrinos and of $\bar{\nu}_e$ during collapse. This is demonstrated in Fig. 1. At high densities of order 10^{13} g/cm^3 also nucleon-nucleon bremsstrahlung becomes a significant source of neutrinos other than ν_e which causes the steep rise of the $\nu_{\mu,\tau}$ and $\bar{\nu}_e$ luminosities at times just before bounce.

3.2. Neutrino nucleosynthesis

When neutrinos, produced in the hot supernova core, pass through the outer shells of the star, they can induce nuclear reactions and in this way contribute to the elementsynthesis (the ν-process, [17]). For example, the nuclides ^{11}B and ^{19}F are produced by $(\nu,\nu'n)$ and $(\nu,\nu'p)$ reactions on the quite abundant nuclei ^{12}C and ^{20}Ne. These reactions are dominantly induced by ν_μ and ν_τ neutrinos and their antiparticles (combined called ν_x neutrinos) which have larger average energies than ν_e and $\bar{\nu}_e$ neutrinos. As found in detailed stellar evolution studies [32] the rare odd-odd nuclides ^{138}La and ^{180}Ta are mainly made by the charged-current reaction ^{138}Ba$(\nu_e,e^-)^{138}$La and ^{180}Hf$(\nu_e,e^-)^{180}$Ta. Hence, the ν-process is potentially sensitive to the spectra and luminosity of ν_e and ν_x neutrinos, which are the neutrino types not observed from SN1987a.

Recent supernova simulations, with improved descriptions of neutrino matter interactions, indicate that the average neutrino energies are smaller than previously assumed [33]. This should result in reduced neutrino-induced cross sections and hence lower elemental production rates. This has been the motivation of recent neutrino nucleosynthesis studies performed for stars with masses between 15 and 40 $M_\odot$ and including neutrino-nucleus cross sections for a large set of nuclei with $Z < 78$. As an additional improvement in comparison to previous calculations these nucleosynthesis studies considered differential cross sections for multi-particle emissions [34]. Mainly due to the change in neutrino spectra, this study finds slightly smaller abundances for ^{7}Li, ^{11}B, ^{138}La and ^{180}Ta, however, it confirms the production of these nuclides by neutrino nucleosynthesis [33]. The study also finds that neutrino-induced reactions, either directly or indirectly by providing an enhanced abundance of light particles, noticeably contribute to the production of the radioactive nuclides ^{22}Na and ^{26}Al, which are both prime candidates for gamma-ray astronomy. However, the studies do not find significant production of two other candidates, ^{44}Ti and ^{60}Fe, due to neutrino-induced reactions.

As a major improvement it has been possible recently to measure the GT strengths on ^{138}Ba and ^{180}Hf below the particle thresholds and to convert these data into the relevant (ν_e,e^-) cross sections [35]. It is found that the new cross sections are slightly larger than the RPA predictions.

Due to the expected hierarchy of average energies for supernova neutrinos ($\langle E_{\nu_e}\rangle < \langle E_{\nu_x}\rangle$, neutrino oscillations are expected to increase the average ν_e energy and consequently also the charged-current cross section induced by supernova neutrinos. As pointed out by Kajino and collaborators, this makes the ratio of ^{7}Li and ^{11}B sensitive to the θ_{13} mixing angle and to the mass hierarchy [36, 37, 38]. Despite this intriguing sensitivity, an accurate derivation of the ^{7}Li/^{11}B abundance ratio requires reliable stellar model calculations and neutrino and nuclear

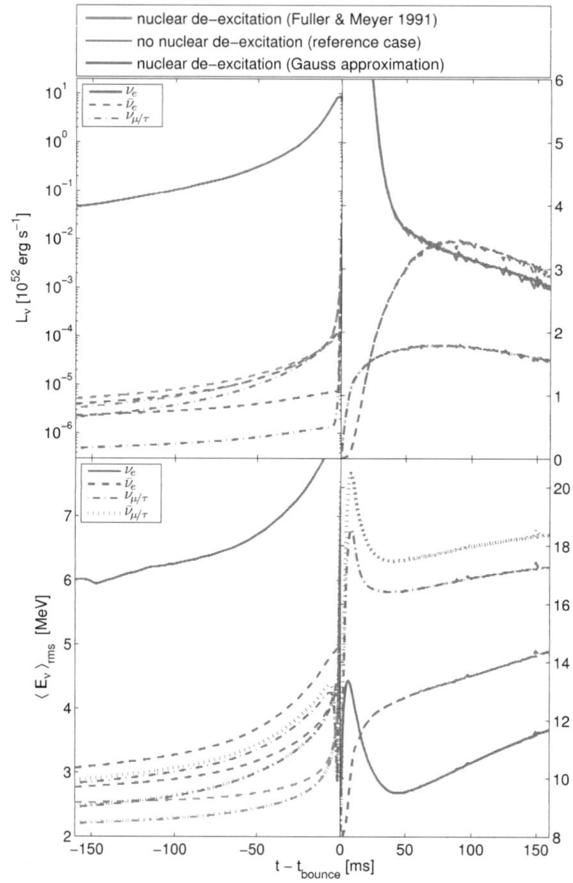

Figure 1. Evolution of neutrino luminosities and average energies from a core-collapse supernova simulation of an 11.2 $M_\odot$ progenitor star including the production of neutrino pairs from heavy-nuclei de-excitations, based on Ref. [31] (green lines) and the Gauss ansatz (see text, red lines), in comparison to a simulation that uses identical input physics, but neglects the nuclear deexcitation process (blue lines). (from [7]).

cross sections, but must also consider the production of the elements from other astrophysical sources; ^{7}Li is, for example, also produced by Big Bang nucleosynthesis [39].

3.3. Detecting supernova neutrinos

The observation of neutrinos from supernova SN1987A by the earthbound detectors Kamiokande [40] and IMB [41] has confirmed and advanced the understanding of core-collapse supernovae. A similar boost is expected from the observation of the next near-by supernova which is likely to test the predictions of supernova models concerning the neutrino spectra for the different flavors, including the noticeable neutrinoburst signal in electron neutrinos [2], originating from electron capture on protons set free by the shock. The prediction for this spectrum has changed recently after inelastic neutrino-nucleus scattering has been included in supernova simulations [42]. These burst neutrinos traverse regions outside the shock where nuclei have yet not been dissociated and, in particular, high-energy neutrinos excite these nuclei. This means that they are down-scattered in energy, in this way significantly reducing the high-

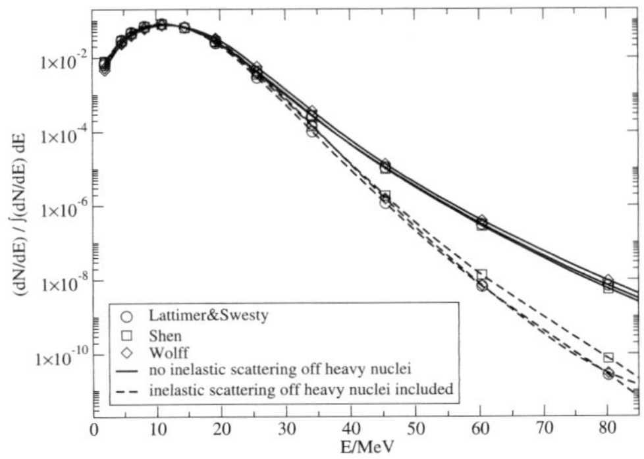

Figure 2. Normalized ν_e number spectra radiated during the shock breakout burst as seen by a distant observer at rest. Results are shown for simulations with three different nuclear EoSs. For better comparison of the strongly time-dependent spectra during this evolution phase, integration in a window of 8 ms around the peak luminosity was performed. Inelastic neutrino scattering off nuclei (dashed lines) leads mostly to energy losses of high-energy neutrinos and thus reduces the high-energy tails of the spectra. The vertical line marks the mean spectral energy. (from [42])

energy tail of the spectrum of emitted supernova neutrinos (see Fig. 2, [42]). This makes the detection of neutrinos, originating from the burst, by earthbound detectors more difficult, as the neutrino detection cross section scales with E_ν^2. Depending on the neutrino reaction threshold the reduction of the expected event rate can be significant, reaching up to more than 50% for ^{16}O which has been proposed, via nucleon emission following neutral current excitation, as a potential detection scheme for supernova ν_x neutrinos in Superkamiokande [43].

For a recent overview on the detectors and their scheme for observing supernova neutrinos the reader is refered to Ref. [44]. Table 2 of this reference lists the present and future supernova neutrino detectors. Their main material are liquid scintillator (C_nH_{2n}), water, lead or liquid argon. Hence translating the event rates of supernova neutrinos observed in the detectors requires a detailed knowledge of the cross sections for the neutrino interaction with the detector materials.

For ^{12}C shell model calculations have been performed for charged- and neutral-current neutrino reactions in a $(0{+}2)\hbar\omega$ model space [45], improving earlier calculations based on the RPA or on more restricted shell model spaces (see [19]). For the double-magic nucleus ^{16}O GT transitions are strongly suppressed and the cross sections are dominated by spin-dipole transitions which have recently been modelled in a shell model calculation [46] considering the p and sd shells. Based on a hybrid model approach calculating the GT contribution within the shell model, the transition to the IAS from the Fermi sum rule and the forbidden transitions within the RPA, Suzuki and Honma have determined the $^{40}\mathrm{Ar}(\nu_e, e^-)^{40}\mathrm{K}$ cross section for neutrino energies up to $E_\nu = 100$ MeV [47]. A similar study of the $(\bar{\nu_e}, e^+)$ cross section on ^{40}Ar has yet not been performed. As the GT contribution to the cross section is strongly suppressed such a study likely requires a shell model calculation performed in the (sd)-(pf) model space for the forbidden transitions. For ^{208}Pb Suzuki and Sagawa presented (ν_e, e^-) cross sections which have been obtained using GT data from a (p, n) experiment and adjusting their Hartree-Fock + Tamm-Dancoff approach for the first-forbidden response to the peaks of the spin-dipole resonances [48]. Furthermore, the spreading and quenching of the GT response has been considered by coupling to 2p-2h configurations. However, the calculations have been performed assuming a muon-decay-at-rest rather than a supernova neutrino spectrum, and are thus not directly relevant to supernova neutrinos.

4. References

[1] H.A. Bethe, Rev. Mod. Phys. **62** (1990) 801
[2] H.-T. Janka, K. Langanke, A. Marek, G. Martínez-Pinedo, B. Müller, Phys. Repts. **442** (2007) 38.
[3] H.A. Bethe, G.E. Brown, J. Applegate and J.M. Lattimer, Nucl. Phys. **A324** (1979) 487
[4] K. Langanke and G. Martinez-Pinedo, Rev. Mod. Phys. **75** (2003) 819
[5] S. Hannestad and G. Raffelt, Astrophys. J. **507** (1998) 339
[6] S. Bacca, K. Hally, M. Liebendörfer, A. Perego, C.J. Pethick and A. Schwenk, Astrophys. J. **758** (2012) 34
[7] T. Fischer, K. Langanke and G. Martinez-Pinedo, Phys. Rev. C **88** (2014) 065804
[8] H. A. Bethe and J. R. Wilson, Astrophys. J. **295** (1985) 14.
[9] C. Fröhlich *et al.*, Astrophys. J. **637** (2005) 415
[10] C. Fröhlich, G. Martinez-Pinedo *et al.*, Phys. Rev. Lett. **96** (2006) 142502
[11] J. Pruet *et al.*, Astrophys. J. **623** (2005) 325
[12] S.E. Woosley *et al.*, Astrophys. J. **433** (1994) 229
[13] C.J. Hansen, F. Montes and A. Arcones, Astr. J. **797** (2014) 123
[14] Y.-Z. Qian *et al.*, Phys. Rev. **C55** (1997) 1532
[15] W.C. Haxton *et al.*, Phys. Rev. Lett. **78** (1997) 2694
[16] M. Terasawa *et al.*, Astrophys. J. **608** (2004) 470
[17] S.E. Woosley *et al.*, Astrophys. J. **356** (1990) 272
[18] K.G. Balasi, K. Langanke and G. Martinez-Pinedo, Prog. Part. Nucl. Phys. **85** (2015) 33
[19] E. Kolbe, K. Langanke, G. Martinez-Pinedo and P. Vogel, J. Phys. **G29** (2003) 1
[20] Y. Fujita, B. Rubio and W. Gelletly, Prog. Part. Nucl. Phys. **66** (2011) 549
[21] E. Caurier, K. Langanke, G. Martínez-Pinedo and F. Nowacki, Nucl. Phys. A **653** (1999) 439
[22] E. Caurier, G. Martinez-Pinedo, F. Nowacki, A. Poves, A.P. Zuker, Rev. Mod. Phys. **77** (2005) 427
[23] K. Langanke and G. Martínez-Pinedo, Nucl. Phys. **A673** (2000) 481
[24] K. Langanke, G. Martínez-Pinedo, At. Data. Nucl. Data Tables **79** (2001) 1.
[25] A. Juodagalvis *et al* 2010, Nucl. Phys. A **848** 454
[26] A.L. Cole *et al.*, Phys. Rev. C **86** (2012) 015809
[27] E. Kolbe, K. Langanke, G. Martinez-Pinedo, Phys. Rev. **C60** (1999) 052801
[28] K. Langanke *et al.*, Phys. Rev. Lett. **93** (2004) 202501
[29] A. Dzhioev, A. Vdovin, V. Ponomarev and J. Wambach, Phys. At. Nucl. **74** (2011) 1162
[30] A.A. Dzhioev, A.I. Volovin, J. Wambach and V. Yu. Ponomarev, Phys. Rev. C **89** (2014) 035805
[31] G.M. Fuller and B.S. Meyer, Astrophys. J. **376** (1991) 701
[32] A. Heger *et al.*, Phys. Lett. **B606** (2005)
[33] A. Sieverding, L. Huther, G. Martinez-Pinedo and K. Langanke, submitted to Phys. Rev. Lett. (2015)
[34] L. Huther, PhD thesis, Technische Universität Darmstadt, (2014)
[35] A. Byelikov *et al.*, Phys. Rev. Lett. **98** (2007) 082501
[36] T.Yoshida *et al.*, Astrophys. J. **686** (2008) 448
[37] M.K. Cheoun *et al.*, Prog. Theor. Phys. Suppl. **196** (2012) 476
[38] T. Kajino, G.J. Mathews and T. Hayakawa, J. Phys. **G41** (2014) 044007
[39] S.G. Ryan, J.E. Norris and T.C. Beers, Astrophys. J. **523** (1999) 654
[40] K. Hirata, T. Kajita, M. Koshiba, M. Nakahata, Y. Oyama, Phys. Rev. Lett. **58** (1987) 1490.
[41] R.M. Bionta*et al.*, Phys. Rev. Lett. **58** (1987) 1494
[42] K. Langanke *et al.*, Phys. Rev. Lett. **100** (2008) 011101
[43] K. Langanke, P. Vogel and E. Kolbe, Phys. Rev. Lett. **76** (1996) 2629
[44] K. Scholberg, Ann. Rev. Nucl. Part. Science **62** (2012) 81
[45] T. Suzuki, S. Chiba, T. Yoshida, T. Kajino and T. Otsuka, Phys. Rev. **C74** (2006) 034307
[46] T. Suzuki, J. Phys. Conf. Ser. **80** (20119 012041
[47] T. Suzuki and M. Honma, Phys. Rev. C **87** (2013) 014607
[48] T. Suzuki and H. Sagawa, Nucl. Phys. **A718** (2003) 446c

Measurement of ^{19}Ne spectroscopic properties via a new method of inelastic scattering to study novae.

F. Boulay[1], B. Bastin[1], F. de Oliveira Santos[1], T. Davinson[2], A. Lemasson[1], P. Ujic[3], J. Mrazek[4], A. M. Sánchez-Benítez[5,1], E. Traykov[1], D. Ramos[6], L. Achouri[7], L. Acosta[5], C. Borcea[8], M. Ciemala[9], P. Delahaye[1], A. Estrade[2], G. de France[1], J. Grinyer[1], B. Jacquot[1], O. Kamalou[1], A. Laird[10], V. Margerin[2], G. Marquinez-Duran[5], F. Negoita[8], M. Rejmund[1], C. Rodriguez-Tajes[1], F. Rotaru[8], C. Schmitt[1], N. de Séréville[11], O. Sorlin[1], M. Stanoiu[8], J.C. Thomas[1], P.J. Woods[2]

[1]GANIL Caen, [2]University of Edinburgh, [3] INS Vinca Belgrade, [4]Nuclear Physics Institute of Rez, [5]University of Huelva, [6]Universidade de Santiago de Compostela, [7]LPC Caen, [8]IFIN-HH Bucharest, [9]Institute of Nuclear Physics Krakòw, [10]University of York, [11]IPN Orsay

E-mail: `florent.boulay@ganil.fr`

Abstract.
The accuracy of the predictions of the γ flux produced by a classical nova during the first hours after the outburst is limited by the uncertainties on several reaction rates, including the ^{18}F(p,α)^{15}O one. Better constraints on this reaction rate can be obtained by determining the spectroscopic properties of the compound nucleus ^{19}Ne. This was achieved in a new inelastic scattering method using a ^{19}Ne radioactive beam (produced by the GANIL-SPIRAL 1 facility) impinging onto a proton target. The experiment was performed at the VAMOS spectrometer. In this article the performances (excitation energy range covered and excitation energy resolution) and limitations of the new technique are discussed. Excitation energy resolution of σ=33 keV and low background were obtained with this inverse kinematics method, which will allow extracting the spectroscopic properties of ^{19}Ne.

1. Introduction

The classical description of a nova consists of an explosion that occurs in a binary system of stars. Typically, the outer layer of a main sequence star is accreted by gravitational attraction onto the surface of a companion white dwarf star (WD), where nuclear reactions are occurring. The surface temperature increases and can reach several hundreds of millions degrees depending on the mass and the type (either ONe or CO) of the WD. At these critical temperatures, the electronic degeneracy of the WD surface matter that prevents the expansion of the gas is lifted giving rise to the explosion and expulsion of matter into the interstellar medium.

Some of the nuclei produced via proton capture reactions are radioactive. The detection of their characteristic γ-lines, by a satellite such as INTEGRAL, would allow to constrain the scenario of the explosion, with the proviso of knowing very well the reaction rates. According to models [1], the ^{18}F nucleus 511 keV decay transition should dominate the low-energy γ spectrum ($\leq$ 2 MeV) during the first hours after the outburst. ^{18}F has a reasonable half-life (110 min) and is predicted to be sufficiently produced to be a good candidate for observation. To date, the evaluation of the ^{18}F(p,α)^{15}O reaction rate at the relevant astrophysics energy range, called Gamow window (GW) is mainly based on the spectroscopic properties of the compound nucleus ^{19}Ne (^{18}F+p). The lack of direct measurements within the GW is explained by the low beam intensities of ^{18}F available at the current radioactive ions beam facilities and the low cross section. The existing uncertainties on the ^{18}F(p,α)^{15}O reaction rate covers several orders of magnitude at typical novae temperatures [2]. To better constrain this reaction rate, a new investigation of the ^{19}Ne spectroscopic properties was performed at GANIL using a novel experimental setup.

2. Method and experimental setup

The inelastic scattering reaction ^{19}Ne(p,p')^{19}Ne* was used to populate the ^{19}Ne states of interest, located around the proton threshold S_p=6.41 MeV (S_α=3.528 MeV). The advantage of such a reaction mechanism is the statistical and non-selective feeding of all states. States that are not accessible via selective reactions like transfer reactions were indeed seen in this experiment. The kinematics of the protons p' allowed to reconstruct the excitation energy spectrum of ^{19}Ne* and to study the states of interest. These states, located in the GW region, can decay following three different channels. The γ decay channel has a very low probability. Thus no dedicated detector was used to measure them. The two main decay channels correspond to charged particle emissions (^{19}Ne $\rightarrow$ ^{18}F + p" or ^{19}Ne $\rightarrow$ ^{15}O + α). The angular distribution of the emitted charged particles (α or p") can be used to reconstruct the spin of the ^{19}Ne states in a model independent way. The coincidence between the proton p' and the emitted particles can be used to extract the individual branching ratios.

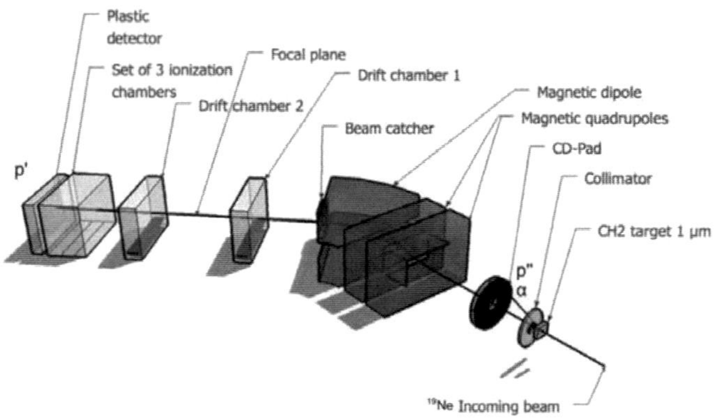

Figure 1. General layout of the experimental setup. The different detectors used in the experiment are indicated.

The ^{19}Ne radioactive beam was produced at the GANIL-SPIRAL 1 facility. It was accelerated to 10 MeV/u by the CIME cyclotron and sent onto a 1 μm thin polyethylene target $(CH_2)_n$. The beam intensity was up to 1×10^8 pps with a purity of 100 %. During the experiment, several beam cuts were applied in the accelerator section to improve as much as possible the beam position and its energy resolution. In those conditions the average incident ^{19}Ne flux on target was of 2×10^7 pps. In inverse kinematics, the proton p' has to be detected at zero degree to get the best energy resolution. The large acceptance VAMOS spectrometer [3] was used to reject the incident beam ions and to ensure the p' energy measurement of the protons with an energy resolution up to $\frac{1}{500}$ (see Fig. 1). VAMOS was used for the first time to detect protons. The α particles and protons p" were identified and characterized with a silicon ΔE-E telescope called CD-PAD [4] located at 10 cm downstream the target.

IOP Conf. Series: Journal of Physics: Conf. Series **940** (2017) 012003 doi:10.1088/1742-6596/940/1/012003

3. Analysis and preliminary results

3.1. Excitation energy

The large magnetic rigidity acceptance of VAMOS allowed measuring coincidences of p' scattered particles with α and p" decay particles in a wide excitation energy range from 4.5 to 8.5 MeV (see Fig. 2) and from 7 to 8.5 MeV respectively. With this setup resolutions of 44 keV at 4.5 MeV to 33 keV at 8.1 MeV were achieved. The amplitude of the excitation energy spectrum was normalized to the beam time and corrected relative to the contribution of the carbon in the target. To perform this correction, measurements with a 0.4 μm thick carbon target (during 15 % of the experiment) were realized (see Fig. 2) and pointed out a background level compatible with zero (observed counts in the spectrum are compatible with an hydrogen contamination usually present in thin carbon targets).

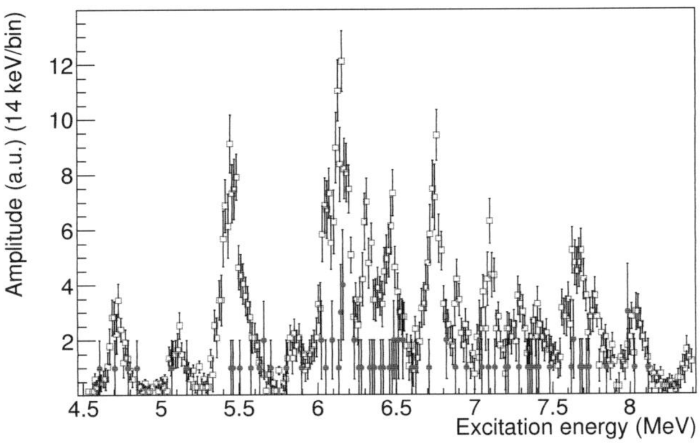

Figure 2. Excitation energy spectrum of ^{19}Ne reconstructed by the VAMOS spectrometer with protons p' detected at 0° in coincidence with alpha particles recorded with the CD-PAD ΔE-E telescope located behind the target (open squares). States are populated through the inelastic reaction mechanism ^{19}Ne(p,p')^{19}Ne*. In circles, the spectrum obtained when the $(CH_2)_n$ target was substituted with a thin carbon target.

3.2. Drawback of the method

A source of contamination was observed in the data, it is related to ^{15}O and ^{18}F coming from the decay of ^{19}Ne*. They are kinematicaly forward focused in the laboratory frame with a magnetic rigidity difference of about 20 % with respect to the dipole nominal value set for the experiment. The large acceptance of VAMOS allows thus a simultaneous detection of some of these particles with the proton p'. The heavy ions have a much higher energy deposit in the VAMOS drift chambers. Therefore, they produced saturation of the signals collected on the strips of the drift chambers leading to the destruction of the proton p' signal of interest.

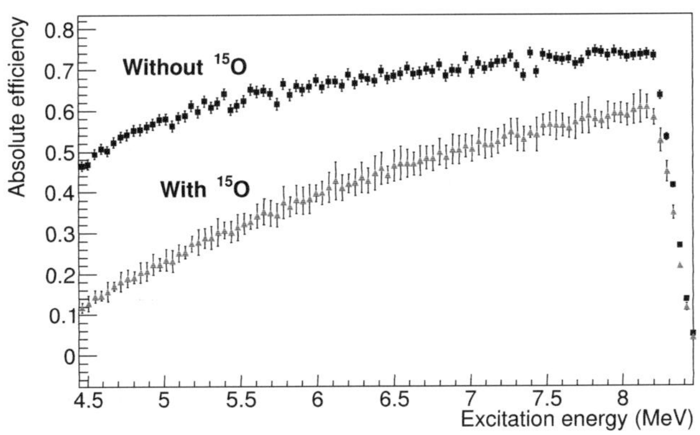

Figure 3. Absolute efficiency as a function of the excitation energy deduced from a simulation in the case the proton p' is detected by VAMOS in coincidence with decaying α particles recorded in the CD-PAD (black curve). The curve in red illustrates the loss of statistics and efficiency due to the heavy ions saturation effect on the drift chamber. Both simulations error bars include the uncertainty on the angular momentum distributions.

In order to correct these losses induced by heavy ions, a simulation that includes the full kinematics, the geometry of the CD-PAD, the angular and the magnetic rigidity acceptances of the VAMOS spectrometer was developped and used. Figure 3 shows the absolute detection efficiency simulated as a function of the excitation energy reconstructed via the p' scattered particles measured in VAMOS in coincidence with an α decay particle detected in the CD-PAD. Same work has been completed in the case of coincidences of p' with p" decay particles.

3.3. Benchmarking of the new technique

Table 1 provides a comparison between the properties (excitation energy, proton to alpha width ratio, total width) extracted in this experiment and the ones from the literature [5] for the well known $\frac{3}{2}^+$ state located at E_x=7.074 MeV. There is a good agreement between these values. The uncertainties in excitation energy given in the literature are better than ours as determined from a compilation of several experiments. Our resolution is however the best achieved up to now for this type of experiment in inverse kinematics. The current limitation in energy resolution is mainly due to the target thickness. The resolution for the partial widths is degraded in this experiment because of the efficiency correction that has to be applied due to the heavy ions contamination. The spin of this state has not been assigned yet because additionnal corrections has to be applied (a precise knowledge of the beam position is needed for instance).

Properties	This experiment	Literature [5]
Excitation energy	7.076(3) MeV	7.074(2) MeV
Γ_{tot}	35(4) keV	39(2) keV
$\frac{\Gamma_p}{\Gamma_\alpha}$	0.64(5)	0.638(5)

Table 1. Summary of the 7.074 MeV state properties measured in this experiment and from the literature [5].

4. Conclusions

Thanks to the performances of the VAMOS spectrometer, we realized a new spectroscopic study of ^{19}Ne with a high energy resolution (σ = 44 to 33 keV) for an inverse kinematics measurement. The heavy ions coming from the decay of ^{19}Ne*, ^{15}O and ^{18}F, induced statistics losses. A correction function was defined through a simulation of this experiment and applied to the data. It allowed us to extract relevant spectroscopic properties such as the total width, as well as the proton and the α partial widths ratio. A comparison of the results obtained from this work for the well-known state at 7.074 MeV with the values given in the literature has been made and pointed out the power and accuracy of our method. New spectroscopic properties will be extracted for all the states located around the GW. For future experiments, a thin layer of aluminium could be used after the VAMOS dipole in order to stop the heavy ions. Its thickness should not impact too much the protons energy and angular resolution.

4.1. Acknowledgements

This work has been supported by the European Community FP7 - Capacities - Integrated Infrastructure Initiative - contract ENSAR n° 262010. This work was also supported by the France-Romanian IN2P3-IFIN-HH collaboration n° 03-33, by the France-Czech LEA NuAG collaboration and the NAVI collaboration. The authors would also like to thank the technical teams of GANIL for their strong implications in this experiment.

References

[1] Jordi José and Margarita Hernanz. Journal of Physics G : Nuclear and Particle Physics, 34(12) :R431, (2007)
[2] C. Iliadis, R. Longland, A.E. Champagne, A. Coc, R. Fitzgerald, Nuclear Physics A, 841, 1, 31-250, (2010)
[3] M. Rejmund, B. Lecornu, A. Navin, et al. Nucl. Instrum. Methods A, 646(1) :184-191, (2011)
[4] A.N Ostrowski, S. Cherubini, T. Davinson, et al. Nucl. Instrum. Methods A, 480(23) :448-455, (2002)
[5] D. W. Bardayan, J. C. Blackmon, W. Bradfield-Smith, et al. Phys. Rev. C, 63 :065802, (2001)

Study of excited states of ^{35}Ar through β-decay of ^{35}K for nucleosynthesis in novae and X-ray bursts

A. Saastamoinen[1], G. J. Lotay[2], A. Kankainen[3], B. T. Roeder[1], R. Chyzh[1], M. Dag[1], E. McCleskey[1], A. Spiridon[1] and R. E. Tribble[1]

[1] Cyclotron Institute, Texas A&M University, College Station, TX-77843, USA
[2] Department of Physics, University of Surrey, Guildford, GU2 7XH, UK
[3] School of Physics and Astronomy, University of Edinburgh, Edinburgh, EH9 3JZ, UK

E-mail: ajsaasta@comp.tamu.edu

Abstract. Excited states in ^{35}Ar have been studied via β-decay of ^{35}K to improve spectroscopic information relevant for the ^{34g,m}Cl(p,γ)^{35}Ar reaction at novae and X-ray burst temperatures. An extensive set of new data has been collected, including several new proton groups, improved half-life, and direct observation of γ-decay from the lowest $T = 3/2$ state in ^{35}Ar. In this contribution selected preliminary results from the experiment are presented.

1. Introduction

The thermonuclear runaway in close binary systems such as novae and X-ray bursts proceeds through proton-rich nuclei, and many of the radiative proton capture reactions (p,γ) involving sd-shell nuclei close to the dripline are dominated by resonant capture. The key parameters in understanding the astrophysical reaction rates are the energies, decay widths and spins of these resonances. One of the reactions for which improved data are needed and which determines the synthesis of nuclei beyond sulfur and chlorine is the radiative proton capture ^{34g,m}Cl(p,γ)^{35}Ar. At the moment the properties of the excited states of ^{35}Ar above the proton separation threshold are rather poorly known and the astrophysical reaction rate is based on statistical Hauser-Feshbach calculations [1, 2, 3].

The short half-life of the ^{34}Cl ground state, 1.53 s, and of the isomer, 32 min [4], prevent fabricating targets for direct reaction measurements. In addition intense and clean enough radioactive beams of ^{34}Cl are not presently available. However, the excited states of ^{35}Ar can be studied through indirect methods, such as β-decay. The excited states above the proton separation threshold, $S_p(^{35}\text{Ar}) = 5896.29(75)$ keV [5], can be populated selectively through the β-decay of the $3/2^+, T = 3/2$ ground state of ^{35}K. These states can de-excite either through γ or proton emission, yielding a probe on the properties of these states. So far there has been only one spectroscopic β-decay study ^{35}K [6, 7], in which the astrophysically interesting region was not covered.

In a recent experiment we have studied the excited states of ^{35}Ar selectively through the β-decay of the ground state of ^{35}K in hopes for providing more detailed data for the astrophysically interesting states in ^{35}Ar. In this contribution we report preliminary results of our experiment, including several new proton groups, the first direct observation of the Isobaric Analogue State (IAS) of ^{35}K ground state in ^{35}Ar in β-decay, and the improved half-life of ^{35}K.

IOP Conf. Series: Journal of Physics: Conf. Series **940** (2017) 012004 doi:10.1088/1742-6596/940/1/012004

2. Experimental method

We have studied β-decay of ^{35}K at the Cyclotron Institute of Texas A&M University. In this experiment the ^{35}K beam was produced in inverse-kinematics reaction ^{1}H(^{36}Ar,^{35}K)2n by bombarding a LN$_2$-cooled hydrogen gas target at 2 atm pressure with a 36-MeV/u ^{36}Ar beam. The reaction products recoiling from the target were separated with the Momentum Achromat Recoil Separator (MARS) [8], resulting a beam of ^{35}K with typical intensity of 500 pps and purity of about 70% as shown in in Fig. 1.

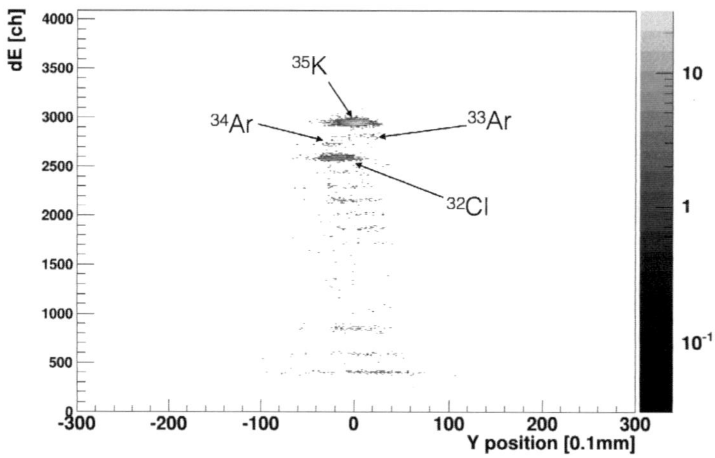

Figure 1. A particle identification of the reaction products at the focal plane of the MARS. The momentum defining slits of MARS were at ±0.5 cm, yielding $\Delta p/p = 0.6\%$. The purity of the resulting ^{35}K beam was about 70%.

Ions of interest were implanted into a Si detector stack consisting of a 140μm Single-Sided Silicon Strip Detector, a 45μm Double-Sided Silicon Strip Detector (DSSSD), and a 1 mm thick Si-pad detector. The Si detectors were surrounded by two 70% high-purity germanium detectors (HPGe) in close geometry for efficient γ-ray detection. The beam implantation depth was controlled by using a rotatable 254 μm thick Al degrader, allowing tuning of the beam into the middle of the DSSSD. The setup is described in more detail in Ref. [9]. During the measurement cycle, the ^{35}K beam was pulsed with implantation period of 400 ms and decay period of 400 ms, and the data was collected only during the decay part, yielding a 50% duty cycle. The setup was calibrated with standard calibration sources and with ^{32}Cl and ^{36}K beams produced with the same primary beam and target as in case of ^{35}K.

3. Results

In the measured γ-data, as shown in Fig. 2, we have the first direct observation of the decay of the 5572 keV Isobaric Analogue State of ^{35}K ground state in ^{35}Ar. In the previous β-decay study reported in Refs. [6, 7], the location of the state was deduced from the observed singles γ-ray spectrum by adding two γ-rays that could be identified to originate only from ^{35}K decay. We have observed not only the previously known transitions, but have the detection in coincidence, and in addition measured the full energy γ-decay from the IAS, thus confirming the earlier assignment.

The astrophysically interesting states are located beyond the proton separation threshold of ^{35}Ar. In our γ-ray data, some of the known states [4] above the proton threshold are observed. With selectivity of the β-decay process, one can constrain the spin-parity assignment of these

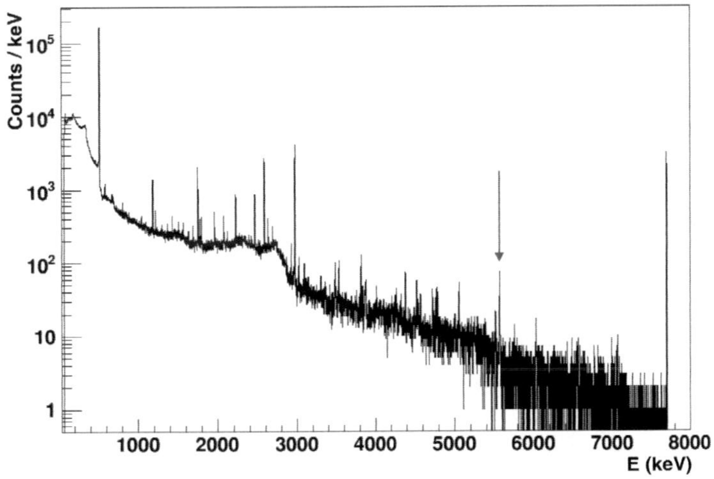

Figure 2. Observed γ-ray spectrum from the β-decay of ^{35}K. The direct γ emission from the IAS of the ^{35}K g.s. in ^{35}Ar is highlighted with an arrow.

levels, as well as improve the resonance energies through the good resolution of the HPGe detectors.

We have observed also several new β-delayed particle decays as demonstrated in Fig. 3. The general features of the spectrum agree with the previous study of Ref. [7] down to the lowest energy decay observed in the previous study at around 1.3 MeV. Most of these new particle groups were likely below the detection threshold of the previous study. However, several of the new particle groups seem to be in coincidence with 461 keV or 665 keV γ-rays from ^{34}Cl, indicating that these decays occur from levels that are well above the Gamow window of the ^{34g,m}Cl(p,γ)^{35}Ar reaction.

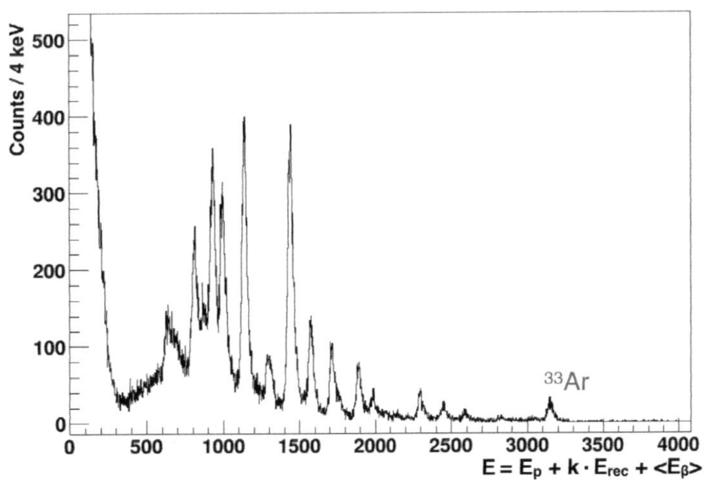

Figure 3. The measured total decay energy spectrum of the charged particles from the β-decay of ^{35}K. The only identifiable impurity present in the spectrum is the known proton group from the β-decay of ^{33}Ar [10].

The high statistics of the study allows decay time analysis of the individual particle and γ decays, yielding a half-life of 175(2) ms, which is in agreement with the known half-life of ^{35}K 178(8) ms [4]. The half-lives of all the new decays are in agreement with the known half-life, supporting the assignment of origin from the decay of ^{35}K rather than any known impurity in the beam. Detailed analysis of the data is an ongoing process.

4. Conclusions and outlook

In conclusion, we have done a high statistics measurement of β-decay of ^{35}K, yielding an extensive new dataset of γ and particle decays from the excited states of ^{35}Ar. In addition to the known decay data, we have observed several new particle groups as well as several γ-rays that were not observed in the previous studies, and improved the half-life value of ^{35}K. The detailed analysis and astrophysical interpretation of the data is under way. In the future, in-beam studies of the states in ^{35}Ar are planned, as well as probing the ^{35}K β-decay with a novel detector system based on Micromegas technology [11], allowing detection of particle decays free of β-background down to about 100 keV.

Acknowledgements

This work has been supported in part by the US DOE under Grants DE-FG02-93ER40773 and DE-NA0001786.

References

[1] Iliadis C, Champagne A, Jose J, Starrfield S and Tupper P 2002 *Astrophys.J. Suppl.* **142** 105
[2] Parikh A, Jose J, Moreno F and Iliadis C 2008 *New Astr. Rev.* **52** 409
[3] Parikh A, Jose J and Sala G 2014 *AIP Advances* **4** 041002
[4] Chen J, Cameron J and Singh B 2011 *Nuclear Data Sheets* **112** 2715
[5] Wang M, Audi G, Wapstra A, Kondev F, MacCormick M, Xu X and Pfeiffer B 2012 *Chin. Phys. C* **36** 1603
[6] Carraz L C, Ewan G T, Hagberg E, Hardy J C, Jonson B, Mattsson S, Ravn H L and Tidemand-Petersson P 1979 *Phys. Lett. B* **85** 212
[7] Ewan G T, Hagberg E, Hardy J C, Jonson B, Mattsson S, Tidemand-Petersson P and Towner I S 1980 *Nucl. Phys. A* **343** 109
[8] Tribble R, Azhari A, Gagliardi C, Hardy J, Mukhamedzhanov A, Tang X, Trache L and Yennello S 2002 *Nucl. Phys. A* **701** 278
[9] McCleskey M, Trache L, Saastamoinen A, Banu A, Simmons E, Roeder B, Tabacaru G, Tribble R, Davinson T, Woods P and Äystö J 2013 *Nucl. Instrum. Meth. in Phys. Res. A* **700** 124
[10] Adimi N, Domínguez-Reyes R, Alcorta M, Bey A, Blank B, Borge M J G, Santos F d O, Dossat C, Fynbo H O U, Giovinazzo J, Knudsen H H, Madurga M, Matea I, Perea A, Sümmerer K, Tengblad O and Thomas J C 2010 *Phys. Rev. C* **81** 024311
[11] Pollacco E, Trache L, Simmons E, Spiridon A, McCleskey M, Roeder B, Saastamoinen A, Tribble R, Pascovici G, Kebbiri M, Mols J and Raillot M 2013 *Nucl. Instrum. Meth. in Phys. Res. A* **723** 102

IOP Conf. Series: Journal of Physics: Conf. Series **940** (2017) 012005 doi:10.1088/1742-6596/940/1/012005

Nuclear physics uncertainties of the astrophysical γ-process studied through the ^{64}Zn(p,α)^{61}Cu and ^{64}Zn(p,γ)^{65}Ga reactions

Gy. Gyürky, Zs. Fülöp, Z. Halász, G. G. Kiss, T. Szücs

Institute for Nuclear Research (Atomki), Bem tér 18/c, H-4026 Debrecen, Hungary

E-mail: gyurky@atomki.mta.hu

Abstract.
In a recent work, the cross section measurement of the ^{64}Zn(p,α)^{61}Cu reaction was used to prove that the standard α-nucleus optical potentials used in astrophysical network calculation fail to reproduce the experimental data at energies relevant for heavy element nucleosynthesis. In the present paper the analysis of the obtained experimental data are continued by comparing the results with the predictions using different parameters. It is shown that the recently suggested modification of the standard optical potential leads to a better description of the data.

1. Introduction

Nature builds up the chemical elements of the universe through nuclear reactions starting from the nucleons that were born after the Big Bang. We need to understand these nuclear reactions in order to explain the abundances and the mere existence of chemical elements. The mechanism of nuclear reactions is therefore a key ingredient in our understanding of nucleosynthesis. Ab initio calculations based on the basic nucleus-nucleus interaction are possible only for the lightest systems of a few nucleons. For the reactions of heavier nuclei simplifications are necessary such as the reduction of the many body problem of a reaction to a two body one. In such a picture the interaction of the two reacting nuclei are represented by a nucleus-nucleus optical potential. The properties of such a potential - among other factors - determine the calculated cross section. It is therefore essential to find a suitable potential in order to obtain reliable cross sections for astrophysical reaction network calculations of the elemental and isotopic abundances.

The appropriate potential acting between interacting nuclei plays an important role in practically all mass regions encountered in nuclear astrophysics from low level density systems (where e.g. direct capture cross sections are calculated by potential models) to high level densities towards the heavier mass region. The motivation of the present work came from the problems related to the synthesis of the heavy proton rich isotopes (p-isotopes) in the astrophysical γ-process [1]. In the mass and energy region of the γ-process nuclear reactions proceed through the formation of compound nuclei at excitation energies where the level density is high. Nuclear reactions in this domain are usually described using the statistical approach of the Hauser-Feshbach theory. An important ingredient of the calculations is the averaged transmission coefficients in all energetically allowed reaction channels. In the case of particle emission or absorption, the transmission coefficients are obtained from the corresponding particle-nucleus optical potential (where the particle can be a single nucleon or a few nucleon system like an α-particle).

Owing to the generally unsatisfactory results of γ-process network calculations, the experimental study of the relevant nuclear reactions became a hot topic of nuclear astrophysics research of recent years [2]. Many experimental results indicate that statistical model calculations are not able to describe well the measured cross sections of reactions involving α-particles. One of the most likely reasons of the failure is the lack of good knowledge of the α-nucleus optical potential at low energies. This was identified typically in measurements of α-induced reactions [3]. The problem with these reactions, however, is the fact that owing to the extremely low cross section the measurements are done at energies higher than the astrophysically relevant ones. This means that the α-nucleus optical potential is not tested in the relevant energy range. Moreover, the cross sections at the studied energies show sensitivity not only to the α-nucleus potential, but to several other parameters. In order to study directly the α-nucleus potential at astrophysical energies, a new approach is followed in the present work.

2. The studied reactions and the experimental technique

In order to study the low energy α-nucleus optical potential at γ-process relevant energies, the cross section of a reaction where the α-particle is in the exit channel was measured. The chosen reaction was ^{64}Zn(p,α)^{61}Cu. The advantages of this reaction and the experimental technique was detailed in our recent publication [4]. Here only the most important aspects are summarized.

Calculations show that in the proton energy range between about 3 and 8 MeV the cross section of this reaction obtained in the statistical model depends solely on the choice of the α-nucleus optical potential [5]. This is indicated as a sensitivity plot shown in figure 1 of ref [4]. Taking into account the Q value of the reaction, the γ-process relevant energy range of the ^{61}Cu + α system in the exit channel of the studied reaction lies in the lower half of this energy interval. Therefore the measurement of the ^{64}Zn(p,α)^{61}Cu cross section provides direct information about the α-nucleus potential.

The experiments were carried out at the cyclotron accelerator of Atomki where ^{64}Zn targets were bombarded by a proton beam in the energy interval between 3.5 and 8 MeV. Since the reaction product is radioactive, the number of reactions was determined by the activation method measuring the γ-radiation following the decay of ^{61}Cu by a HPGe detector.

The most important conclusion of the measurements was already drawn in ref [4]. Namely, the experiments proved for the first time at astrophysical energies that the commonly used optical potential strongly overestimates the measured cross sections towards the lowest energies. This overestimation is a general finding of earlier experiments, which were however carried out at much higher α + nucleus relative energies. In this paper some further comparison of the measured cross sections with model calculations is shown. For the comparisons two statistical model codes are used. The SMARAGD code [6] is a successor of the NON-SMOKER code [7] which is often applied in nuclear astrophysics network calculations. Calculations with the widely used TALYS [8] code are also carried out.

3. The ^{64}Zn(p,γ)^{65}Ga reaction, γ-strength functions and level densities

Although the primary aim of the present work was the measurement of the ^{64}Zn(p,α)^{61}Cu cross section, the ^{64}Zn(p,γ)^{65}Ga reaction was also studied as this is the other open reaction channel on ^{64}Zn in the studied energy range. Similarly to figure 1 of ref [4], the left panel of figure 1 here shows the sensitivity of the ^{64}Zn(p,γ)^{65}Ga calculated cross section to the variation of different partial widths. As one can see, the (p,γ) cross section in the studied energy range (shaded area) is sensitive solely to the γ-width.

The right panel of figure 1 shows the measured ^{64}Zn(p,γ)^{65}Ga cross section and the predictions of the statistical model calculations. The standard SMARAGD calculation overestimates the

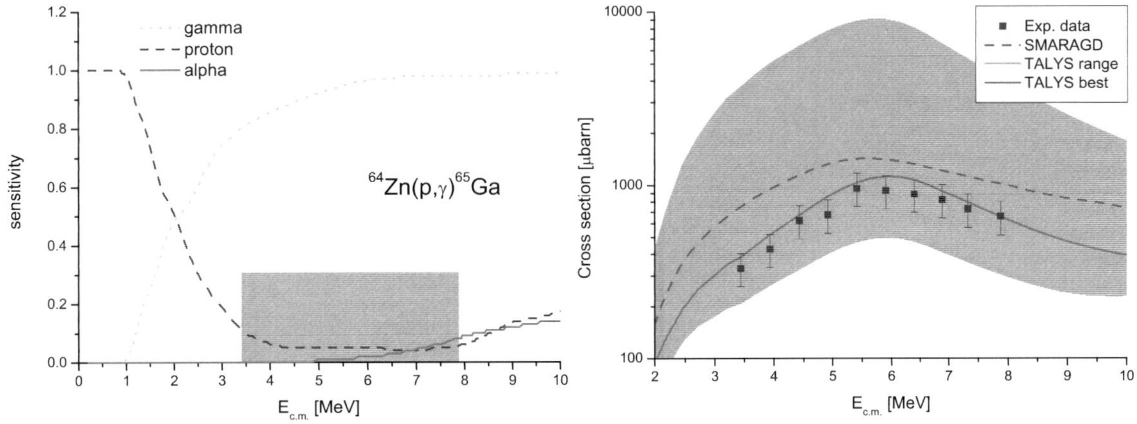

Figure 1. Left: sensitivity of the calculated ^{64}Zn(p,γ)^{65}Ga cross section to the various partial widths. Right: measured and calculated ^{64}Zn(p,γ)^{65}Ga cross section. See text for details.

measured data by about a factor of two but describes relatively well the energy dependence. The γ-width is determined by the level density and γ-ray strength function used in the statistical model. TALYS calculations were therefore carried out with varying these input parameters. For both parameters TALYS version 1.4 used in this work offers five different built-in options. Calculations with all possible combinations of these parameters were done and the cross section range covered by the obtained results is shown in the figure as the gray band. As one can see, differences of up to a factor a 50 can be obtained with the different input parameters. It is worth noting that there is no a priori information about which of the possible input parameters are correct. The wide range of possible calculated cross sections show that the not correct nuclear reaction rates entering the γ-process models can indeed be the reason of the poor reproduction of the p-isotope abundances. Only experiments like the present one can help find the best input parameters for the calculations. In the case of the ^{64}Zn(p,γ)^{65}Ga reaction the best TALYS result can be obtained using the microscopic level density of Goriely *et al.* [9] and the γ-strength function of Kopecky and Uhl (generalized Lorentzian form of the giant dipole resonance for the dominant E1 transition) [10]. This is shown as the red curve in figure 1.

4. The ^{64}Zn(p,α)^{61}Cu reaction and the modification of the α-nucleus potential

The most important conclusion of our measurement presented in ref [4] was that the standard statistical model calculations overestimate the measured ^{64}Zn(p,α)^{61}Cu at the lowest studied energies while the agreement becomes fairly good when the energy increases. In the case of the SMARAGD code (and in its predecessor NON-SMOKER) the standard calculations are carried out with the alpha-nucleus optical potential of McFadden and Satchler [11]. This potential was developed based on 25 MeV experimental data and does not have any energy dependence. It is therefore not surprising that experimental evidence is gathering that the potential fails at low energies where an energy dependence of at least the imaginary part of the potential seems to be required.

In order to cure the low energy problem of the McFadden–Satchler potential, an energy dependent modification was recently recommended [12]. The depth of the imaginary part was modified using a Fermi-type function with a diffuseness parameter a_E. Several recent experiments indicate that this modification leads to a better reproduction of the measured data (see e.g. [13] and references therein). In the present paper this modification is tested for the ^{64}Zn(p,α)^{65}Cu reaction. Figure 2 shows the comparison of the experimental data with SMARAGD calculation in the form of astrophysical S-factor. (Calculations with the TALYS code

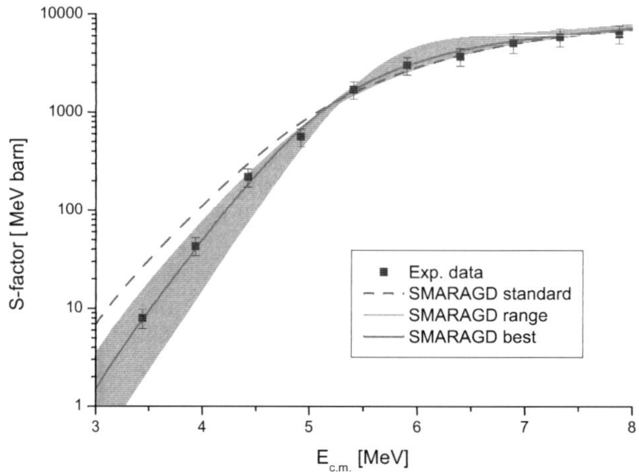

Figure 2. Measured and calculated astrophysical S-factor of the $^{64}Zn(p,\alpha)^{61}Cu$ cross section. See text for details.

are omitted here as they were presented in [4].) The green line shows the standard calculation using the original McFadden–Satchler potential. The grey band indicates the obtained S-factors by the modified potential with the a_E parameter in the range between 1.5 and 5 MeV. It is evident that the modification leads to a much better reproduction of the experimental data. The best agreement is achieved with $a_E = 2.5$ MeV as indicated by the red curve.

5. Summary and conclusions

The present experimental data prove directly at astrophysical energies that the standard low energy α-nucleus optical potential must be modified. The new data support the recently suggested modification of the potential. The implementation of this modification in the reaction rate calculations for astrophysical models is recommended. The comparison of the measured and calculated $^{64}Zn(p,\gamma)^{65}Ga$ cross section data indicate that further experimental effort is needed as only measured data can help find the correct nuclear input parameters of calculations which show highly divergent results depending on the parameters.

Acknowledgments

The authors thank T. Rauscher for providing the results of the SMARAGD code. This work was supported by OTKA grants No. K101328 and K108459.

References

[1] Rauscher T *et al.* 2013 *Rep. Prog. Phys.* **76** 066201
[2] Gyürky Gy *et al.* 2015 *Proceeding of the Nuclear Physics in Astrophysics VI conference, in press*
[3] Szücs T, Dillmann I, Plag R, Fülöp Zs 2014 Nuclear Data Sheets **120** 191
[4] Gyürky Gy *et al.* 2014 *Phys. Rev. C* **90** 052801(R)
[5] Rauscher T 2012 *The Astrophysical Journal Supplement Series* **201** 26
[6] T. Rauscher, code SMARAGD, http://nucastro.org/smaragd.html
[7] Rauscher T and Thielemann F-K 2001 *At. Data Nucl. Data Tables* **79** 47; www.nucastro.org
[8] Koning A J, Hilaire S and Duijvestijn M C 2008, TALYS version 1.4 *Proceedings of the International Conference on Nuclear Data for Science and Technology - ND2007*, p.211,
[9] Goriely S, Tondeur F, Pearson J M 2001 *Atom. Data Nucl. Data Tables* **77** 311
[10] Kopecky J and Uhl M 1990 *Phys. Rev. C* **42** 1941
[11] McFadden L, Satchler G R 1966 *Nucl. Phys.* **84** 177
[12] Sauerwein A *et al.* 2011 *Phys. Rev. C* **84** 045808
[13] Yalçın C *et al.* 2015 *Phys. Rev. C* **91** 034610

Online tools for nucleosynthesis studies

K Göbel[1,2], J Glorius[1,3], A Koloczek[1,2], M Pignatari[2,4], R Plag[3], R Reifarth[1,2], C Ritter[1,2,5], S Schmidt[1], K Sonnabend[1], B Thomas[1] and C Travaglio[2,6]

[1] Goethe University Frankfurt, Germany
[2] http://www.nugridstars.org
[3] GSI Helmholtzzentrum für Schwerionenforschung, Darmstadt, Germany
[4] Konkoly Observatory of the Hungarian Academy of Sciences, Hungary
[5] University of Victoria, Canada
[6] INAF - Astrophysical Observatory Turin, Italy

E-mail: goebel@physik.uni-frankfurt.de

Abstract. The nucleosynthesis of the elements between iron and uranium involves many different astrophysical scenarios covering wide ranges of temperatures and densities. Thousands of nuclei and ten thousands of reaction rates have to be included in the corresponding simulations. We investigate the impact of single rates on the predicted abundance distributions with post-processing nucleosynthesis simulations. We present online tools, which allow the investigation of sensitivities and integrated mass fluxes in different astrophysical scenarios.

1. Introduction

1.1. Neutron capture processes

The heavy elements from iron to uranium are mainly produced by the rapid and slow neutron capture processes, r- and s-process. In stars, the **s-process** takes place in helium and carbon burning phases with neutron number densities between 10^8 and 10^{10} cm^{-3} [1]. The reaction path closely follows the valley of stability since neutron capture time scales of typically ten years are much slower than most of the β-decay time scales of the involved nuclei. Three components contribute to the s-process: the weak, main and strong components. The main differences lie in the neutron-to-seed ratio, the temperature, as well as the neutron densities [2].

The **main** and the **strong components of the s-process** take place in low-mass Asymptotic Giant Branch (AGB) stars and produce the nuclei with mass numbers above $A \approx 90$ [3]. The low-mass AGB star (1.5 to 3 $M_\odot$) consists of a C/O core surrounded by alternately burning He and H shells, which are separated by a thin He-rich layer. The H envelope is fully convective. He is produced during the H shell burning, accumulating He-rich material on top of the He intershell. The temperature rises and triggers He burning, which generates the flash and the following thermal pulse (TP) [4]. The intershell becomes convective, the envelope expands and the H burning can be temporarily extinguished [5]. These flashes reoccur until the AGB envelope is entirely lost by mass loss. The number of TPs depends on the initial mass of the star, the mass loss description, the opacities etc. [6]. After each TP, if the H burning shell is inactive, the so-called third dredge-up (TDU) occurs and mixes H into the He intershell [7, 2]. The available protons are captured by ^{12}C producing ^{13}N, which β-decays to ^{13}C, forming the

IOP Conf. Series: Journal of Physics: Conf. Series **940** (2017) 012006 doi:10.1088/1742-6596/940/1/012006

^{13}C pocket [8]. The s-process starts when neutrons are released by the ^{13}C(α,n)^{16}O reaction. A second neutron source ^{22}Ne(α,n)^{25}Mg is partially activated after subsequent thermal pulses when the temperature exceeds $2.5 \cdot 10^8$ K [6].

The **weak component of the s-process** takes place in massive stars with $M_{\rm star} > 8\ M_\odot$. The time-integrated neutron flux is lower compared to the main component producing nuclei with mass numbers $A \leq 90$. The reaction ^{22}Ne(α,n)^{25}Mg provides neutrons during convective He burning at temperatures of about $3 \cdot 10^8$ K and during convective shell C burning at about 10^9 K. The star explodes as a core-collapse supernova (ccSN) at the end of its evolution and enriches the interstellar medium [2, 9, 10, 11].

In addition to the s- and r-processes an intermediate neutron capture process with neutron number densities around 10^{15}cm^{-3} has been proposed, the **i-process** [12]. Very late thermal pulses in post-AGB stars are a possible astrophysical site [13]. When the He burning convective zone brings some hydrogen from the H-rich layers outward, protons can be captured by ^{12}C. The produced ^{13}N decays to ^{13}C which reaches the bottom of the convective He-burning layers. Neutrons are produced via ^{13}C(α,n) at high temperatures resulting in neutron densities typical for the i-process [14, 15].

1.2. p-process

About 35 proton-rich isotopes heavier than iron cannot be produced by neutron-capture reactions since they are bypassed by s- and r-process paths [16]. The origin of the p-nuclei is, to date, little understood and many possible production sites are under discussion [17]. The widely accepted model of the **γ-process** assumes that the p-nuclei are produced in sequences of photodisintegration reactions starting at r- and s-process nuclei as seed. The necessary temperatures are 2.0 to $3.5 \cdot 10^9$ K, which implies explosive conditions. Neutron-, proton- and α-dissociation reactions on more neutron-rich nuclei drive the masses towards the proton-rich side of the valley of stability, reaching the p-nuclei directly or via β-decays of unstable nuclei. The explosive burning in the shock-heated Ne/O shell of a core-collapse supernova was identified as a possible production site [18, 19, 20, 21].

2. Post-processing network

Post-processing nucleosynthesis (PPN) studies were carried out within the Nucleosynthesis Grid (NuGrid) research platform [22]. NuGrid offers a software framework for nucleosynthesis simulations in astrophysical environments [23, 24]. Temperature and density profiles (so-called "trajectories") are obtained from simulations of stellar evolution that include only reactions relevant for the energy production. PPN processes the trajectories and calculates the stellar yields with a large network including up to 5,000 isotopes and 60,000 reactions. Different nuclear data sets can be chosen as input for the reaction network.

3. Sensitivity studies

The impact of nuclear reaction rates on the final abundance of the isotopes was determined by sensitivity studies for different s-process scenarios and an i-process scenario. In each nucleosynthesis simulation one reaction rate was changed compared to the default rates. The sensitivity was calculated as the ratio of the relative change in abundance $\Delta N_j/N_j$ of isotope j and the relative change of the rate $\Delta r_i/r_i$ [15, 25]:

$$s_{ij} = \frac{\Delta N_j/N_j}{\Delta r_i/r_i}$$

IOP Conf. Series: Journal of Physics: Conf. Series **940** (2017) 012006 doi:10.1088/1742-6596/940/1/012006

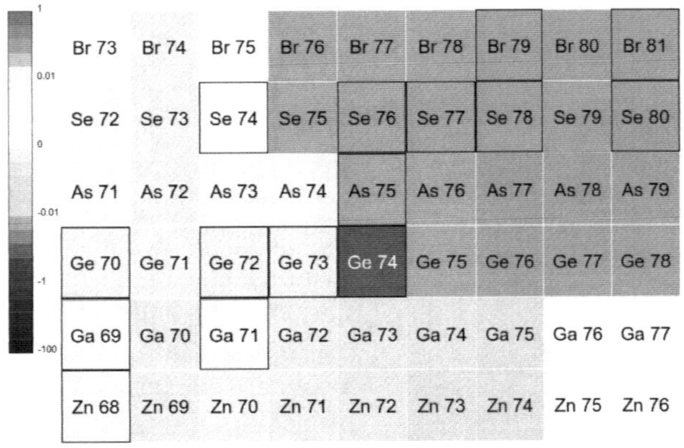

Figure 1. Sensitivity of isotopes on the reaction rate of ^{74}Ge(n,γ).

4. Nucleosynthesis fluxes during a core-collapse supernova

The post-processing nucleosynthesis simulations for the γ-process used a stellar progenitor with an initial mass of 25 $M_\odot$ and metallicity $Z = 0.02$ [26]. The nucleosynthesis was calculated by using classic ccSN trajectories [27].

Nucleosynthesis fluxes show the main reaction paths for a certain time range in the simulation. The fluxes give important information about the relevance of a reaction for the isotopic abundances. The flux leading to the product nucleus via a certain reaction is defined by [24]

$$\frac{f_{\text{Reaction}}}{\Delta t} = \frac{\Delta Y_{\text{Product,Reaction}}}{\Delta t}.$$

The reaction rate and the abundance of the parent species determine the flux [23]. Relative nucleosynthesis fluxes for a single isotope represent the relative net abundance yield in the simulation. They also illustrate the most relevant reaction paths for the nucleus.

5. Online tools

We provide the results of the sensitivity studies via the web interface `http://exp-astro.physik.uni-frankfurt.de/sensitivities/`. The user may choose the scenario ("i-process", "C-shell", "C13 pocket", "Thermal pulse", "weak s process"; the trajectories are described in detail on the webpage) and the reaction of interest. The results are shown color-coded for a part of the chart of nuclei (Figure 1). Other parameters may be displayed in addition to the sensitivities. The results are shown in different color schemes and layouts according to the chosen parameter ("Sensitivity", "Decay", "(n,g) MACS", "Abundance", "Halflife" or "custom").

We show the integrated nucleosynthesis fluxes of the classic ccSN model via the web interface `http://exp-astro.physik.uni-frankfurt.de/fluxes/`. The pictures show the relative time-integrated production and destruction fluxes (Figure 2). The sum of all production fluxes is normalized to 100% for each isotope, and the destruction fluxes are scaled with the same factor.

6. Outlook

The online tools are under development. Our goal is to provide sensitivity studies for a large variety of astrophysical scenarios, and nucleosynthesis flux studies to investigate in detail the production and destruction of isotopes in different stellar conditions.

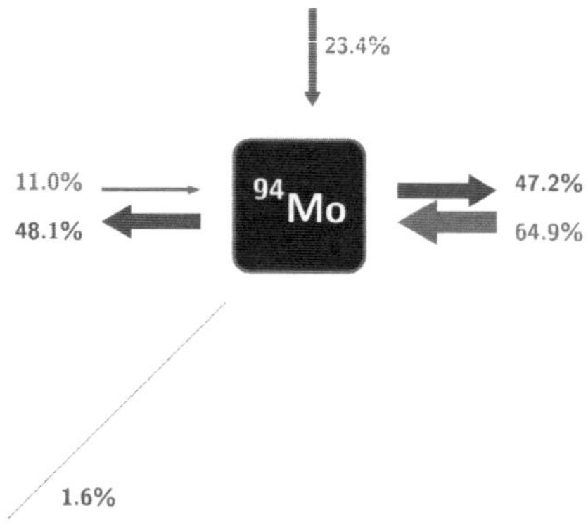

Figure 2. Relative time-integrated fluxes producing (blue) and destroying (red) the p-isotope ^{94}Mo in the post-processing nucleosynthesis simulation of the 25 $M_{\odot}$ ccSN model. The sum of all production fluxes is normalized to 100%, and the destruction fluxes are scaled with the same factor. Fluxes smaller than 1% are not shown.

Acknowledgments
This project was supported by the Helmholtz International Center for FAIR, DFG (SO907/2-1) and HGS-HIRe. NuGrid acknowledges significant support from NSF grants PHY 02-16783 and PHY 09-22648 (JINA), EU MIRG-CT-2006-046520, from Lendület-2014 Programme of the Hungarian Academy of Sciences (Hungary) and from SNF (Switzerland).

References
[1] Käppeler F, Gallino R, Bisterzo S and Aoki W 2011 *Reviews of Modern Physics* **83** 157–194
[2] Reifarth R, Lederer C and Käppeler F 2014 *Journal of Physics G Nuclear Physics* **41** 053101
[3] Gallino R, Arlandini C, Busso M, Lugaro M, Travaglio C, Straniero O, Chieffi A and Limongi M 1998 *ApJ* **497** 388–403
[4] Herwig F 2005 *Annual Review of Astronomy & Astrophysics* **43** 435–479
[5] Herwig F, Freytag B, Hueckstaedt R M and Timmes F X 2006 *ApJ* **642** 1057–1074
[6] Karakas A I and Lattanzio J C 2014 *Publications of the Astronomical Society of Australia* **31** e030
[7] Straniero O, Gallino R and Cristallo S 2006 *Nuclear Physics A* **777** 311–339
[8] Straniero O, Chieffi A, Limongi M, Busso M, Gallino R and Arlandini C 1997 *ApJ* **478** 332–339
[9] Rauscher T, Heger A, Hoffman R D and Woosley S E 2002 *ApJ* **576** 323–348
[10] The L S, El Eid M F and Meyer B S 2007 *ApJ* **655** 1058–1078
[11] Pignatari M, Gallino R, Heil M, Wiescher M, Käppeler F, Herwig F and Bisterzo S 2010 *ApJ* **710** 1557–1577
[12] Cowan J J and Rose W K 1977 *ApJ* **212** 149–158
[13] Herwig F, Pignatari M, Woodward P R, Porter D H, Rockefeller G, Fryer C L, Bennett M and Hirschi R 2011 *ApJ* **727** 89
[14] Bertolli M G, Herwig F, Pignatari M and Kawano T 2013 *ArXiv e-prints* (*Preprint* 1310.4578)
[15] Koloczek A 2015 *Systematic s-process sensitivity studies* Ph.D. thesis Goethe University Frankfurt, Germany
[16] Burbidge E M, Burbidge G R, Fowler W A and Hoyle F 1957 *Reviews of Modern Physics* **29** 547–650
[17] Rauscher T, Dauphas N, Dillmann I, Fröhlich C, Fülöp Z and Gyürky G 2013 *Reports on Progress in Physics* **76** 066201
[18] Woosley S E and Howard W M 1978 *ApJS* **36** 285–304
[19] Rayet M, Arnould M and Prantzos N 1990 *A&A* **227** 271–281
[20] Lambert D L 1992 *A&A Rev.* **3** 201–256
[21] Rayet M, Arnould M, Hashimoto M, Prantzos N and Nomoto K 1995 *A&A* **298** 517
[22] The NuGrid collaboration http://www.nugridstars.org
[23] Herwig F, Diehl S, Fryer C L, Hirschi R, Hungerford A, Magkotsios G, Pignatari M, Rockefeller G, Timmes F X, Young P and Bennet M E 2008 *PoS(NIC X)023*
[24] Pignatari M and Herwig F 2012 *NPN* **22** 18–23
[25] Koloczek A, Thomas B, Glorius J, Plag R, Pignatari M, Reifarth R, Ritter C, Schmidt S and Sonnabend K 2015 *Atomic Data and Nuclear Data Tables* Status: submitted
[26] Pignatari M, Herwig F, Hirschi R, Bennett M, Rockefeller G, Fryer C, Timmes F X, Heger A, Jones S, Battino U, Ritter C, Dotter A, Trappitsch R, Diehl S, Frischknecht U, Hungerford A, Magkotsios G, Travaglio C and Young P 2013 *ArXiv e-prints* (*Preprint* 1307.6961)
[27] Rapp W, Görres J, Wiescher M, Schatz H and Käppeler F 2006 *ApJ* **653** 474–489

IOP Conf. Series: Journal of Physics: Conf. Series **940** (2017) 012008 doi:10.1088/1742-6596/940/1/012008

Gamma-ray line diagnostics of supernova explosions - SN2014J and Cas A

Thomas Siegert
Max-Planck-Institute for extraterrestrial Physics, Gießenbachstraße 1, D-85748, Garching bei München, Germany

E-mail: tsiegert@mpe.mpg.de

Abstract. Gamma-rays from nuclear de-excitation of newly procuced isotopes during supernovae (SNe) provide a unique window to the explosion mechanisms. SNe interiors are accessible only by γ-rays as they are energetic enough to penetrate the SN cloud. Both thermonuclear explosions (type Ia) and core-collapse SNe (CCSN, type II) are key producers of heavy elements in the Universe. In SNe Ia, a white dwarf (WD) is disrupted by ignition from inside or by triggering the explosive event from outside, producing major amounts of ^{56}Ni. Type II SNe are powered by the gravitational collapse of a massive star, having burnt all its nuclear fuel.

In this work, we present a diagnostic study of γ-ray lines from SN2014J and Cassiopeia A (Cas A). INTEGRAL observed SN2014J for several months and for the first time, it was possible to measure the characteristic lines from the ^{56}Ni-decay chain in a SN Ia event. Surprisingly, ^{56}Ni was seen only 20 days after the explosion which indicates that some ^{56}Ni must be located outside the WD and not deeply embedded. We provide a ^{56}Co γ-ray line light curve and estimate a visible ^{56}Ni mass of 0.5 $M_\odot$ from a comparison to 1D model light curves. Cas A observations have been revisited and we detect both, the characteristic hard X-ray line from the decay of ^{44}Ti at 78 keV, and the subsequent γ-ray line from the decay of ^{44}Sc at 1157 keV in one coherent data set. Expansion velocities in the range of 2000 – 5000 km s^{-1} and an initially synthesised ^{44}Ti mass of 1.37×10^{-4} $M_\odot$ are found.

1. Introduction

A SN is the final fate of a star, producing major amounts of iron-peak and also heavier elements. When a massive star ($\gtrsim 8$ $M_\odot$) starts to lack of nuclear fuel in its interior, the radiation pressure cannot sustain gravitation any more and the Fe/Ni-core of the star collapses. The outer layers, consisting of lighter elements than Fe and Ni, are free-falling towards the centre and interact with the supernova shock wave, backbouncing from the newly build proto-neutron star. If this shock wave is energetic enough, it can accelerate the outer layers of stellar material up to the escape velocity which is called a CCSN. In the interaction region of shock wave and outer layers, explosive nucleosynthesis occurs and procudes predominantly ^{56}Ni as the most stable nuclear configuration. This is described by nuclear statistical equilibrium for the short time of explosive burning. However, as the explosion site dilutes, other nuclear reactions are in favour, falling out of the equilibrium. In this case, an overabundance of α-particles easily produces α-multiple nuclei, e.g. ^{44}Ti. This is known as α-rich freeze-out [1, 2, 3, 4, 5, 6].

In contrast to massive stars, a SN Ia involves a WD that is triggered to explode by various mechanisms. There are two commonly accepted scenarios to disrupt the electron-degenerate matter of which a WD consists of: The single degenerate branch describes a WD and a giant star orbiting each other. Binary mass transfer due to Roche-lobe overflow gradually accretes mass from the giant companion star onto the

WD, building another layer of material. If C and/or O is accreted, the WD can reach its Chandrasekhar limiting mass and ignites in one more spots of its interior. A thin flame is rushing through the WD, totally disrupting it and procuding new elements of which ^{56}Ni has the largest fraction with $0.1 - 1.0$ M$_\odot$. If there is a He layer accreted, it can ignite in a flash and trigger the WD explosion from outside. The double-degenerate scenario explains a type Ia SN as the merger of two WDs, if they are not fusing to a neutron star [7, 8, 9, 10, 11, 12, 13, 14].

In both cases, the temporal and the spectral behaviour of γ-ray lines from various decaying isotopes can give deeper insights into the kinematics of the explosion and provide hints for the progenitor channels. Kinematic characteristics are reflected by the γ-ray line parameters, such as the line centroid and width. The line centroid is interpreted as the bulk (line-of-sight) motion, while the width represents the spread of expansion velocities. The amount of produced elements is most directly measured through the total γ-ray line flux of a particular isotope. Tracing the history of the parameters can tighten interpretations when comparing to models. For example, ^{56}Ni and its daughter nucleus ^{56}Co have comparatively short half-life times of 6 d and 77 d, respectively, and are therefore suitable follow the very early phases of a supernova explosion. On the other hand, ^{44}Ti has a half-life time of 60 a which makes studies of the interiors of considerably older SN remnants (SNRs) possible [15, 16, 17, 18, 19].

Utilizing two examples, one CCSN (Cas A [20]) and one SN Ia (SN2014J [21, 22]), we illustrate the necessity and capabilities of γ-ray line diagnostics for explosive stellar events. Gamma-rays originating in nuclear decay reaction open up a unique window into the deep interiors of SNe by "X-raying it with γ-rays". For SN2014J, we focus on the decay chain ^{56}Ni$\rightarrow^{56}$Co$\rightarrow^{56}$Fe, in which ^{56}Ni decays to an excited state of ^{56}Co which is predominantly emitting γ-rays of energies 158 keV and 812 keV when de-exciting to its ground state. Similarly, ^{56}Co is decaying to stable ^{56}Fe, producing γ-rays at 847 keV and 1238 keV. In Cas A, most of the ^{56}Ni and ^{56}Co already decayed and the dominant energy provider are γ-rays from the decay chain ^{44}Ti$\rightarrow^{44}$Sc$\rightarrow^{44}$Ca. ^{44}Ti is decaying to ^{44}Sc, emitting X-rays at 68 keV and 78 keV, followed by a β^+-decay of ^{44}Sc to stable ^{44}Ca only 3.9 h after, emitting γ-rays at 1157 keV. Radioactive decay γ-rays from the ^{56}Ni-decay chain are what powers the light-curves in SNe Ia as they are absorbed in the SN could and re-emitted at lower energies. Likewise, γ-rays from the ^{44}Ti-decay chain (and positrons) power the late light curve of CCSNe [23, 16].

2. Data Description and Analysis

Our observations have been made with the γ-ray spectrometer SPI aboard ESA's INTEGRAL satellite. SPI is a coded-mask telescope, consisting of a honeycomb shaped array of 19 high-purity Ge detectors, measuring γ-rays with energies between 20 keV and 8 MeV at a resolution of 2.2 keV at 662 keV, and a segmented W mask 1.71 m above the array with partly opaque and partly transparent elements. Depending on the SPI pointing direction, a celestial γ-ray source imprints a certain shadowgram on the detector array which has to be discriminated against a large instrumental background, making typically 99% of the total counts. In order to create distinct sequences of patterns from celestial sources, the telescope is redirected around the target position, typically every half an hour by $\sim 2^\circ$ (dithering). The background patterns needed to build a self-consistent description of the measured data are essentially constant for particular, distinguishable physical sources in the spacecraft. They are determined by an elaborate spectral fitting procedure on time scales long enough to obtain statistical accuracy but at the same time short enough to trace detector degradation effects due to cosmic ray bombardment. Each instrumental background line is characterised by four line-shape parameters for each detector and for each time interval that is chosen (three days). The background continuum is treated separately as it exhibits a congeries of physical processes. Therefore, on longer time-scales, the detector patterns for the two background components *lines* and *continuum*, and also for proposed (modelled) celestial sources, are fixed and need to be renormalised to the actual data. This is done by a maximum likelihood fit for each energy bin, estimating intensity scaling paramters for sky and background models simultanously [24, 25, 26, 27, 28, 29]. Details for the whole procedure can be found in [21], [22], [20], and [30], for example; see also [31].

IOP Conf. Series: Journal of Physics: Conf. Series **940** (2017) 012008 doi:10.1088/1742-6596/940/1/012008

SN2014J occured on January 2014, UT 14.75 ± 0.21 in the starburst galaxy M82 at a distance of 3.3 Mpc, $(l/b) = (141.43°/40.56°)$, and was recognised the closest type Ia event since four decades. It provided a unique opportunity to study such an event by its emitted γ-rays. INTEGRAL observed SN2014J between January 31 and June 26 2014, corresponding to days 16.3 and 164 after the explosion, with a total exposure time of ~7 Ms [32, 33, 34, 35].

Cas A is a young CCSNR at a distance of 3.4 kpc in the Galactic plane, $(l/b) = (111.74°/-2.13°)$, and was targeted several times by INTEGRAL during its ongoing mission. Our data set formally comprises ten years of data, in which Cas A was seen with an exposure of 10.8 Ms in SPI's field of view $(16° \times 16°)$ [36, 37, 38, 39].

Given the angular resolution of SPI of ~ 2.7°, both sources are treated as point-sources in our analysis.

3. Results and Interpretation

3.1. SN2014J

Figure 1 shows the measurements of the early phase after the explosion of SN2014J, i.e. from days 16 to 34. The two most prominent γ-ray lines from the ^{56}Ni-decay at 158 keV and 812 keV, are detected only 18 days after the explosion (Figs. 1a and 1b). The measured line fluxes are $(1.10 \pm 0.42) \times 10^{-4}$ ph cm^{-2} s^{-1} (158 keV) and $(1.90 \pm 0.66) \times 10^{-4}$ ph cm^{-2} s^{-1} (812 keV). The line centroids are 159.43 ± 0.43 keV and 811.84 ± 0.42 keV, consistent with the laboratory energies (2σ). The lines are not broadened which suggests velocity spreads (expansion velocities) of less than 2000 km s^{-1} (2σ). Within measurements uncertainties, the two lines are identical. Following the two γ-ray lines over a longer period substantiates the finding of ^{56}Ni measured early in SN2014J as the measured light curves for both lines (see Fig. 1c) are consistent with the half-life time of ~ 6 days for ^{56}Ni.

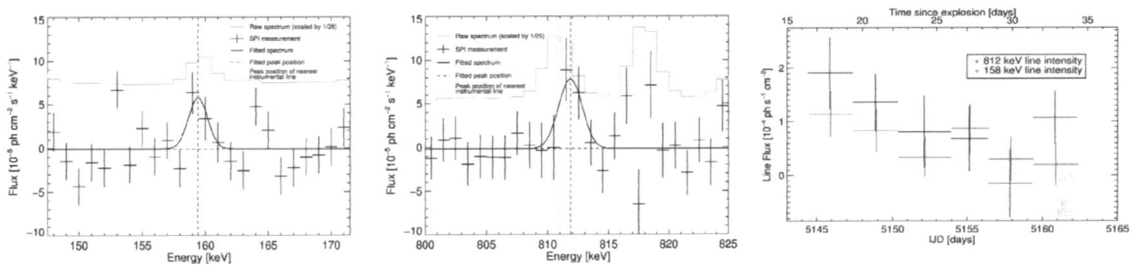

(a) 158 keV line. (b) 812 keV line. (c) ^{56}Ni γ-ray light curves.

Figure 1: ^{56}Ni-measurements. Left and middle panel show the spectra (black crosses) from the position of SN2014J about 18 days after the explosion with a Gaussian line fitted to the data points. Superimposed is a scaled raw spectrum (gray histogram) to compare peak positions of nearest intrumental lines (gray/blue dashed). The right panel shows the ^{56}Ni γ-ray line light curve for both lines in time intervals of three days between days 16 and 34 after the explosion.

Measuring decaying ^{56}Ni that early for a type Ia is surprising as ^{56}Ni is expected to be buried deeply in the SN cloud during that time. The visible mass of ^{56}Ni 18 days after the explosion is estimated to be (0.06 ± 0.03) M$_\odot$. Our interpretation is a that some fraction of the ^{56}Ni must be located near the surface of the WD. A possible though speculative configuration to explain this amount of ^{56}Ni and to meet the kinematic constraints from the γ-ray measurements, which is unexpected in most explosion models, is as follows: If a He belt is accreted from a donor star, exploding, and triggering the explosion of the WD, the γ-ray data constraints can be met if this scenario is essentially seen pole-on. A He belt rather than a layer can be conserved if the accretion rate is of the order 10^{-5} M$_\odot$ a^{-1} and/or if the WD is rapidly rotating so that the accretion is faster than it loses angular momentum. Furthermore, radiation transfer simulations produce almost indistinguishable spectra in UVOIR wavelengths, when comparing standard scenarios to our proposed configuration. This makes SN2014J appear rather normal in these wavelength observations, although there are also hints for irregularities [21, 40, 35, 41, 42, 43, 44].

Figure 2 shows the measurements of decaying ^{56}Co from the position of SN2014J. During the γ-ray maximum around days 60-100, both major γ-ray lines originating in the decay of ^{56}Co at 847 keV and 1238 keV are clearly detected (Figs. 1a and 1b). Obtained flux values are $(3.7 \pm 1.2) \times 10^{-4}$ ph cm^{-2} s^{-1} (847 keV) and $(2.3 \pm 0.7) \times 10^{-4}$ ph cm^{-2} s^{-1} (1238 keV), respectively. This can directly be converted to an initially synthesised ^{56}Ni mass seen to decay of (0.49 ± 0.12) M$_\odot$. The lines are Doppler-broadened and suggest expansion velocities of 7250 ± 1500 km s^{-1}. This is in concordance with most explosion models.

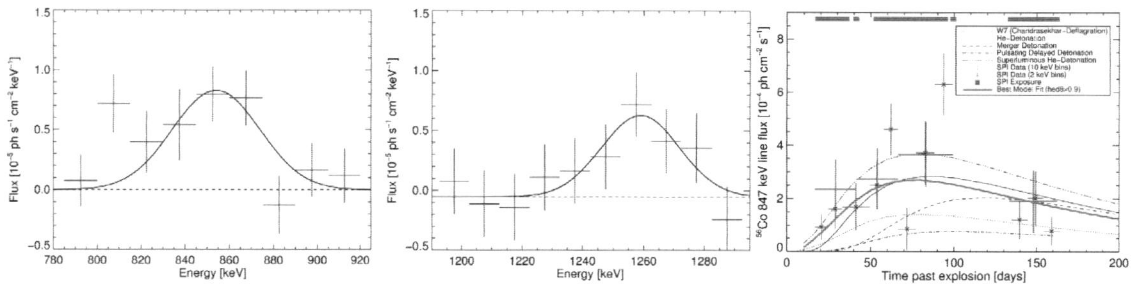

(a) 847 keV line. (b) 1238 keV line. (c) ^{56}Co γ-ray light curve.

Figure 2: ^{56}Co-measurements. Left and middle panel show the spectra (black crosses) from the position of SN2014J about 80 days after the explosion with a Gaussian line fitted to the data points. The right panel shows the 847 keV ^{56}Co γ-ray line light curve with time resolutions of $\sim$ 30 days (red) and $\sim$ 10 days (blue). There are theoretical 1D γ-ray line light curves exemplified, taken from [45].

In Fig. 2c, the γ-ray line light curve for the 847 keV line during INTEGRAL observations is illustrated. The line flux shows a rise in intensity towards the γ-ray maximum and a decline consistent with the half-life time of ^{56}Co, as expected. However, utilizing the fine energy resolution of SPI and a smaller time binning reveals discrepancies between the theoretical, smooth behaviour during the rise to the maximum, which may indicate asymmetries in the explosion or shadowing effects in the morphology (blue data points). The spectra for all epochs, except after day 134, indicate aspherical ^{56}Ni distributions, perhaps caused by clumps or co-moving volume elements at certain times, occulting the γ-ray signal. By testing several 1D models from [45], we independently estimate the synthesised ^{56}Ni mass to (0.49 ± 0.12) M$_\odot$, formally favouring models with He near the surface of the SN. Acceptable fits provide ^{56}Ni masses between 0.4 and 0.8 M$_\odot$ and we want to emphasise that these estimates are based on 1D models which assume single and smoothly broadened gaussian lines. This may not be reflected by our data and estimates should be taken with caution. Despite the clear detection of ^{56}Ni and ^{56}Co γ-ray lines, a thorough statistical discrimination among progenitor models is inadequate and most models are in agreement with our ^{56}Ni measurements if the ^{56}Ni mass is a free parameter [22, 45, 46, 47].

3.2. Cas A

SPI allows one to measure the whole ^{44}Ti-decay chain using the same instrument in one coherent data set. Figures 3a and 3b show the spectra from the position of Cas A about 340 years after the estimated explosion date in AD 1681. The 78 keV X-ray line form the decay of ^{44}Ti has a total flux of $(2.1 \pm 0.4) \times 10^{-5}$ ph cm^{-2} s^{-1}, is centred at 78.9 ± 1.5 keV and broadened 1.9 ± 0.6 keV above the instrumental resolution. These values lead to a ^{44}Ti mass estimate of $(1.5 \pm 0.4) \times 10^{-4}$ M$_\odot$ seen to decay, and an estimate for the velocity spread of 4300 ± 1600 km s^{-1} at essentially no bulk motion. Due to

IOP Conf. Series: Journal of Physics: Conf. Series **940** (2017) 012008 doi:10.1088/1742-6596/940/1/012008

complications in modelling background features below ~ 73 keV, increasing the background level by two orders of magnitude, the 68 keV line is not detected. The flux from γ-ray line at 1157 keV from the subsequent decay of ^{44}Sc is $(3.5 \pm 1.2) \times 10^{-5}$ ph cm^{-2} s^{-1}, and is centred at 1158.0 ± 3.6 keV and broadened by 8.9 ± 1.4 keV. Although the ^{44}Ti mass estimate is considerably larger, $(2.4 \pm 0.9) \times 10^{-4}$ M$_{\odot}$, the expansion velocity estimate of the ejecta is only 2200 ± 400 km s^{-1}.

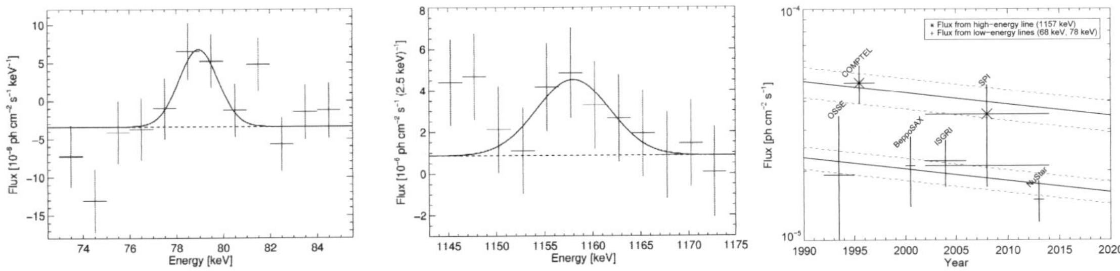

(a) 78 keV line. (b) 1157 keV line. (c) ^{44}Ti and ^{44}Sc γ-ray light curves.

Figure 3: ^{44}Ti- and ^{44}Sc-measurements. Left and middle panel show the spectra (black crosses) from the position of Cas A measured with SPI with a Gaussian line fitted to the data points. The right panel shows ^{44}Ti and ^{44}Sc measurements from this work (SPI) as well as previous measurements with other instruments. The data points for each γ-ray line have been fitted with an exponential decay function, separately, revealing a flux discrepancy between γ-rays originating in excited ^{44}Sc and excited ^{44}Ca.

This apparent discrepancy becomes even more significant when measurements from earlier instruments are included in the analysis (see Fig. 3c). The measurements for the high-energy line at 1157 keV (COMPTEL and SPI) show a systematically higher flux value, compared to the low-energy line measurements from OSSE, BeppoSAX, ISGRI, SPI, and NuSTAR, although the values should be almost equivalent 340 years after the explosion. We argue that the true kinematics of the Cas A SNR are reflected in the line shapes originating in the decay of ^{44}Ti to ^{44}Sc*, emitting 68 keV and 78 keV X-rays. The second step in the decay chain, from ^{44}Sc to ^{44}Ca*, may not be the only process that leads to excited ^{44}Ca because low-energy cosmic rays are believed to be able to excite ambient ^{44}Ca which promtply emits 1157 keV γ-rays. As ^{44}Sc is decaying with a half-life time of only 3.9 h, no ^{44}Sc is available to be excited efficiently enough to procude additional flux in the low-energy lines. Under this assumption, i.e. by combining only the low-energy line measurements, the ^{44}Ti-mass estimate is $(1.28 \pm 0.14) \times 10^{-4}$ M$_{\odot}$, and an additional flux level of $(2.3 \pm 0.6) \times 10^{-5}$ ph cm^{-2} s^{-1} is revealed for the 1157 keV line. Although this effect is expected, the measured magnitude is 10-100 times larger than from theoretical estimates. Combining all previous measurements without any assumptions obtains an estimate for the initially synthesised ^{44}Ti-mass of $(1.37 \pm 0.19) \times 10^{-4}$ M$_{\odot}$ [20, 48, 49, 50, 51, 52, 53, 54, 55, 56].

References

[1] D. Bodansky, D. D. Clayton, et al. Nov. 1968, ApJS, 16, 299
[2] S. E. Woosley, W. D. Arnett, et al. Nov. 1973, ApJS, 26, 231
[3] F. X. Timmes, S. E. Woosley, et al. Jun. 1996, ApJ, 464, 332
[4] S. Woosley & T. Janka. Dec. 2005, Nature Physics, 1, 147
[5] L.-S. The, D. D. Clayton, et al. May 2006, A&A, 450, 1037
[6] G. Magkotsios, F. X. Timmes, et al. Nov. 2010, ApJS, 191, 66
[7] K. Nomoto, K. Iwamoto, et al. Feb. 1997, Nuclear Physics A, 621, 467
[8] M. Stritzinger, B. Leibundgut, et al. Apr. 2006, A&A, 450, 241
[9] P. A. Mazzali, F. K. Röpke, et al. Feb. 2007, Science, 315, 825
[10] S. Rosswog, D. Kasen, et al. Nov. 2009, ApJ, 705, L128
[11] M. Fink, F. K. Röpke, et al. May 2010, A&A, 514, A53

[12] J. Guillochon, M. Dan, et al. Jan. 2010, ApJ, 709, L64
[13] R. Pakmor, M. Kromer, et al. Mar. 2012, ApJ, 747, L10
[14] W. Hillebrandt, M. Kromer, et al. Apr. 2013, Frontiers of Physics, 8, 116
[15] J. Isern, J. Gómez-Gomar, et al. 1997, The Transparent Universe, eds. C. Winkler, T. J.-L. Courvoisier, & P. Durouchoux, vol. 382 of ESA Special Publication, 89
[16] G. Audi, O. Bersillon, et al. Dec. 2003, Nuclear Physics A, 729, 3
[17] J. Isern, E. Bravo, et al. Oct. 2008, New A Rev., 52, 377
[18] J. Isern, M. Hernanz, et al. 2011, Lecture Notes in Physics, Berlin Springer Verlag, eds. R. Diehl, D. H. Hartmann, & N. Prantzos, vol. 812 of Lecture Notes in Physics, Berlin Springer Verlag, 233–308
[19] I. R. Seitenzahl, F. X. Timmes, et al. Sep. 2014, ApJ, 792, 10
[20] T. Siegert, R. Diehl, et al. Jul. 2015, A&A, 579, A124
[21] R. Diehl, T. Siegert, et al. Sep. 2014, Science, 345, 1162
[22] R. Diehl, T. Siegert, et al. Feb. 2015, A&A, 574, A72
[23] R.B. Firestone, V.S. Shirley, et al. 1996, Table of Isotopes (Wiley)
[24] C. Winkler, T. J.-L. Courvoisier, et al. Nov. 2003, A&A, 411, L1
[25] G. Vedrenne, J.-P. Roques, et al. Nov. 2003, A&A, 411, L63
[26] S. J. Sturner, C. R. Shrader, et al. Nov. 2003, A&A, 411, L81
[27] D. Attié, B. Cordier, et al. Nov. 2003, A&A, 411, L71
[28] G. Weidenspointner, J. Kiener, et al. Nov. 2003, A&A, 411, L113
[29] S. Schanne, B. Cordier, et al. Oct. 2003, ArXiv Astrophysics e-prints
[30] T. Siegert & R. Diehl. Jan. 2015, ArXiv e-prints, 1501.05648
[31] A. W. Strong, R. Diehl, et al. Dec. 2005, A&A, 444, 495
[32] Y. Cao, M. M. Kasliwal, et al. Jan. 2014, The Astronomer's Telegram, 5786, 1
[33] J. Fossey, B. Cooke, et al. Jan. 2014, Central Bureau Electronic Telegrams, 3792, 1
[34] E. Kuulkers. Jan. 2014, The Astronomer's Telegram, 5835, 1
[35] W. Zheng, I. Shivvers, et al. Mar. 2014, ApJ, 783, L24
[36] W. B. Ashworth, Jr. 1980, Journal for the History of Astronomy, 11, 1
[37] J. E. Reed, J. J. Hester, et al. Feb. 1995, ApJ, 440, 706
[38] J. R. Thorstensen, R. A. Fesen, et al. Jul. 2001, AJ, 122, 297
[39] D. A. Green. 2002, Highlights of Astronomy, 12, 350
[40] A. Goobar, J. Johansson, et al. Mar. 2014, ApJ, 784, L12
[41] P. E. Nugent, M. Sullivan, et al. Dec. 2011, Nature, 480, 344
[42] R. Kippenhahn & H.-C. Thomas. Feb. 1978, A&A, 63, 265
[43] W. Y. Law & H. Ritter. Jun. 1983, A&A, 123, 33
[44] A. L. Piro & L. Bildsten. Aug. 2004, ApJ, 610, 977
[45] L.-S. The & A. Burrows. May 2014, ApJ, 786, 141
[46] E. Churazov, R. Sunyaev, et al. Aug. 2014, Nature, 512, 406
[47] E. Churazov, R. Sunyaev, et al. Feb. 2015, ArXiv e-prints, 1502.00255
[48] L.-S. The, M. D. Leising, et al. Dec. 1996, A&AS, 120, C357
[49] A. F. Iyudin, R. Diehl, et al. 1997, The Transparent Universe, eds. C. Winkler, T. J.-L. Courvoisier, & P. Durouchoux, vol. 382 of ESA Special Publication, 37
[50] J. Vink, J. M. Laming, et al. Oct. 2001, ApJ, 560, L79
[51] M. Renaud, J. Vink, et al. Aug. 2006, ApJ, 647, L41
[52] B. W. Grefenstette, F. A. Harrison, et al. Feb. 2014, Nature, 506, 339
[53] W. Baade & F. Zwicky. 1934, Contributions from the Mount Wilson Observatory, vol. 3, pp.79-83, 3, 79
[54] R. Ramaty, B. Kozlovsky, et al. Jul. 1979, ApJS, 40, 487
[55] E. G. Berezhko & H. J. Völk. May 2004, A&A, 419, L27
[56] A. Summa, D. Elsässer, et al. Sep. 2011, A&A, 533, A13

Using stellar observations to trace the formation processes of Mo, Ru, Pd, and Ag

Camilla J. Hansen, Terese T. Hansen, Almudena Arcones and Anja C. Andersen

Dark Cosmology Centre, University of Copenhagen, Juliane Maries Vej 30, DK-2100 Copenhagen, Denmark

E-mail: cjhansen@dark-cosmology.dk

Abstract. The exact formation mechanism of many heavy elements remains unknown. Models of the formation site and environment have greatly improved over the past decades, and experiments have provided new data for many of the heavy isotopes. However, much information is still missing to fully describe the neutron-capture formation processes and their sites. Stellar observations combined with mass spectroscopy of meteorites can help to place some of the needed constraints. This, in turn, will help to improve the models and our knowledge on the formation channels creating these heavy elements. Recent studies of Mo, Ru, Pd, and Ag showed that different processes or environments other than the main r-process (forming, e.g., Eu) are needed to explain the production (and observations) of elements with $40 < Z < 50$. An observational study of Mo–Ag is presented, where stellar abundances are compared to meteoritic isotopic abundances (at metallicities [Fe/H] > -1.5) to extract information on differences or similarities in their production. Finally, the number of formation processes needed to describe the chemical composition in low metallicity stars ([Fe/H] < -2.5) is discussed.

1. Introduction and Motivation

To understand the heavy element production in the early (high z) universe old, unevolved, unmixed stars are crucial study cases. These have preserved the chemical signatures of the processes and sites that created the stellar abundances. At these early times only primary formation channels will have had time to operate, most likely in the form of a main rapid neutron-capture (r-)process.

Observations and theoretical predictions indicate that the r-process has two components and/or take place in different astrophysical environments giving rise to a main and a weak formation channel. The slow n-capture (s-)process also has a weak component associated with massive stars which creates the lighter elements with $Z > 30$ [1, 2] and a main component that takes place in asymptotic giant branch (AGB) stars of typically less than $4M_\odot$ and forms more massive elements like Ba [3, 4].

The presence of more than one r-process was observationally deduced by, e.g., a large observational star-to-star abundance scatter seen among n-capture elements in stars with $-4 \lesssim$[Fe/H]< -2.5 but not in $\alpha-$elements ([5] and see Figure 1) which are mainly formed in type II supernovae (SNe II). A long standing quest is therefore to unveil how many n-capture formation processes exist at low metallicity and understand the nature of these processes.

IOP Conf. Series: Journal of Physics: Conf. Series **940** (2017) 012009 doi:10.1088/1742-6596/940/1/012009

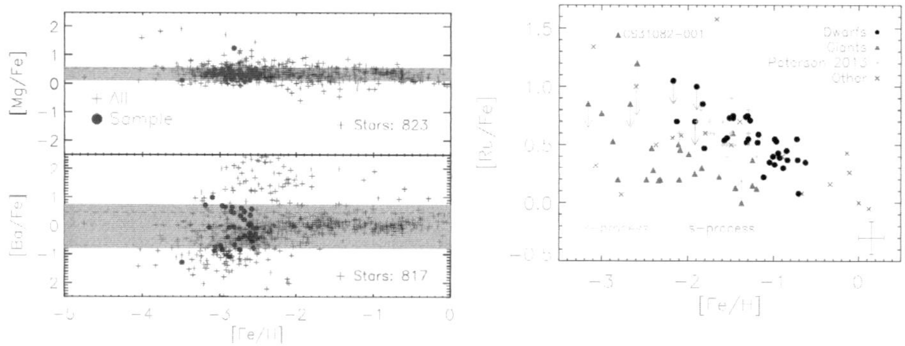

Figure 1. A large star-to-star scatter in n-capture elements but not in α-elements [5].

2. Observational indications of a second, weak r-process

The integrated area of the absorption lines (equivalent width, W) relate to the stellar abundances in the following way,

$$\log \mathrm{W} = \log(const) + \log(A) + \log(gf\lambda) - \theta\chi - \log(\kappa) \tag{1}$$

where A is the abundance, $gf\lambda$ is the product of the statistical weight, oscillator strength, and wavelength, $\theta = 5040\mathrm{K/Temperature}$, χ the excitation potential, and κ the absorption coefficient. When two elements X and Y are formed by the same process their [X/Y][1] ratio will grow at the same rate and we find a flat trend. However, if we consider the absolute logarithmic abundances of two elements co-produced by the same process a linear (1:1) trend is expected. Comparing a well known main r-process element like Eu to Ag shows a clear 'anti-correlation'. This indicates that Eu and Ag are not produced by the same process, while Ru and Ag clearly correlate which indicates co-production by a weak r-process (Figure 2).

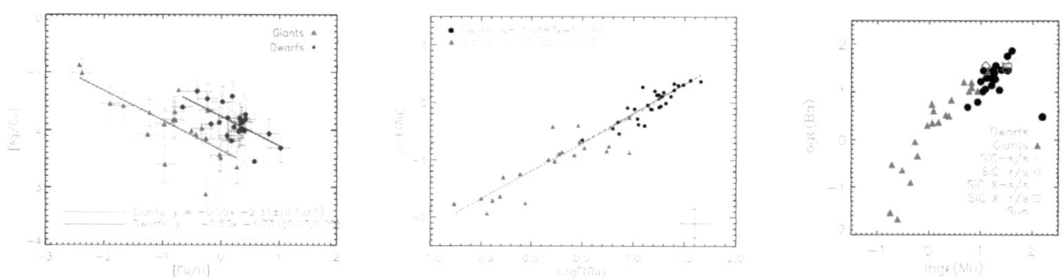

Figure 2. Left: Ag vs. Eu (from [7]) and middle: Ag vs. Ru ([6]). Giant stars are shown as red triangles, and dwarf stars as blue or black circles. Right: Comparing isotopic r/s and s/s fractions in two different SiC grains to elemental stellar abundances [6].

[1] where X and Y are the abundances of two elements scaled to the solar ($\odot$) abundances. The metallicity is typically based on the stellar (*) Fe abundances, in which case we have:

$$[\mathrm{Fe/H}] \equiv \log(N_{\mathrm{Fe}}/N_{\mathrm{H}})_* - \log(N_{\mathrm{Fe}}/N_{\mathrm{H}})_\odot \tag{2}$$

By intercomparing Sr, Y, Zr, Mo, Ru, Pd, Ag, Ba, and Eu we find a number of correlations and anti-correlations. At solar metallicity, the elements are produced by the following processes:

- Weak s-process: $Z \lesssim 40$ (or 42)
- Main s-process: a broad atomic range - typically Ba ($Z = 56$) and heavier
- Weak r-process: $40 < Z < 50$ - in particular Ru, Pd, and Ag
- Main r-process: possibly the full range - or $Z > 50$

Isotopic abundances can be measured in presolar grains. These provide more information on the underlying nuclear formation processes than elemental abundances. By comparing the isotope fractions of r-/s-process and s-/s-process in Silicon Carbide (SiC) grains to the elemental stellar abundances we can extract more direct information on the nuclear formation process than from the elemental abundances ([6]). From this comparison we find that the r/s ratio of SiC X grains (formed in SNe [8]) are larger than the same ratio in mainstream SiC grains. One explanation is that mainstream grains are enriched by AGB stars where the s-isotopes are created at the expense of the r-isotopes. Moreover, the r/s fraction is both types of grains agree with the elemental Zr, Mo, and Ba abundances in stars with a metallicity of $-1.1 <$[Fe/H]< -1.5 confirming the formation processes extracted from elemental stellar abundances. The SiC grains contain more s-process material than stars, which could indicate that the gases stars form from are more diluted or differently mixed than the gases in presolar grains. Below log(Mo) $=$ log(Ba) $= 0$ the tight correlation breaks down and more scatter is found. This point coincides with [Fe/H]~ -2.5 (see Figure 1).

3. Discussion: The neutron-capture processes at low metallicity

Many recent observations indicate that some of the most Fe-poor stars are strongly enhanced in carbon ([C/Fe]> 0.7). These stars might be amongst the first stars that ever formed [9, 10] and are generally referred to as carbon enhanced metal-poor (CEMP) stars. They are subclassified by their heavy element content, namely CEMP-s, CEMP-r, CEMP-r/s and CEMP-no (with little or no heavy elements; see [11]). The CEMP-s stars are thought to reside in binary systems, where an AGB star enriches the low-mass, less evolved companion in C and s-process elements, while CEMP-no stars can possibly be explained by fall-back SNe [12]. However, the origin of CEMP-r and -r/s stars is less straightforward to disentangle. Figure 3 shows that enrichment

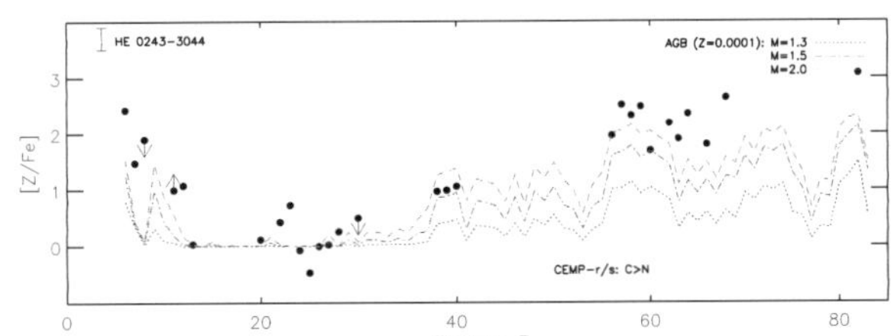

Figure 3. A CEMP-r/s star compared to AGB yields (from [10]).

by a low-mass AGB star cannot account for the full stellar abundance pattern in a CEMP-r/s star. Instead an intermediate process (between r and s) is required – the so-called i–process [13], while the CEMP-r stars are thought to form from a medium that was pre-enriched in C and r-process elements.

Numerous nuclear processes have been suggested as early chemical enrichment channels. Hence we explore if a linear superposition of only two primary processes is sufficient to describe

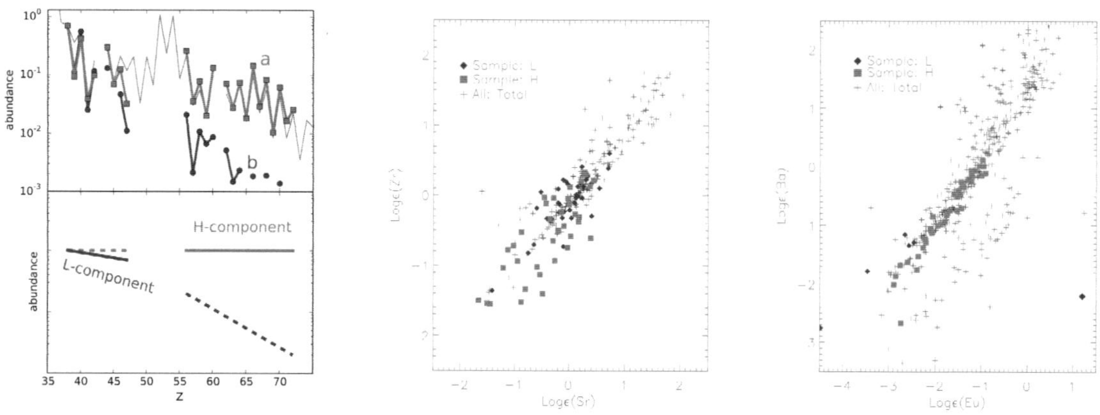

Figure 4. Left: Illustation of the H and L components. Mid and right panel: Scatter in total observed abundances (+) vs. calculated L (blue circles) and H (red squares) abundances [5].

the composition in most metal-poor stars. This test is site independent and follows [14] in using two components: H and L (Figure 4) which are characterised by the two well-studied stars, CS22892-052 and HD122563. Moreover we assume that:

1) All metal-poor stars ([Fe/H< -2.5) are enriched by only two primary processes (H and L);
2) Each of these processes produce a robust abundance pattern in every event.

This leaves us with a calculated abundance pattern that for every element, Z, can be accounted for by a weight (C) and a component abundance (Y) for each of the H and L components:

$$Y_{calc}(Z) = (C_H Y_H(Z) + C_L Y_L(Z)) * 10^{[Fe/H]} \tag{3}$$

By separating every elemental abundance into its L and H fraction we find that the H component is robust and can explain the primary production of the heavy elements very well (this agrees with the main r-process pattern), while L may be less robust or only explain the formation of elements lighter than $Z \sim 50$ (Figure 4).

We used the calculated L component to show the potential of using observations to constrain the extreme astrophysical conditions (e.g., $\nu-$driven winds) where heavy elements are produced.

4. References

[1] Kaeppeler, F., Wiescher, M., Giesen, U., et al. 1994, *The Astrophysical Journal*, **437**, 396
[2] Frischknecht, U., Hirschi, R., & Thielemann, F.-K. 2012, *Astronomy & Astrophysics*, **538**, L2
[3] Busso, M., Picchio, G., Gallino, R., & Chieffi, A. 1988, *The Astrophysical Journal*, **326**, 196
[4] Bisterzo, S., Gallino, R., Straniero, O., Cristallo, S., Käppeler, F. 2011, *MNRAS*, **418**, 284
[5] Hansen, C. J., Montes, F., & Arcones, A. 2014, *The Astrophysical Journal*, **797**, 123
[6] Hansen, C. J., Andersen, A. C., & Christlieb, N. 2014, *Astronomy & Astrophysics*, **568**, A47
[7] Hansen, C. J., Primas, F., Hartman, H., et al. 2012, *Astronomy & Astrophysics*, **545**, A31
[8] Hoppe, P., Leitner, J., Gröner, E., et al. 2010, *The Astrophysical Journal*, **719**, 1370
[9] Keller, S. C., Bessell, M. S., Frebel, A., et al. 2014, *Nature*, **506**, 463
[10] Hansen, T., Hansen, C. J., Christlieb, N., et al. 2015, *The Astrophysical Journal*, **807**, 173
[11] Beers, T. C., & Christlieb, N. 2005, *Annual Review of Astronomy & Astrophysics*, **43**, 531
[12] Tominaga, N., Iwamoto, N., & Nomoto, K. 2014, *The Astrophysical Journal*, **785**, 98
[13] Cowan, J. J., & Rose, W. K. 1977, *The Astrophysical Journal*, **212**, 149
[14] Qian, Y.-Z., & Wasserburg, G. J. 2001, *The Astrophysical Journal Letters*, **552**, L55
[15] Arcones, A., & Montes, F. 2011, *The Astrophysical Journal*, **731**, 5

IOP Conf. Series: Journal of Physics: Conf. Series **940** (2017) 012010 doi:10.1088/1742-6596/940/1/012010

Electron capture rates in stars studied with heavy ion charge exchange reactions

C.A. Bertulani

Department of Physics and Astronomy, Texas A&M University-Commerce, Commerce, TX 75429, USA

E-mail: carlos.bertulani@tamuc.edu

Abstract. Indirect methods using nucleus-nucleus reactions at high energies (here, high energies mean $\sim$ 50 MeV/nucleon and higher) are now routinely used to extract information of interest for nuclear astrophysics. This is of extreme relevance as many of the nuclei involved in stellar evolution are short-lived. Therefore, indirect methods became the focus of recent studies carried out in major nuclear physics facilities. Among such methods, heavy ion charge exchange is thought to be a useful tool to infer Gamow-Teller matrix elements needed to describe electron capture rates in stars and also double beta-decay experiments. In this short review, I provide a theoretical guidance based on a simple reaction model for charge exchange reactions.

1. Introduction

Supernovae explosions may provide a scenario for the production of heavy elements in the Universe, although it is uncertain what are the mechanisms driving such explosions [1]. The uncertainties arise from several possibilities, including how fast nuclear reactions occur in their environments. E.g., electron-capture on neutron-rich nuclei beyond mass number 60 is thought to become important as the density increases during core-collapse supernovae [2]. One cannot access these reaction rates directly in laboratory experiments. Additional problems are, e.g., the stellar enhancement due to thermal population of excited states. Moreover, in the medium nuclear mass range, accurate shell-model calculations are difficult and theoretical methods employing mean-field techniques have been introduced which include large uncertainties. To check their validity, theoretical calculations must be tested against experiment. Another even more difficult problem arises because laboratory-based experiments do not reproduce the conditions (density and temperature) present in stellar environments [2, 3]. Thus the numerous electron capture reactions occurring in stars need coordinated efforts involving theory and experiments.

To fill in parts of our knowledge gap, (p,n), (^{3}He,t) and heavy ion charge-exchange reactions have been used to extract Gamow-Teller matrix elements inherent to electron capture rates. Experimental analysis often use the formalism popularized in Refs. [4, 5, 6, 7], which relates the forward scattering charge-exchange formula with the reduced matrix elements for non spin-flip Fermi, $B(F)$, and spin-flip Gamow-Teller, $B(GT)$, matrix elements, i.e.,

$$\frac{d\sigma}{d\Omega}(\theta = 0^\circ) = \left(\frac{\mu}{2\pi\hbar}\right)^2 \frac{k_f}{k_i} N_D |J_{\sigma\tau}|^2 \left[B(GT) + \mathcal{C}B(F)\right], \qquad (1)$$

IOP Conf. Series: Journal of Physics: Conf. Series **940** (2017) 012010 doi:10.1088/1742-6596/940/1/012010

where N_D is a distortion factor (accounting for initial and final state interactions), $J_{\sigma\tau}$ is the Fourier transform of the GT part of the effective nucleon-nucleon interaction, $\mathcal{C} = |J_\tau/J_{\sigma\tau}|^2$, and $B(F) = A_J|\langle f|| \sum_k \tau_k^{(\pm)}||i\rangle|^2$, and $B(GT) = A_J|\langle f|| \sum_k \sigma_k\tau_k^{(\pm)}||i\rangle|^2$ with $A_J = (2J_f+1)/(2J_i+1)$. Eq. 1 is purely empirical and there are strong evidences that it works reasonably well for most reactions of interest. It has been used indiscriminately in the analysis of charge exchange reactions, although it has been also shown that it fails in few cases. It lacks a solid theoretical basis and should be used with caution to reach the accuracy needed for the electron capture response functions [8, 9]. In view of the great relevance of the ongoing experimental campaign in radioactive beam facilities, I will show how Eq. 1 arises from a very simple model and what are its limitations.

2. Charge-exchange processes with heavy ion reactions

As shown in Fig. 1 left panel, charge-exchange reactions can proceed via virtual pion and rho exchange in one or two step processes (a, b), whereas a physical nucleon exchange always requires at least a two step process (c). Therefore, pion and rho exchange is a more efficient way to induce charge exchange reactions with heavy ions, unless the reaction occurs at low energies of about 50 MeV/nucleon and lower, as demonstrated by Lenske et al. [10]. This fact also allows the reaction mechanism to be much simpler to handle because nucleon exchange is much harder to describe theoretically than pion or rho exchange.

2.1. General formalism

The DWBA scattering amplitude for inelastic processes in nucleus-nucleus collisions is given by [11]

$$f_{aA\to bB}(\theta) = -\frac{m_0}{2\pi\hbar}\left\langle \chi_{\mathbf{k}'}^{(-)}(\mathbf{R})\phi_f(\mathbf{r})|V(\mathbf{R},\mathbf{r})|\chi_{\mathbf{k}}^{(+)}(\mathbf{R})\phi_i(\mathbf{r})\right\rangle \tag{2}$$

where $\chi_{\mathbf{k}',\mathbf{k}}^{(\pm)}$ are the (outgoing/incoming) distorted scattered waves for the c.m. motion of the two nuclei with reduced mass m_0 and $\phi_{i,f}(\mathbf{r})$ are the initial and final wavefunctions for the intrinsic nuclear motion, respectively. The c.m. of relative motion between the nuclei is described by the coordinate $\mathbf{R}$ and $\mathbf{r}$ is the intrinsic coordinate of the wavefunction for the nucleus of interest (usually the projectile in heavy ion charge exchange with radioactive beams).

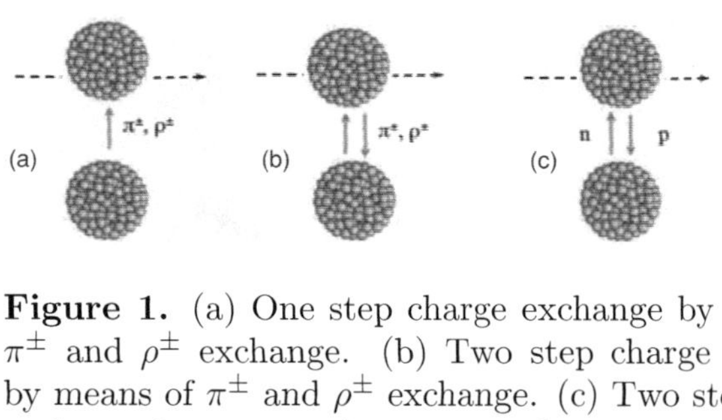

Figure 1. (a) One step charge exchange by means of $\pi^\pm$ and $\rho^\pm$ exchange. (b) Two step charge exchange by means of $\pi^\pm$ and $\rho^\pm$ exchange. (c) Two step charge exchange by means of neutron and proton exchange.

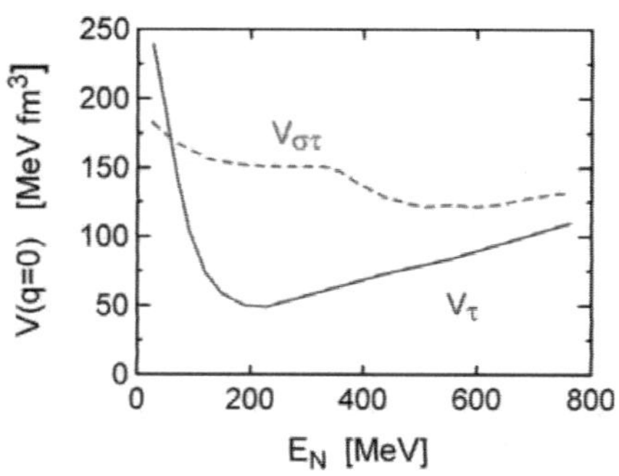

Figure 2. Spin and spin-isospin dependence of the effective nucleon-nucleon interaction used in direct reactions [12].

In Eq. 2, $V(\mathbf{R},\mathbf{r})$ is the interaction potential between the two nuclei which leads to the inelastic process $aA \to bB$. It is taken as the interaction potential between a nucleon in nucleus

t (A or B) at position $\mathbf{r}_t$, with a nucleon in nucleus p (a or b) at position $\mathbf{r}_p$. The meson exchange between these nucleons is responsible for the charge exchange process in nucleus-nucleus collisions at high energies ($E_{lab} > 50$ MeV/nucleon), for which the physical exchange of nucleons is negligible due to severe phase-space limitations [8].

We will now describe a method appropriate for inelastic scattering in high energy collisions, along the lines of Ref. [7, 8, 13]. Even without details of the calculations, it is very easy to follow. Using eikonal wavefunctions, the DWBA amplitude becomes

$$f_{aA\to bB}(\theta) = -\frac{m_0}{2\pi^-\hbar}\sum_{p,t}\int d^3R\left\langle \phi^{(b)}(\mathbf{r}_p)\phi^{(B)}(\mathbf{r}_t)\,|V_{pt}(\mathbf{r})\,S(b)\exp\left[-i\mathbf{q}\cdot\mathbf{R}\right]|\,\phi^{(a)}(\mathbf{r}_p)\phi^{(A)}(\mathbf{r}_t)\right\rangle, \tag{3}$$

where the sum includes all pairs of nucleons, with one from nucleus p and another from nucleus t. The relative coordinates of the pair are related by $\mathbf{r} = \mathbf{R} + \mathbf{r}_p - \mathbf{r}_t$, $\mathbf{q} = \mathbf{k}' - \mathbf{k}$ is the momentum transfer, $\mathbf{k}$ and $\mathbf{k}'$ are the initial and final momentum in the center of mass, and $b = |\mathbf{R}\times\hat{\mathbf{k}}|$ is often interpreted as the impact parameter.

In Ref. [8] the charge exchange cross sections were studied in terms of $\pi + \rho$ exchange interactions based on the OPEP potential. In-medium corrections of the nucleon-nucleon interaction were not taken into account. These effects have been extensively studied by Franey and Love [12] and other authors. The important parts of the interaction for the charge-exchange reaction are the central and the tensor components. In general, the spin-orbit part of the interaction produces a negligible contribution to the cross sections. The (in-medium) nucleon-nucleon interaction may be written as

$$\begin{aligned} V_{pt}(\mathbf{r}) &= V^C(r) + V^C_\sigma(r)(\boldsymbol{\sigma}_p\cdot\boldsymbol{\sigma}_t) + \left[V^C_\tau(r) + V^C_{\sigma\tau}(r)(\boldsymbol{\sigma}_p\cdot\boldsymbol{\sigma}_t)\right](\boldsymbol{\tau}_t\cdot\boldsymbol{\tau}_p) \\ &\quad + \left[V^T(r) + V^T_\tau(r)(\boldsymbol{\tau}_t\cdot\boldsymbol{\tau}_p)\right]S_{pt}(\hat{\mathbf{r}}) + V^{LS}(r)\,\mathbf{l}\cdot(\boldsymbol{\sigma}_t + \boldsymbol{\sigma}_p), \end{aligned} \tag{4}$$

where the tensor operator is given by $S_{pt}(\hat{\mathbf{r}}) = 3(\boldsymbol{\sigma}_p\cdot\hat{\mathbf{r}})(\boldsymbol{\sigma}_t\cdot\hat{\mathbf{r}}) - (\boldsymbol{\sigma}_p\cdot\boldsymbol{\sigma}_t)$, and the spin-orbit term contains the angular momentum operator $\mathbf{l} = (\mathbf{r}_t - \mathbf{r}_p)\times(\mathbf{p}_t - \mathbf{p}_p)/\hbar$, often yielding a negligible contribution to the cross sections. The effective nucleon-nucleon potential has also to include an exchange operator to fully account for anti-symmetrization. Thus, the interaction (4) should be understood as $V_{pt}(\mathbf{r}) \to V_{pt}(\mathbf{r})\left[1 + (-1)^l P^x\right]$, where P^x is the space exchange operator which changes $\mathbf{r}\to-\mathbf{r}$ on the right and $(-)^l$ (with l = relative angular momentum in the nucleon-nucleon system) ensures anti-symmetrization.

Using a spherical basis ($\mu = 0, \pm1$), the scalar products can be written as

$$\boldsymbol{\tau}_t\cdot\boldsymbol{\tau}_p = -\textstyle\sum_\mu\sqrt{3}\,\langle 1\mu 1 -\mu|00\rangle\,\tau_t^{(\mu)}\tau_p^{(-\mu)} = \sum_\mu(-1)^\mu\,\tau_t^{(\mu)}\tau_p^{(-\mu)} \tag{5}$$

$$\boldsymbol{\sigma}_t\cdot\boldsymbol{\sigma}_p = -\textstyle\sum_\mu\sqrt{3}\,\langle 1\mu 1 -\mu|00\rangle\,\sigma_t^{(\mu)}\sigma_p^{(-\mu)} = \sum_\mu(-1)^\mu\,\sigma_t^{(\mu)}\sigma_p^{(-\mu)}, \tag{6}$$

where

$$\tau^{(\pm1)} = \mp\frac{1}{\sqrt{2}}(\tau_x \pm i\tau_y), \qquad \tau^{(0)} = \tau_z .$$

The $\boldsymbol{\tau}_t\cdot\boldsymbol{\tau}_p$ operator is responsible for isospin exchange via the combination of $\tau^\pm$ operators in Eq. (5). Likewise, the $\boldsymbol{\sigma}_t\cdot\boldsymbol{\sigma}_p$ is responsible for spin-flip interactions. Further, introducing $V^C(r)\equiv V^0_{00}(r)$, $V^C_\tau(r)\equiv V^0_{01}(r)$, $V^C_\sigma(r)\equiv V^0_{10}(r)$ $V^C_{\sigma\tau}(r)\equiv V^0_{11}(r)$, $V^T(r)\equiv V^2_{10}(r)$, $V^T_\tau(r)\equiv V^2_{11}(r)$, Eq. (4) can be written in the following more compact form [14]

$$V_{pt}(\mathbf{r}) = \sum_{(K=0,2)ST} V^K_{ST}(r)\,C^K_S\,Y_K(\hat{\mathbf{r}})\cdot\left[\boldsymbol{\sigma}^S_t\otimes\boldsymbol{\sigma}^S_p\right]^K\left[\boldsymbol{\tau}^T_t\cdot\boldsymbol{\tau}^T_p\right], \tag{7}$$

where $K = 0$ corresponds to the central force, $K = 2$ to the tensor force, and S, T to the spin and isospin labels of the force. The constants C_S^K have values $C_0^0 = (4\pi)^{1/2}$, $C_1^0 = -(12\pi)^{1/2}$, $C_0^2 = 0$, and $C_1^2 = (24\pi/5)^{1/2}$. When $S = 0$, $\boldsymbol{\sigma}^S$ is the unit operator, and when $S = 1$, it becomes the Pauli spin operator $\boldsymbol{\sigma}$. Likewise for the isospin operator $\boldsymbol{\tau}^T$. In Eq. 7 the following notation has been used

$$Y_K(\widehat{\mathbf{r}}) \cdot \left[\boldsymbol{\sigma}_t^S \otimes \boldsymbol{\sigma}_p^S\right]^K = \sum_\mu (-1)^\mu Y_{K,-\mu}(\widehat{\mathbf{r}}) \left[\boldsymbol{\sigma}_t^S \otimes \boldsymbol{\sigma}_p^S\right]^{K\mu} , \tag{8}$$

and

$$\left[\boldsymbol{\sigma}_t^S \otimes \boldsymbol{\sigma}_p^S\right]^{K\mu} = \sum_m \langle SmS(\mu - m)|K\mu\rangle \, \sigma_t^{S(m)} \sigma_p^{S(\mu-m)} . \tag{9}$$

Now, defining

$$\widetilde{V}_{ST}^K(p) = 4\pi \int V_{ST}^K(r) \; j_K(pr) \; r^2 \, dr , \tag{10}$$

it is straightforward to show that

$$\begin{aligned} f_{aA\to bB}(\theta) &= -\frac{m_0}{(2\pi)^4 \hbar^2} \int d^3p \, d^3R \, S(b) \; \exp\left[-i(\mathbf{q}+\mathbf{p})\cdot\mathbf{R}\right] \\ &\times \left\langle \phi^{(b)}(\mathbf{r}_p)\phi^{(B)}(\mathbf{r}_t) \left| \exp\left[-i\mathbf{p}\cdot(\mathbf{r}_p - \mathbf{r}_t)\right] \widetilde{V}_{pt}(\mathbf{p}) \right| \phi^{(a)}(\mathbf{r}_p)\phi^{(A)}(\mathbf{r}_t) \right\rangle \\ &= -\frac{m_0}{(2\pi)^4 \hbar^2} \sum_{(K=0,2)ST} i^K C_S^K \int d^3p \, d^3R \, S(b) \; \exp\left[-i(\mathbf{q}+\mathbf{p})\cdot\mathbf{R}\right] \widetilde{V}_{ST}^K(p) \, Y_K(\widehat{\mathbf{p}}) \\ &\times \left\langle \phi^{(b)}(\mathbf{r}_p)\phi^{(B)}(\mathbf{r}_t) \left| \exp\left[-i\mathbf{p}\cdot(\mathbf{r}_p - \mathbf{r}_t)\right] \left[\boldsymbol{\sigma}_t^S \otimes \boldsymbol{\sigma}_p^S\right]^K \left[\boldsymbol{\tau}_t^T \cdot \boldsymbol{\tau}_p^T\right] \right| \phi^{(a)}(\mathbf{r}_p)\phi^{(A)}(\mathbf{r}_t) \right\rangle . \end{aligned} \tag{11}$$

This result is simple and is also *exact*, depending on the following assumptions: (a) the validity of DWBA, (b) the use of eikonal wavefunctions and (c) an effective nucleon-nucleon interaction obeying all symmetries allowed. Let us them under scrutiny one by one:

a - *Validity of DWBA*. This is not questionable, as the reaction is a genuine perturbative process. In fact, DWBA is used in all experimental analysis of charge-exchange reactions.

b - *Validity of eikonal waves.* This also leaves for little room for discussion. Eikonal waves are known to describe accurately the distorted waves in high energy collisions involving low energy transfers and forward angles [11].

c- *Effective nucleon-nucleon interactions.* This is not a problem either. Effective interactions are the basis of all approaches to nucleus-nucleus collisions. Earlier calculations used a single Yukawa interaction with a range of 1 fm [15]. A review of existing data [(p,p') and (p,n) reactions] gave $V_{11}^0 = 12 \pm 2.5$ MeV [16] for such a force. Later, Petrovich *et al.* [17, 18] proposed to include a pseudopotential, i.e., a $\delta(\mathbf{r})$ force, to it. Also, effective interactions derived by fitting matrix elements of a sum of Yukawas to G-matrix elements of phenomenological nucleon-nucleon potentials were introduced [19, 20, 21]. These interactions involve more than one Yukawa potential in an effort to consider different meson exchanges. A popular interaction, developed by Love and Franey [12] uses the antisymmetrized matrix elements of an effective potential to fit the nucleon-nucleon t-matrix in heavy ion collisions at high energies. The right-hand side of Figure 1 shows the isospin and the spin-isospin dependence of the interaction separately as a function of the bombarding energy. One sees that in the region around 200 MeV/nucleon the spin-isospin part clearly

IOP Conf. Series: Journal of Physics: Conf. Series **940** (2017) 012010 doi:10.1088/1742-6596/940/1/012010

dominates over the isospin one. In fact, most of charge-exchange experiments are carried out at about this energy. The high energy also validates the use of the eikonal distorted waves.

The differential cross section for charge exchange is obtained by an average of initial spins and sum over final spins, i.e.,

$$\frac{d\sigma}{d\Omega} = \frac{1}{(2J_a+1)(2J_A+1)} \sum_{\text{spins}} \left| f_{aA\to bB}(\theta) \right|^2 . \tag{12}$$

Obviously, at forward angles $q = 2k\sin(\theta/2) \simeq 0$ and the integral in Eq. 11 becomes even simpler. But the replacement of $q = 0$ in Eq. 11 only yields Eq. 1 with additional simplifications. A closer inspection of Eq. 11 shows that, neglecting the p-dependence of the nucleon-nucleon interaction, $V_{ST}^{K}(p)$, and of the angular momentum dependence in $Y_K(\widehat{\mathbf{p}})$, will lead to the result that $\mathbf{R} \simeq \mathbf{r}_p - \mathbf{r}_t$. Due to the strong absorption at small values of b (small R), the reaction is peripheral and only rather large values of R will contribute in the integral. This means that only the tail of the nuclear wavefunctions will be probed in heavy ion charge exchange reactions, as in any other direct reaction. Eq. 1 is thus justified if we can neglect the momentum dependence in the prodcut $V_{ST}^{K}(p)\, Y_K(\widehat{\mathbf{p}})$. A much stronger approximation assumes the absence of refraction and absorption, such that $S(\mathbf{b}) \simeq 1$. In this case the integral over $\mathbf{R}$ yields $\mathbf{q} \simeq -\mathbf{p}$, and for $q \simeq 0$, we again recover Eq. 1.

A detailed study of the theoretical validity of Eq. 1 was carried out in Ref. [9]. It was found that the extraction of Gamow-Teller matrix elements from heavy ion charge exchange experiments can be misleading if a bettter treatment of the reaction mechanism is not taken into account. At forward angles a clear difference between exact calculations based on Eq. 11 and Eq. 1 have been found [9]. This conclusion might seem surprising, because weak interaction strengths have routinely been extracted from (p,n) and heavy ion exchange reactions using the simple formulation based on the empirical Eq. 1.

3. Double-beta decay studies with heavy ion charge exchange reactions

Double beta decay in nuclei can occur in nuclei without emission of a neutrino, if the neutrino is a Majorana particle (i.e., its own antiparticle). Theorists believe that if the neutrino is a Majorana particle then it is easier to explain the origins of its small, but non-zero, mass. If neutrinoless double beta decay is observed, it will also provide the neutrino mass, since its decay rate is proportional to the square of the neutrino mass. Ordinary double beta decay with neutrino emission has been observed (see Ref. [22] and references therein), but up to now no experiment has reported positive results on neutrinoless beta-decay. Based on the V-A theory of weak interactions it is possible to show that the half-life of neutrinoless double beta decay can exceed by large that of ordinary double beta decay. For all such reasons, there is a large ongoing effort in the nuclear theory community to calculate reliably, or measure directly or indirectly, matrix elements relevant for double beta-decay. As with the case of electron capture matrix elements, spin and spin-isospin nuclear response needs to be assessed.

Charge exchange reactions can play a vital role in extracting matrix elements for double beta-decay by means of the two-step process displayed in part (b) on the left-hand side of Figure 1. The physical exchange of two nucleons is ruled out because the cross section is simply too small at the relevant heavy ion energies. Besides, the theoretical formalism for it is much more involved than that for pion and rho exchange. In Ref. [8] it was estimated that double charge exchange total cross sections in heavy ion collisions are smaller by a factor $10^{-4} - 10^{-5}$ compared to single charge-exchange. If the peak of the differential cross section at forward angles for single charge exchange is of the order of millibarns, it was estimated that the equivalent value for double charge exchange is of the order of microbarns. This requires intense beams and excellent detection efficiency. With recent advances on both fronts, it is expected that a new line of research will focus on these reactions in the near future.

4. Conclusions

Indirect techniques for nuclear astrophysics using new radioactive beam facilities worldwide consist indisputably of one of the most interesting work in present nuclear physics research. In this short review, I have focused on charge exchange reactions and the extraction of Gamow-Teller matrix elements of relevance for electron capture in stars and for neutrino induced reactions. The purpose is to illustrate the level of understanding in nuclear reaction theory for this particular case. We know that empirical formulas are very useful in physics. And it seems that the empirical Eq. 1 works reasonably well for charge exchange reactions. But is lacks a solid physics basis in some situations. This leaves us in an uncomfortable position because physics is driven by the desire to understand physical phenomena from first principles.

5. Acknowledgments

This work was partially supported by the U.S. NSF Grant No. 1415656 and the U.S. DOE Grant No. DE-FG02-08ER41533.

References

[1] Heger A, Fryer C L., Woosley S E, Langer N and Hartmann D H 2003 *Astrophys. J.* **591** 288
[2] Hix W R et al. 2003 *Phys. Rev. Lett.* **91** 201102
[3] Martnez-Pinedo G, Langanke and K, Dean D 2000 *Astrop. J. Sup.* **126** 493
[4] Goodman C D, Austin Sam M, Bloom S D, Rapaport J and Satchler G R 1980 *The (p,n) Reaction and the Nucleon-Nucleon Force* (Plenum, New York)
[5] Petrovich F 1980 ibid.
[6] Taddeucci T N et al. 1987 *Nucl. Phys.* **A469** 125
[7] Osterfeld F 1992 *Rev. Mod. Phys.* **64** 491
[8] Bertulani C A 1993 *Nucl. Phys.* **A554** 493
[9] Bertulani C A and Lotti P 1997 *Phys. Lett.* **B402** 237
[10] Lenske H, Wolter H H, Bohlen H G 1989 *Phys. Rev. Lett.* **62** 1457
[11] Bertulani C A and Danielewicz D 2004 *Introduction to Nuclear Reactions* (IOP Publishing, London)
[12] Franey M A and Love W G 1985 *Phys. Rev.* **C31**, 481
[13] Dmitrev V F et al. 2001 *Phys. Rev.* **C65** 015803
[14] Cook J et al. 1984 *Phys. Rev.* **C30** 1538
[15] Wharton W R and Debevec P T 1975 *Phys. Rev.* **C11** 1963
[16] Austin Sam M and Crawley G M 1972 *The Two-Body Force on Nuclei* (Plenum, New York)
[17] Golin M, Petrovich F and Robson D, 1976 *Phys. Lett.* **B64** 253
[18] Petrovich F and Stanley D 1977 *Nucl. Phys.* **A275** 487
[19] Bertsch G, Borysowich J, McManus H, and Love W G 1977 *Nucl. Phys.* **A284** 399
[20] Hosaka A, Kubo K I, and Toki H 1986 *Nucl. Phys.* **A444** 76
[21] Anantaraman N, Toki H, and Bertsch G F 1983 *Nucl. Phys.* **A398** 269
[22] Beringer J et al. 2012 *Phys. Rev.* **D86** 010001

IOP Conf. Series: Journal of Physics: Conf. Series **940** (2017) 012011 doi:10.1088/1742-6596/940/1/012011

New direct investigation of the ^{19}F(p,α_0)^{16}O down to 0.2 MeV

I. Lombardo[1,2], D. Dell'Aquila[1,2], A. Di Leva[1,2], I. Indelicato[3,4], M. La Cognata[3,4], M. La Commara[3,4], A. Ordine[2], V. Rigato[5], M. Romoli[2], E. Rosato[1,2], G. Spadaccini[1,2], C. Spitaleri[3,4], A. Tumino[4,6], M. Vigilante[1,2]

[1] Dipartimento di Fisica, Università di Napoli Federico II, Via Cinthia, 80126, Napoli, Italy
[2] INFN - Sezione di Napoli, Via Cinthia, 80126, Napoli, Italy
[3] Dipartimento di Fisica ed Astronomia, Università di Catania, Via S. Sofia, 95125, Catania, Italy
[4] INFN - Laboratori Nazionali del Sud, Via S. Sofia, 95125, Catania, Italy
[5] INFN - Laboratori Nazionali di Legnaro, Legnaro (Padova), Italy
[6] Facoltà di Ingegneria e Architettura, Univ. Kore di Enna, Enna, Italy

E-mail: ivlombardo@na.infn.it

Abstract. We discuss new results concerning the investigation of the ^{19}F(p,α_0)^{16}O reaction at very low energies (0.6-0.2 MeV). This reaction is important both in Nuclear Astrophysics, because it closes the CNOF cycle and it is an important fluorine destruction channel in hydrogen rich stellar environments, and in Nuclear Structure, because it allows to study the spectroscopy of the self-conjugated ^{20}Ne compound nucleus. Despite its importance, very few direct data have been published at low energies in the literature. Our work allowed to deduce the *S*-factor down to $\approx$ 0.2 MeV and to point out the role played by broad resonances at low energies. The experimental *S*-factor is $\approx$1.5-2 times larger than non-resonant extrapolations reported by NACRE with potentially important astrophysical consequences.

1. Introduction

The study of low energy ^{19}F(p,α_0) and ^{19}F(p,α_π) reactions represents a powerful tool to probe the structure of the self-conjugated ^{20}Ne compound nucleus [1, 2, 3, 4, 5], that can show important α-cluster structures [6, 7] near some N-α disintegration thresholds (e.g. 11.89 MeV for the ^{12}C+2α and 19.17 MeV for the 5α disintegration mode [8]). For this reason in past times several experimental efforts were done on the study of this reaction in order to highlight the existence of quartet excitations in ^{20}Ne [9]. Nevertheless, some ambiguities in the ^{20}Ne spectroscopy (mainly concerning the J$^\pi$ assignment of states and the determination of branching ratios $\frac{\gamma^2_{\alpha\pi}}{\gamma^2_{\alpha_0}}$) are still persisting [8], especially in the lowermost energy region accessible with direct measurements (E_p <0.5 MeV).

Furthermore, the $p+^{19}F \rightarrow \alpha+^{16}O$ reaction plays an important role in Nuclear Astrophysics for two reasons. First, this reaction closes the CNOF cycle in the hydrogen-burning phase of massive stars [10]; second, it can be an important fluorine destruction channel in hydrogen-rich stellar environments [11, 12, 13]. Unfortunately, very few direct data-sets of the *S*-factor at

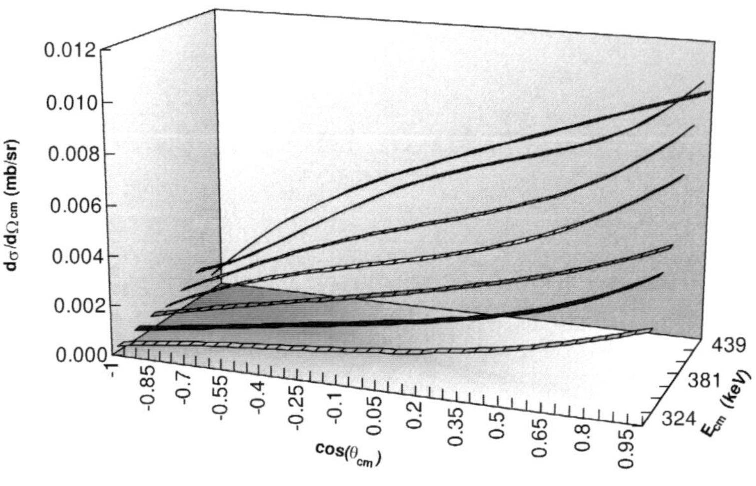

Figure 1. Evolution with the energy of the shape of angular distributions of the $^{19}F(p,\alpha_0)^{16}O$ reaction. Curves are the results of Legendre Polynomial fits of experimental data in the 0.45-0.3 MeV center of mass energy domain. A noticeable forward anisotropy is observed.

$E_{cm} < 1$ MeV have been reported in the literature [1, 3, 14], and the reaction rate estimate at astrophysical energies is mainly based on extrapolation from high energy data [15]. Very interestingly, recent indirect data obtained with the Trojan Horse Method [16, 17] pointed out the important role played by a low energy resonance at 113 keV [12, 13]. Considering the importance of this reaction and taking into account the lack of data reported in the literature, we performed a new experiment at INFN - Laboratori Nazionali di Legnaro aimed at measuring excitation functions and angular distributions of the $^{19}F(p,\alpha_0)$ reaction in the bombarding energy region $E_p \approx$0.6-0.2 MeV. The details of this experiment are widely discussed in Ref. [18]. In the following sections we report the main experimental findings obtained in this experiment.

2. Experimental apparatus

Low energy proton beams have been delivered by the AN2000 Van de Graaf accelerator of Laboratori Nazionali di Legnaro (Padua, Italy). The calibration procedure used in this experiment is described in details in Ref. [18]. The energy uncertainty was of the order of $\approx \pm$ 2.5 keV; the proton beam intensity was of the order of 0.3-0.9 μA. The used targets were CaF_2 layers 30 μg/cm^2 thick, evaporated onto a 20 μg/cm^2 carbon backing, and were frequently changed during the experiment. The detection array was constituted by 12 silicon detectors (300 μm thick) mounted at polar angles ranging from 20° to 160°. 8 μm aluminium absorbers were placed in front of the silicon detectors to stop scattered protons [1, 19, 20, 21, 22]. Beam current was measured by means of a Faraday cup (-300 V suppression voltage). A copper rod cooled to liquid nitrogen temperature was used inside the scattering chamber to reduce carbon build-up effects. The vacuum in the scattering chamber was better than 10^{-6} mbar. Due to the high Q-value (8.114 MeV), experimental spectra show clean peaks due to the $^{19}F(p,\alpha_0)^{16}O$ reaction (see, e.g. Figure 1 of Ref. [18]). From α_0 particle yields it is possible to deduce angular distributions and excitation functions in the 0.2-0.6 MeV bombarding energy range.

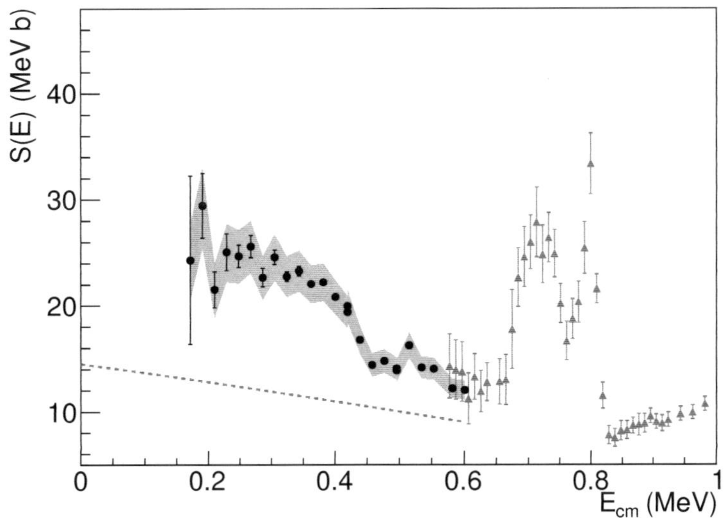

Figure 2. S-factor of the ^{19}F(p,α_0)^{16}O reaction at low energies ($\lesssim$ 1 MeV). Black full dots: experimental data obtained in this work (error bars: statistical errors; light grey band: non-statistical errors). Green triangles: data reported in Ref. [19]. Dashed line: non-resonant extrapolation reported by NACRE in Ref. [15].

3. Experimental results

Figure 1 shows the shape of the present experimental angular distributions as a function of the energy in the $E_{cm} \approx 0.3$-0.45 MeV domain. A pronounced forward asymmetry can be observed; this effect characterizes the whole E_p=0.2-0.5 MeV region and could be attributed either to the interference between adjacent resonances with opposite parities [1, 3, 14] (in fact, in this case, the R-matrix theory predicts non vanishing contributions due to odd terms in the Legendre polynomial expansion of angular distributions [23]) or to the onset of direct processes at very low energies, that could reflect some cluster structure in the target nucleus [24, 25]. The analysis of experimental angular distributions in terms of cosine polynomials [1, 3], discussed in details in Ref. [18], reveals that both processes co-exist. In particular, the overall trend of angular distribution seems to be consistent with the presence of a direct contribution in the E_p=0.3-0.5 MeV region, together with the excitation of a broad 2^+ state in ^{20}Ne (E_{cm}=0.251 MeV, E_x=13.095 MeV, Γ=162 keV [8]).

The ^{19}F(p,α_0)^{16}O cross section $\sigma(E)$ has been obtained by integrating experimental angular distributions over 4π. From the $\sigma(E)$ data we deduced the astrophysical S-factor, that is shown in Figure 2 as black dots. Vertical bars represent statistical errors, while the band represents non-statistical errors, estimated as discussed in Ref. [18]. In Figure 2 we also report the recent data of Ref. [19] as green triangles, obtained in a previous experiment performed at the TTT3 tandem accelerator in Naples at higher energies, and the non-resonant S-factor extrapolation by the NACRE collaboration [15] (red dashed line). Results obtained in the present experiment agree well with the lowermost energy data of Ref. [19]. The analysis of the low energy part of Figure 2 reveals a behaviour of the S-factor that can be attributed to the sum of a direct component and of various resonances of the compound nucleus. For example, we can see a resonant contribution due to the broad 2^+ state at 0.251 MeV (E_x=13.095 MeV, Γ=162 keV), in agreement with the considerations on the angular distributions made before. This broad resonance leads to an increase of the low energy S-factor with respect to the non-resonant case.

At low energy, the S-factor is $\approx$1.5-2 times larger than the corresponding NACRE evaluation based on the non-resonant extrapolation of high energy data. As a consequence, also the reaction rate is well larger than the NACRE predictions (for further details on the reaction rate calculation see Ref. [18]). This larger reaction rate could lead to a more efficient ^{19}F destruction in extra-mixing phenomena. This seems in agreement with recent experimental observations of fluorine abundance in metal-poor AGB stars [11, 26] and could contribute to improve the description of fluorine nucleosyntesis in stars [27].

4. Summary

In these proceedings we briefly discuss new direct experimental measurements of the ^{19}F(p,α_0)^{16}O S-factor at $E_{cm} \approx 0.2$ - 0.6 MeV. More details on the experimental procedures and results can be found in Ref. [18]. The investigated reaction is interesting for (a) the study of the spectroscopy of natural parity states in ^{20}Ne and (b) the astrophysical implication of the ^{19}F(p,α)^{16}O reaction rate on fluorine nucleosynthesis in AGB stars. We point out for the first time the contribute of the low-energy 0.251 MeV resonance in ^{20}Ne to the experimental S-factor, that is well larger than the currently adopted extrapolations reported by NACRE. The presence of low-energy resonant contributions is in agreement with results of recent experiments with the Trojan Horse method. These findings could help to improve our knowledge of ^{19}F nucleosynthesis by reaction rate reassessments [28, 29].

References

[1] Isoya A, Ohmura H and Momota T 1958, *Nucl. Phys.* **7** 116
[2] Isoya A 1958, *Nucl. Phys.* **7** 126
[3] Caracciolo R et al 1974, *Lett. Nuovo Cim.* **11** 33
[4] De Rosa A, Perillo E et al 1978, *Nuovo Cim.* **44** 433
[5] Cuzzocrea P et al 1980, *Lett. Nuovo Cim.* **28** 515
[6] Beck C (ed.) 2010, *Lect. Not. Phys.* 818, Vol. 1
[7] Dell'Aquila D 2016, *Nuovo Cim.* **C 39** 272
[8] Tilley DR et al 1998, *Nucl. Phys.* **A 636** 249
[9] Arima A, Gillet V and Ginocchio J 1970, *Phys. Rev. Lett.* **25** 1043; Middleton R, Garret JD and Fortune HT 1971, *Phys. Rev. Lett.* **27** 950
[10] Rolfs C and Rodney WS 1988, *Cauldrons in the Cosmos*, Chicago Univ. Press
[11] Lucatello S et al 2011, *Astr. Jour.* **729** 40
[12] La Cognata M et al 2011, *Astr. J. Lett.* **739** L54
[13] La Cognata M et al 2015, *Astr. J.* **805** 128
[14] Breuer G 1959, *Zeit. Phys.* **154** 339
[15] Angulo C et al 1999, *Nucl. Phys.* **A 656** 3
[16] Spitaleri C et al 1999, *Phys. Rev.* **C 60** 055802
[17] Spitaleri C et al 2017, *Phys. Rev.* **C 95** 035801
[18] Lombardo I et al 2015, *Phys. Lett.* **B 748** 178
[19] Lombardo I et al 2013, *J. Phys.* **G 40** 125102
[20] Lombardo I et al 2014, *J. Phys. Conf. Ser.* **569** 012068
[21] Lombardo I et al 2014, *Bull. Rus. Acad. Sci.* **78** 1093
[22] Lombardo I et al 2016, *J. Phys. Conf. Ser.* **730** 012016
[23] Kuperus J et al 1965, *Physica* **31** 1603
[24] Herndl H et al 1991, *Phys. Rev.* **C 44** R952
[25] Yamashita Y and Kudo Y 1993, *Prog. Theor. Phys.* **90** 1303
[26] Abia C et al 2011, *Astr. J. Lett.* **737** L8
[27] Cristallo S et al 2014, *Astr. Astr.* **A 46** 570
[28] Indelicato I et al 2017, *Astr. Jour.* **845** 19
[29] He JJ et al 2018, *Chin. Phys.* **C 42** 015001

IOP Conf. Series: Journal of Physics: Conf. Series **940** (2017) 012012 doi:10.1088/1742-6596/940/1/012012

Theoretical cross sections in the ^{12}C(α,γ)^{16}O reaction

M. Katsuma

Advanced Mathematical Institute, Osaka City University, 558-8585 Osaka, Japan

E-mail: mkatsuma@sci.osaka-cu.ac.jp

Abstract. The ^{12}C(α,γ)^{16}O reaction seems to be explained by the simple reaction mechanism of the direct-capture potential model. The theoretical cross section at $E_{c.m.} = 0.3$ MeV is found to be enhanced with the $E2$ transition. This is caused by the tail of the subthreshold 2_1^+ state ($E_x = 6.92$ MeV). The total S-factor is consistent with the previous studies. The photodisintegration of ^{16}O is also found to be dominated by the $E2$ excitation in the vicinity of the α-particle threshold.

1. Introduction

The ^{12}C(α,γ)^{16}O reaction is considered as a key reaction for the carbon-oxygen ratio in the universe, and it has a great influence on the stellar evolution of massive stars and nucleosynthesis of elements [1]. However, because of the Coulomb barrier, the cross section is too small to measure at the centre-of-mass energy $E_{c.m.} = 0.3$ MeV, corresponding to the helium burning temperature. To understand the reaction rates, the experimental efforts have been made with the various methods, e.g. [2, 3, 4, 5], including indirect measurements. The cross section of the inverse reaction is expected to be larger than that of ^{12}C(α,γ)^{16}O. The experimental projects with the high-intensity laser are in progress to reveal the tiny cross section, e.g. [6].

Under the circumstances, I have recently calculated the low-energy cross section of ^{12}C(α,γ)^{16}O by using the direct-capture potential model (PM) [7], and have converted them into the reaction rates [8]. In this presentation, I report the theoretical low-energy extrapolation of the cross section and examine the additional resonant contribution to PM [9]. In addition, I discuss the photodisintegration of ^{16}O for the future experiments by the same model [10].

In the next section, the PM is described, briefly. The low-energy cross section and the reaction rates are discussed in § 3. The photoelectric cross section of ^{16}O(γ,α)^{12}C is also illustrated. The summary is in § 4.

2. Potential model (PM)

The PM describes the fundamental process of radiative capture reactions, and it works in the direct reaction mechanism, where only a few degrees of freedom of motion are activated.

The γ-ray angular distribution for ^{12}C(α,γ_0)^{16}O at low energies is given by $d\sigma/d\Omega_\gamma = \sum_{j=0}^{4} c_j P_j(\cos\theta_\gamma)$, where P_j is the Legendre polynomials. c_j is the coefficient of the expansion, which is calculated from the matrix element of the electromagnetic operator between the initial wave and the final bound state. To generate the initial wave, i.e. the α+^{12}C continuum states, I use the internuclear potential to reproduce the phase shifts of elastic scattering [11] and the α-particle binding energy of 2_1^+ ($E_x = 6.92$ MeV). E_x denotes the excitation energy of ^{16}O. The

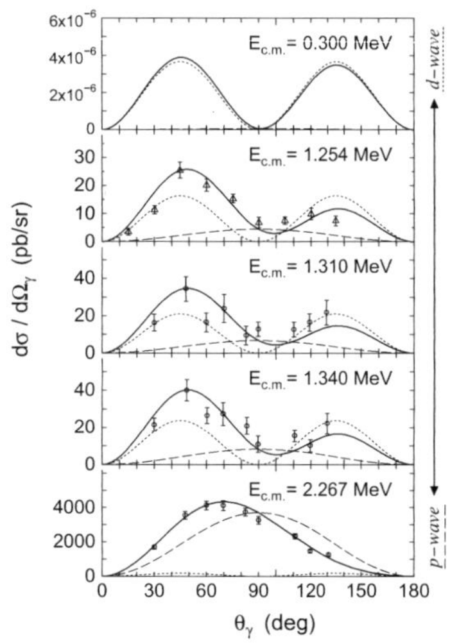

Figure 1. The γ-ray angular distribution of $^{12}C(\alpha,\gamma_0)^{16}O$. The solid curves are the results of PM. The dotted and dashed curves are the pure $E2$ and $E1$ components, respectively. The experimental data are taken from [2, 3].

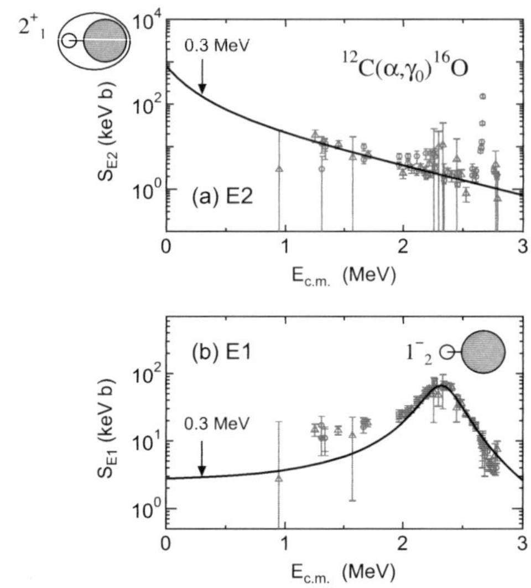

Figure 2. Astrophysical S-factors of the $^{12}C(\alpha,\gamma_0)^{16}O$ reaction: (a) $E2$ and (b) $E1$. The solid curves are the results of PM. The experimental data are taken from [2, 3].

appropriate strength of the potential is determined from the nuclear rainbow at high energies [12]. For the bound state, the separation energy method is used to generate the appropriate wavefunctions in the peripheral region that nuclear reactions are sensitive to [13]. The derived state works as a doorway from the continuum state to the fused nuclei.

The radiative capture cross section is given by $\sigma_{\alpha\gamma} = 4\pi c_0$. To compensate for the rapid variation in $\sigma_{\alpha\gamma}$ below the Coulomb barrier, the astrophysical S-factor is defined in $S_{E\lambda} = E_{c.m.} \exp(2\pi\eta)\sigma_{\alpha\gamma}^{E\lambda}$, where λ denotes the multipolarity of the transition; $\eta = 12e^2/(\hbar v)$; v is the relative velocity. The effective charge in the electromagnetic operator is used as the adjustable parameter to minimise the difference between the theoretical cross sections and the experimental data. The violation of the isospin selection rule is included in the effective charge. The theoretical cross section is converted into the Maxwellian averaged reaction rates [1, 14].

From the statistical equilibrium of $\alpha+{}^{12}C \leftrightarrow \gamma+{}^{16}O$, the cross section of $^{16}O(\gamma,\alpha)^{12}C$ is given as $\sigma_{\gamma\alpha} = [\,k_c^2/(2k_\gamma^2)\,]\sigma_{\alpha\gamma}$, where k_c is the wavenumber of relative motion between the α-particle and ^{12}C; k_γ is the wavenumber of photons $k_\gamma = E_\gamma/(\hbar c)$; $E_\gamma = E_{c.m.} + 7.162$ MeV [10].

3. Low-energy cross sections of $^{12}C(\alpha,\gamma_0)^{16}O$

In this section, I show the calculated results of the cross section from PM. After discussing the additional contribution, I also illustrate the predicted cross section of the inverse reaction.

The calculated γ-ray angular distribution is compared with the experimental data [2, 3] in Fig. 1. The trend of the peak and valley appears to be reproduced by the solid curve obtained from PM. At $E_{c.m.} = 2.267$ MeV, the γ-ray angular distribution has a single peak, that means the p-wave (dashed curve) is predominant. At low energies, the d-wave (dotted curve) seems to become more important because the angular distribution has two peaks. The $E2$ and $E1$ components of the S-factor are shown in Figs. 2(a) and (b) as a function of the incident energy. From Fig. 2(a), the $E2$ transition is found to be enhanced at low energies. This is caused by the subthreshold 2_1^+ state that has the $\alpha+{}^{12}C$ cluster structure [9, 15]. For the $E1$ transition, the 1_2^- molecular resonance is found at $E_{c.m.} \approx 2.4$ MeV ($E_x \approx 9.6$ MeV) in Fig. 2(b).

Table 1. Astrophysical S-factors at $E_{c.m.} = 0.3$ MeV, listed in keV b unit.

	PM [7, 8]	[3]	[4]	[14]
$E1$	3	76 ± 20	79 ± 21	79 ± 21
$E2$	150^{+41}_{-17}	85 ± 30	70 ± 70	120 ± 60
Cascade	$18^{+4.5}_{-4.5}$	4 ± 4	16 ± 16	
Total	171^{+46}_{-22}	165 ± 54	165 ± 107	199 ± 81

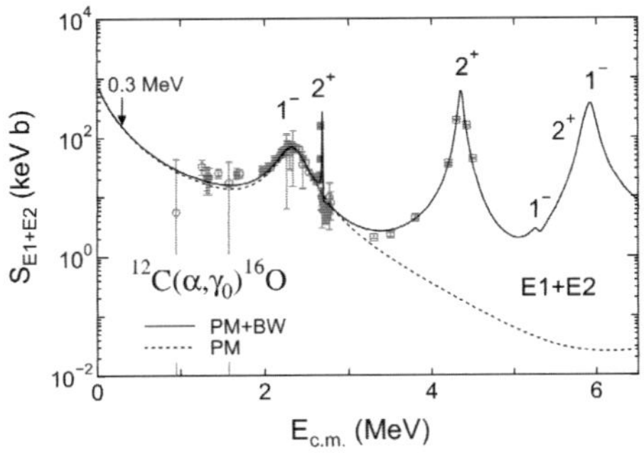

Figure 3. $E1$+$E2$ S-factor of $^{12}C(\alpha,\gamma_0)^{16}O$. The additional resonant contributions are included in the solid curve (PM+BW). The dotted curve is from PM. (Exp.:[2, 3, 5])

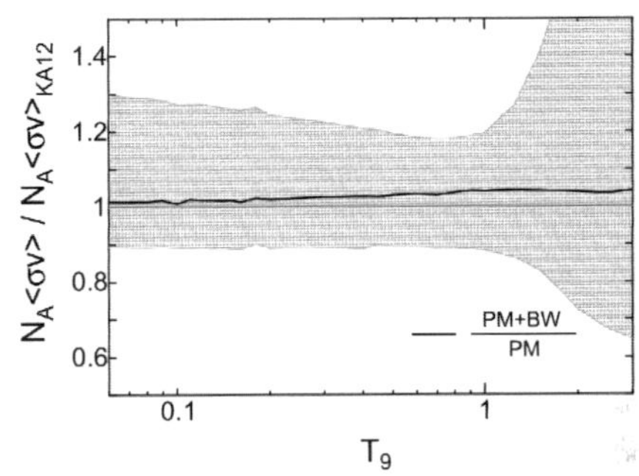

Figure 4. Comparison of the reaction rates between PM and PM+BW. The shade area is the uncertainty of the rates estimated from PM [8]. The result is shown in ratio.

The α+^{12}C cluster states play a crucial role in the discussion about the reaction mechanism of $^{12}C(\alpha,\gamma)^{16}O$. The potential used here reproduces the elastic scattering data for $E_{c.m.} < 5$ MeV [7], and it describes the α+^{12}C rotational bands very well [15]. The γ-ray angular distribution shown in Fig. 1 seems to be made by the interference between 2_1^+ and 1_2^-.

The extrapolated S-factor at $E_{c.m.} = 0.3$ MeV is listed in Table 1. As comparison, the results from [3, 4, 14] are also listed. The total S-factor from PM is consistent with that from [3, 4, 14]. In contrast, the $E1$ and $E2$ components are different. It should be, however, noted that the present result is based on the analysis of the γ-ray angular distribution, that looks good in Fig. 1. The resulting reaction rate resembles the previous results because the total S-factor is comparable. The deduced reaction rate is available in [8].

To examine the contribution from other resonances, the Breit-Wigner (BW) type of the experimental resonances [16] is temporarily appended to PM. Figure 3 shows the sum of the $E1$ and $E2$ S-factors for $^{12}C(\alpha,\gamma_0)^{16}O$. The dotted curve is the result of PM. The solid curve is the result with the experimental resonances (PM+BW). The contribution from the 1_1^- state ($E_x = 7.12$ MeV) is hindered by the isospin selection rule. Under the weak coupling [7], 1_1^- does not have the large α-particle width. The large difference between two curves can be seen above $E_{c.m.} = 3$ MeV. However, the difference between the derived reaction rates is very small, as shown in Fig. 4. To compare the models, the resulting reaction rate is displayed in ratio. From Figs. 3 and 4, I confirm that the additional resonances do not have the important contribution. I, therefore, consider that the $^{12}C(\alpha,\gamma)^{16}O$ reaction is described by PM successfully enough.

Finally, I illustrate the photoelectric cross section of $^{16}O(\gamma,\alpha)^{12}C$ in Fig. 5. The solid curve is the result from PM. The dashed and dotted curves are the $E1$ and $E2$ components, respectively. From Fig. 5, the $E2$ excitation is found to dominate the reaction in the vicinity of the α-particle

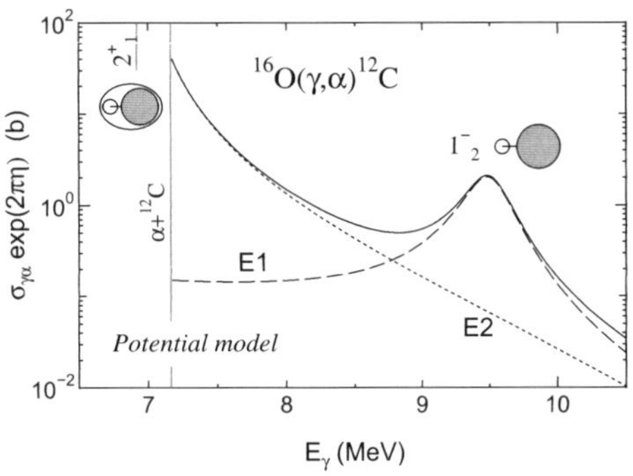

Figure 5. Photoelectric cross sections for ^{16}O(γ,α)^{12}C as a function of the γ-ray energy. The solid curve is the result obtained from PM. The dashed and dotted curves are the $E1$ and $E2$ components, respectively. The cross section is multiplied by $\exp(2\pi\eta)$ to compensate for the rapid drop in the vicinity of the α-particle threshold.

threshold. I also predict that the coupling between the 1^-_1 and 1^-_2 states is weak, and that the angular distribution of the emitted α-particle at $E_\gamma \approx 8.5$ MeV is described by the interference between 2^+_1 and 1^-_2 as well as the counterpart of ^{12}C(α,γ)^{16}O shown in Fig. 1.

4. Summary

I have discussed the theoretical cross section in the ^{12}C(α,γ)^{16}O reaction obtained from PM. The potential used here is concordant with the optical potential reproducing the elastic scattering data over the wide energy region. In addition, I have shown the calculated result of ^{16}O(γ,α)^{12}C.

The ^{12}C(α,γ_0)^{16}O reaction seems to be explained by the direct reaction mechanism of PM. The theoretical cross section at $E_{c.m.} = 0.3$ MeV is found to be enhanced with the $E2$ transition. This is caused by the tail of the subthreshold 2^+_1 state having the α+^{12}C cluster structure. The PM makes the relatively large $E1$ transition from the 1^-_2 molecular resonance at $E_{c.m.} \approx 2.4$ MeV. The γ-ray angular distribution at $E_{c.m.} \approx 1.3$ MeV appears to be made by the interference between 2^+_1 and 1^-_2. The total S-factor at $E_{c.m.} = 0.3$ MeV is consistent with the previous studies. The contribution from other resonances is negligible. The photodisintegration of ^{16}O is also found to be dominated by the $E2$ excitation at the interest energies.

Acknowledgments

I express my gratitude to Professors K. Langanke, P. Descouvemont, and I.J. Thompson for their comments and encouragement. I also thank Y. Kondō for early days of collaboration, and Y. Ohnita and Y. Sakuragi for their hospitality. I am grateful to M. Arnould, A. Jorissen, K. Takahashi and H. Utsunomiya for their encouragement during my stay in Bruxelles.

References

[1] Rolfs C E and Rodney W S 1988 *Cauldrons in the Cosmos* (Chicago IL: Univ. Chicago Press)
[2] Assunção M *et al* 2006 *Phys. Rev. C* **73** 055801
[3] Kunz R *et al* 2001 *Phys. Rev. Lett.* **86** 3244; Kunz R *et al* 2002 *Astrophys. J.* **567** 643
[4] Buchmann L *et al* 1996 *Phys. Rev. C* **54** 393; Buchmann L 1996 *Astrophys. J.* **468** L127
[5] Schürmann D *et al* 2012 *Phys. Lett. B* **711** 35
[6] Matei C *et al* 2015 *This conference talk*; Tesileanu O and Zamfir N V 2015 *PoS(NIC XIII)* 056
[7] Katsuma M 2008 *Phys. Rev. C* **78** 034606; 2010 *Phys. Rev. C* **81** 029804; 2010 *Phys. Rev. C* **81** 067603
[8] Katsuma M 2012 *Astrophys. J.* **745** 192; Katsuma M 2015 *PoS(NIC XIII)* 106
[9] Katsuma M 2014 *Preprint* arXiv:1404.3966 [nucl-th]
[10] Katsuma M 2014 *Phys. Rev. C* **90** 068801
[11] Plaga R *et al* 1987 *Nucl. Phys. A* **465** 291; Tischhauser P *et al* 2009 *Phys. Rev. C* **79** 055803
[12] Brandan M E and Satchler G R 1997 *Phys. Rep.* **285** 143; Ingemarsson A *et al* 1994 *Phys. Rev. C* **49** 1609
[13] Satchler G 1983 *Direct Nuclear Reactions* (New York: Oxford Univ. Press)
[14] Angulo C *et al* 1999 *Nucl. Phys. A* **656** 3
[15] Katsuma M 2013 *J Phys. G* **40** 025107; Katsuma M 2014 *EPJ Web of Conf.* **66** 03041
[16] Tilley D R *et al* 1993 *Nucl. Phys. A* **564** 1

IOP Conf. Series: Journal of Physics: Conf. Series **940** (2017) 012013 doi:10.1088/1742-6596/940/1/012013

Microscopic description of fission properties for r-process nuclei

S A Giuliani[1], G Martínez-Pinedo[1] and L M Robledo[2]

[1] Institut für Kernphysik (Theoriezentrum), Technische Universität Darmstadt, Schlossgartenstraße 2, 64289 Darmstadt, Germany.

[2] Departamento de Física Teórica, Universidad Autónoma de Madrid, E-28049 Madrid, Spain.

E-mail: giuliani@theorie.ikp.physik.tu-darmstadt.de

Abstract. Fission properties of 886 even-even nuclei in the region $84 \leq Z \leq 120$ and $118 \leq Z \leq 250$ were computed using the Barcelona-Catania-Paris-Madrid energy density functional. An extensive study of both the potential energy surfaces and collectives inertias was performed. Spontaneous fission half-lives are computed using the semiclassical Wentzel-Kramers-Brillouin formalism. By comparing these three quantities we found that the stability of the nucleus against the fission process is driven by the interplay between both the potential energy and the collective inertias. In our calculations, nuclei with relative long half-lives were found in two regions around $Z = 120$, $N = 182$ and $Z = 104$, $N = 222$.

1. Introduction and formalism

Fission is a crucial phenomenon to understand r-process nucleosynthesis [1, 2, 3]. Results coming from r-process simulations for matter ejected dynamically in neutron star mergers have been recently published [4, 5, 6]. In this specific astrophysical site the fission process determines how matter is recycled during the neutron irradiation and the production of superheavy nuclei. In ref. [5] was showed that fission fragments distributions have a strong impact in the final r-process distributions, specially in the region around the second r-process peak and the rare-earth sub-peak. On the other hand, the accumulation of material in the fissioning region leads to a robust r-process pattern, as required by metal-poor star observations [6].

These simulations involve nuclei located far from the stability valley, where experimental data is not available. Despite of the great success in the synthesis of heavy nuclei, we are still far from having an exhaustive experimental data set able to provide all the required information for such kind of studies. Due to this, in the last decades a great number of studies were devoted to the description of the fission process within the Energy Density Functional model (EDF). This mean-field approach based in the Hartree-Fock-Bogoliubov (HFB) theory requires for a phenomenological effective interaction fitted to reproduce nuclear matter properties and/or equations of state (see [7] for a recent review and comparison with macroscopic-microscopic models). In this paper, we present the result of the fission properties obtained with the Barcelona-Catania-Paris-Madrid (BCPM) EDF. BCPM is a newly proposed EDF inspired by the Kohn-Sham theory [8]. Its free parameters were adjusted in order to reproduce the binding energy of the 518 even-even nuclei of the Audi and Wapstra 2003 mass table evaluation. In recent papers [9, 10] we explored the fission properties of this interaction, comparing our results with the available experimental data and the predictions obtained from other theoretical models. Following the same procedure described in [9, 10], we extended the calculations to nuclei with protons number $84 \leq Z \leq 120$ and neutrons number $118 \leq N \leq 250$. These calculations are an initial step towards the provision of a global and complete data set covering the major range of interest for the r-process nucleosynthesis.

2. Results

Spontaneous fission lifetimes are obtained using the semiclassical approach given by the WKB formalism. Within this approach, the half-lives (in seconds) can be written as:

$$t_{\rm sf} = 2.86 \times 10^{-21}(1 + \exp(2S)), \tag{1}$$

where S is the action integral computed along the fission path s:

$$S = \int_a^b ds \sqrt{2B(s)\left[(V(s) - E_0)\right]}. \tag{2}$$

and E_0 is the zero-point energy. As it is shown in the expression above, the (spontaneous) fission process is driven by the interplay of two major quantities, namely the collective inertias $B(s)$ and the potential energy $V(s)$. Both quantities must be evaluated along the fission path s, where a and b are the classical turning points fulfilling the condition $V(s) = E_0$. In this work collective inertias were calculated with the Adiabatic Time Dependent HFB (ATDHFB) method [11]. The potential energy $V(s)$ is given by the HFB theory corrected by both zero-point energy and rotational corrections, subtracting the energy of the ground-state [10].

The fission path is determined by minimizing the total energy as a function of the quadrupole moment operator $s = Q_{20}$. This approach, also known as static approximation, is required in order to reduce the computational time and allows the systematic calculation of the fission properties for a large set of nuclei. We should mention here that in recent papers extensive calculations studying the relevance of a multidimensional space with a dynamic determination of the fission path s were performed [12, 13]. However, despite the great interest that these studies deserve, their extension to a systematic calculation of the fission properties is nowadays unfeasible from the computational point of view.

From eq. (2) it is clear that the height of the fission barrier by itself does not provide a complete characterization of the fission process. A more complete picture of the fission properties can be obtained by studying the mean value of the potential energy $\overline{V}(Q_{20})$ and the collective inertias $\overline{B}(Q_{20})$. These quantities are obtained by averaging arithmetically the magnitude of $V(Q_{20})$ and $B(Q_{20})$ along the fission path, between the turning points a and b:

$$\overline{\mathcal{O}}(Q_{20}) \equiv \frac{1}{n} \sum_{i=a}^{b} \mathcal{O}(i), \tag{3}$$

being n the number of Q_{20} points along the fission path.

In Fig.1 the average values $\overline{V}(Q_{20})$ and $\overline{B}(Q_{20})$, depicted as a function of the protons and neutrons number, are compared with the spontaneous fission lifetimes $t_{\rm sf}$. Comparing these three panels we can conclude that for nuclei with $1.7 \leq \overline{V} \leq 4.5$ MeV collective inertias determine whether the nucleus will be stable, or not, against the spontaneous fission process. A clear example is given by the results obtained in the region around $Z = 104$ and $N = 220$. Nuclei close to this region show a similar value of $\overline{V} \sim 3.0$ MeV, while the lifetimes span a range of 15 orders of magnitude. The main reason of such a large variation in the lifetimes is the strong change in the inertias values $\overline{B}$ between neighbor nuclei. Of course, the sensitivity with the collective inertias is reduced in regions where the $\overline{V}$ is extremely low.

IOP Conf. Series: Journal of Physics: Conf. Series **940** (2017) 012013 doi:10.1088/1742-6596/940/1/012013

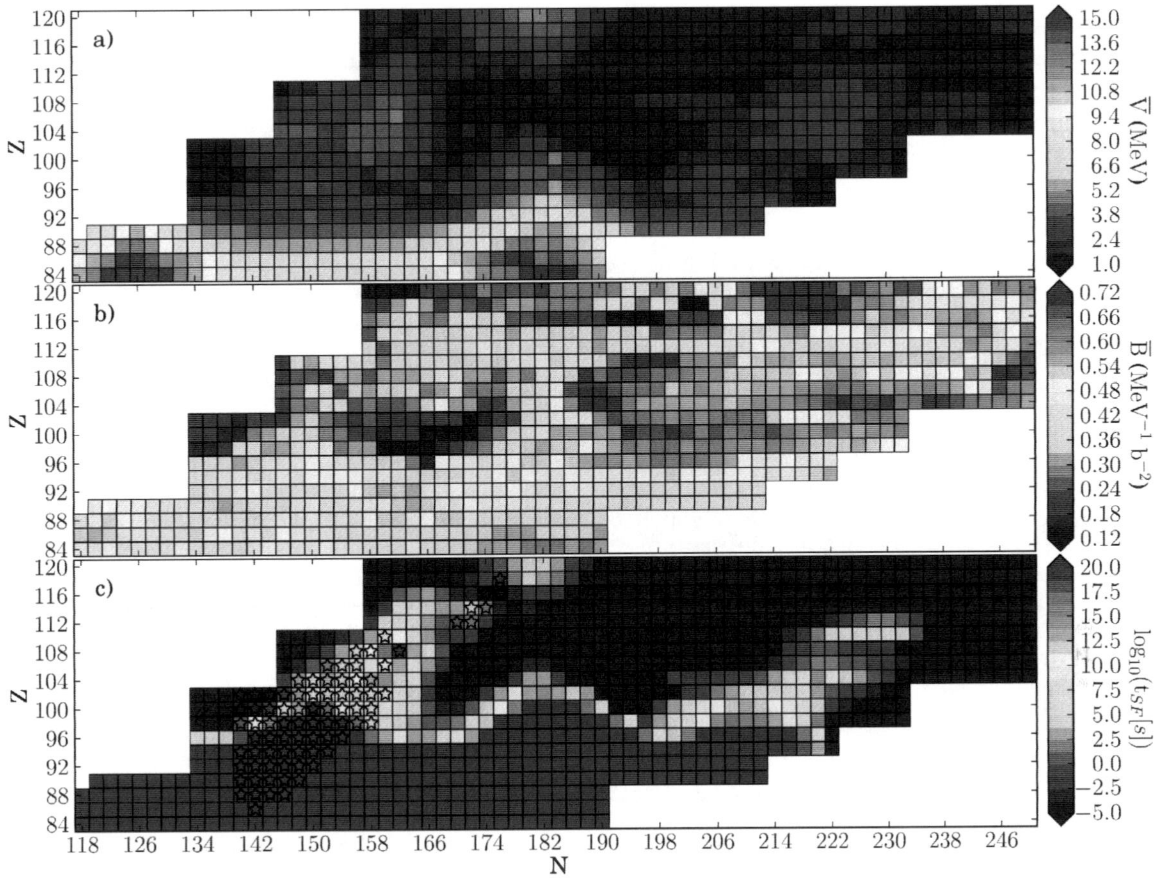

Figure 1. Scatter plot of the mean potential energies $\overline{V}(Q_{20})$ (panel a), mean collective inertias $\overline{B}(Q_{20})$ (panel b) and spontaneous fission lifetimes $t_{\rm sf}$ (panel c) as a function of proton (Z) and neutron number (N). Stars in panel c) show nuclei that have been experimentally observed.

While it is not possible to extract a comprehensive information by just analyzing the fission barrier height, the quantity $\overline{V}$ seems to provide a more general description of the fission properties of single nuclei. For example, comparing the values of $\overline{V}$ of panel a) with the half-lives of panel c), it is possible to conclude that almost all the nuclei with an average fission barrier below 1 MeV have fission lifetimes below 10^{-5} s. It is worth to mention that some of these nuclei show a fission barrier height well above this average value of 1 MeV, leading to a possible overestimation of the hindrance of the nucleus against the fission process. Finally, our calculations predict nuclei with relative long half-lives in the region around the protons number $Z = 120$ and the neutrons magic number $N = 120$. It should be noticed that this region is close to nuclei that were experimentally observed.

3. Conclusions

Using the BCPM EDF, we performed an extensive calculation of the potential energy surface, collective inertias and spontaneous fission lifetimes of 886 even-even nuclei in the region $84 \leq Z \leq 120$ and $118 \leq N \leq 250$. By comparing the half lives with the average value of $V(Q_{20})$ and $B(Q_{20})$ along the fission path, we found that the stability against the fission process is driven by the interplay of these two quantities. This result go against the common description, were the fission characteristics are determined only by the height of the fission barriers. The strong sensitivity of the fission lifetimes with the collective inertias point out the extreme importance of a proper calculation of $B(s)$. In order to obtain a more realistic description of fission dynamics, both studies using a non perturbative calculation of the collective masses as well as a theory beyond the actual cranking approximation are required. Work in this direction is already in progress.

Acknowledgments

The work of SAG and GMP was supported by the Helmholtz Association through the Nuclear Astrophysics Virtual Institute (VH–VI–417) and the BMBF–Verbundforschungsprojekt number 06DA7047I. The work of LMR was supported by the Spanish MICINN Grants No. FPA2012–4694 and No. FIS2012–34479 and by the Consolider–Ingenio 2010 MULTIDARK CSD2009-00064.

References

[1] Martinez-Pinedo G, Mocelj D, Zinner N T, Kelić A, Langanke K, Panov I V, Pfeiffer B, Rauscher T, Schmidt K H and Thielemann F K 2007 **59** 199–205
[2] Petermann I, Langanke K, Martinez-Pinedo G, Panov I V, Reinhard P G and Thielemann F K 2012 *Eur. Phys. J. A* **48** 122
[3] Panov I V, Korneev I Y and Thielemann F K 2009 *Phys. At. Nucl.* **72** 1026–1033
[4] Goriely S 2015 *Eur. Phys. J. A* **51**
[5] Eichler M, Arcones A, Kelic A, Korobkin O, Langanke K, Marketin T, Martinez-Pinedo G, Panov I, Rauscher T, Rosswog S, Winteler C, Zinner N T and Thielemann F K 2015 *Astrophys. J.* **808** 30
[6] Mendoza-Temis J J, Martinez-Pinedo G, Langanke K, Bauswein A and Janka H T 2014 1–12 (*Preprint* 1409.6135)
[7] Baran A, Kowal M, Reinhard P G, Robledo L M, Staszczak A and Warda M 2015 *Nucl. Phys. A* **1503.01608** 1–28 (*Preprint* 1503.01608)
[8] Baldo M, Robledo L M, Schuck P and Viñas X 2013 *Phys. Rev. C* **87** 1–23 (*Preprint* 1210.1321)
[9] Giuliani S A, Martinez-Pinedo G and Robledo L M 2015 Fission properties of neutron rich Uranium isotopes *Proc. Sci. (NIC XIII)* vol 095
[10] Giuliani S A and Robledo L M 2013 *Phys. Rev. C* **88** 54325 (*Preprint* 1305.0293)
[11] Ring P and Schuck P 1980 *The Nuclear Many-Body Problem*
[12] Giuliani S A, Robledo L M and Rodríguez-Guzmán R 2014 *Phys. Rev. C* **90** 054311 (*Preprint* 1312.7229)
[13] Sadhukhan J, Dobaczewski J, Nazarewicz W, Sheikh J A and Baran A 2014 *Phys. Rev. C* **90** 1–5

IOP Conf. Series: Journal of Physics: Conf. Series **940** (2017) 012014 doi:10.1088/1742-6596/940/1/012014

Beyond BCS pairing in high-density neutron matter

A. Rios[1], D. Ding[2], H. Dussan[2], W. H. Dickhoff[2], S. J. Witte[2,3] and A. Polls[4]

[1]Department of Physics, Faculty of Engineering and Physical Sciences, University of Surrey, Guildford, Surrey GU2 7XH, United Kingdom

[2]Department of Physics, Washington University, St. Louis, Missouri 63130, USA

[3]Department of Physics and Astronomy, University of California, Los Angeles, CA 90095, USA

[4]Departament d'Estructura i Constituents de la Matèria and Institut de Ciènces del Cosmos, Universitat de Barcelona, Avinguda Diagonal 647, E-8028 Barcelona, Spain

E-mail: a.rios@surrey.ac.uk

Abstract. Pairing gaps in neutron matter need to be computed in a wide range of densities to address open questions in neutron star phenomenology. Traditionally, the Bardeen-Cooper-Schrieffer approach has been used to compute gaps from bare nucleon-nucleon interactions. Here, we incorporate the influence of short- and long-range correlations into pairing properties. Short-range correlations are treated including the appropriate fragmentation of single-particle states, and they suppress the gaps substantially. Long-range correlations dress the pairing interaction via density and spin modes, and provide a relatively small correction. We use three different interactions as a starting point to control for any systematic effects. Results are relevant for neutron-star cooling scenarios, in particular in view of the recent observational data on Cassiopeia A.

Nucleonic superfluids play an important role in a variety of nuclear physics aspects, from nuclear structure to collective dynamics [1, 2]. Pairing phenomena affect and, at times, substantially alter the properties of nuclear systems. Modern models of neutron stars assume the existence of a neutron superfluid, extending from the core to the crust, as well as proton superconductors [3, 4]. In terms of observations, the theory of pulsar glitches requires a substantial part of the star to be in a superfluid state [5]. Neutrino cooling is also sensitive to the superfluid content of the neutron star [6]. Continuous observations of the cooling curve in Cassiopeia A place more and more stringent constraints on cooling mechanisms [7, 8]. In some models, observations of very rapid cooling can be explained by the onset of the pair formation process in the triplet channel [9].

All in all, neutron star astrophysics would benefit from reliable theoretical calculations of neutron matter pairing. Traditionally, these have been computed within a lowest-order Bardeen-Cooper-Schrieffer (BCS) framework [1]. This well understood methodology works best for uncorrelated systems, and one can put into question its applicability in nuclear systems [2]. In particular, BCS predicts large, of order ≈ 10 MeV, gaps in the isospin $T = 0$ singlet channel, in stark contrast to experimental evidence, which shows a lack of strong nuclear np pairing correlations [10, 11, 12]. An explanation of this puzzle requires an extension of the pairing problem beyond the simplest BCS approach, and should incorporate physically motivated correlation effects.

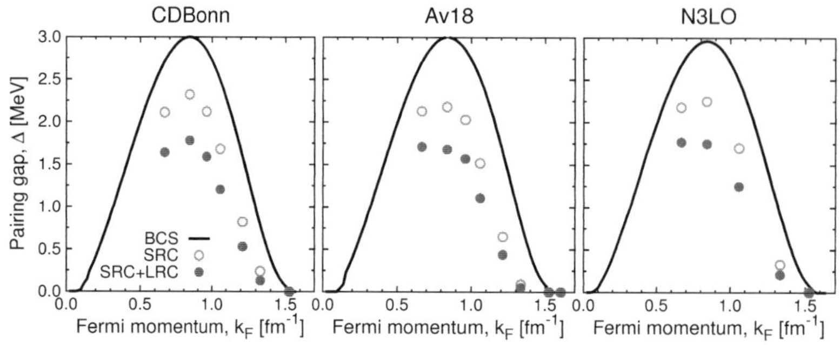

Figure 1. Pairing gaps at the Fermi surface as a function of Fermi momentum for three different NN interactions in the $^1S_0$ channel. Results for three different approximations are presented: BCS (solid lines), beyond BCS with short-range correlations (empty circles) and beyond BCS with both short- and long-range correlations included (solid circles).

On the one hand, the traditional BCS approach does not account for the well-known fragmentation of single-particle (sp) strength firmly established by electron scattering experiments [13, 14]. About a third of the approximately 35 % missing sp strength is due to short-range correlations (SRC), which deplete the nucleon Fermi sea and populate high-momentum components [15, 16]. SRC are ubiquitous in nuclei, and present universal features which are now well understood at the theoretical level [14, 17]. On the other hand, long-range correlations (LRC) in nuclei are responsible for the remaining 20 % of fragmentation. LRC in finite nuclei and infinite systems could be substantially different [18]. At the LRC level, however, nuclear matter can be safely treated as a Fermi liquid and Landau theory can be applied to determine the qualitative features of the induced interaction [19, 20]. In particular, one expects substantial corrections at the level of the effective pairing interaction, which does not necessarily resemble the free space nucleon-nucleon (NN) force [21].

The extension of Green's functions techniques to superfluid systems provides a good starting point to incorporate SRC into pairing calculations [11, 2]. The generalised gap equation,

$$\Delta_L(k) = -\sum_{L'} \int_0^\infty \frac{dk'\, k'^2}{\pi} \frac{\langle k|V_{LL'}|k'\rangle}{E(k')} \Delta_{L'}(k') \tag{1}$$

is used to determine the pairing gap, $\Delta_L(k)$, for a given partial wave channel, L. The squared energy denominator, $E^2(k) = \chi^2(k) + \bar{\Delta}^2(k)$, is the sum of an effective sp energy, $\chi(k)$, and a gap average. For the latter, we consider a sum over coupled channels, but assume that different partial-wave do not overlap for a given momentum. In the simplest BCS approximation, the effective sp energy corresponds to a kinetic spectrum, $\chi(k) = \frac{k^2}{2m} - \mu$, with μ the chemical potential of the system. When SRC are considered for the calculation of the superfluid Green's function, in contrast, $\chi(k)$ becomes a double energy convolution of spectral functions,

$$\frac{1}{2\chi(k)} = \int_{-\infty}^{\infty} \frac{d\omega}{2\pi} \frac{d\omega'}{2\pi} \frac{1 - f(\omega) - f(\omega')}{\omega + \omega'} A(k,\omega) A(k,\omega') \,. \tag{2}$$

The normal spectral function, $A(k,\omega)$, encodes all information related to sp fragmentation. For a given momentum, k, it can be interpreted as a distribution of strength over energies. We compute these at different densities within a ladder self-consistent Green's functions approach at finite temperature [15, 16], hence the Fermi-Dirac distribution, $f(\omega)$ in the previous

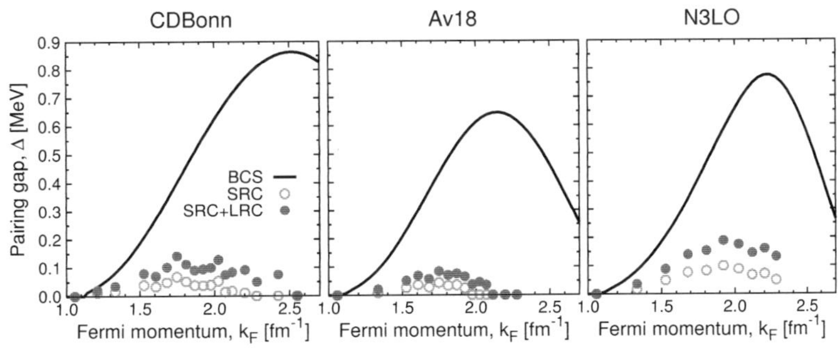

Figure 2. The same as Fig. 1 for the coupled $^3PF_2$ channel.

equation. The pairing results presented below are obtained by extrapolating finite temperature spectral functions to zero temperature. The extrapolation procedure is constrained by imposing thermodynamical consistency [22]. We compute an effective sp energy from Eq. (2), and use it to find the gap via Eq. (1).

LRC primarily influence the pairing effective interaction, due to the exchange of collective spin and density modes. Following Refs. [19, 20], we dress the interaction in our BCS equation, $V_{LL'}$, so that it accounts for fluctuations in a physically motivated way. On the one hand, we consider excitations that couple to the dressed nucleons by means of medium-modified interactions (G-matrices). On the other, the excitations are aware of the Fermi liquid surrounding them and hence experience a screening effect, which we parametrize by means of standard Landau parameters [22].

Figure 1 provides the pairing gaps in the $^1S_0$ channel in neutron matter for three different approximations. The gaps have been computed at the corresponding Fermi surfaces, and are shown as a function of Fermi momentum. Each panel represents a different NN interaction: CD-Bonn [23], Argonne v18 [24] and the Entem-Machledit N3LO potential [25]. The BCS results (solid line) are very similar, which confirms that phase-shift equivalence is enough to fix the value of the gap in this channel [1]. The BCS gap peaks at about 3 MeV around $k_F = 0.7$-0.8 fm^{-1}, and closes at $k_F \approx 1.5$ fm^{-1}. Including SRC within the Green's function formalism outlined above (empty circles), we find that the overall gap is reduced, with a maximum that now sits just below 2 MeV, and a similar density dependence. Screening effects in this channel are repulsive, and as a consequence the corresponding LRC+SRC gaps (solid circles) decrease in size by about 25%. The results for three different NN interactions are robust, and provide useful constraints on their model dependence.

Whereas singlet pairing is active at relatively low densities and affects the dynamics of both the core and the crust, triplet pairing, concerning the coupled $^3PF_2$ channel, takes place within the neutron star core, at Fermi momenta $k_F > 1$ fm^{-1} [8, 9]. Higher Fermi momenta imply that higher relative momenta are explored in the bare (or the effective) interaction. Because phase-shift equivalent interactions are constrained only at low energies and relative momenta, it is not surprising that the corresponding gaps show a larger dependence on the NN force at the BCS level. All gaps open around $k_F \approx 1$ fm^{-1}, and they peak around 2-2.5 fm^{-1}. CDBonn provides the largest maximum ($\Delta_{\text{max}} \approx 0.86$ MeV), followed by N3LO ($\Delta_{\text{max}} \approx 0.77$ MeV) and Av18 ($\Delta_{\text{max}} \approx 0.64$ MeV). In turn, the gap closure happens at larger densities for interactions with larger gaps. Note also that N3LO results are sensitive to the cut-off regularization of the NN force at large Fermi momenta [12].

When SRC are considered, all $^3PF_2$ gaps become strongly suppressed, with maxima below 0.1 MeV: 0.04 MeV for Av18), 0.07 for CDBonn and about 0.1 MeV for N3LO. The gap closure also happens substantially earlier, well below 2.5 fm^{-1}. LRC do not qualitatively change the picture. In contrast to the singlet case, however, spin-density fluctuations lead to a more attractive pairing interaction, and hence LRC increase the pairing gap. The maximum gap that is produced, however, is of the order of 0.2 MeV (N3LO). All in all, these gaps are in general smaller than those found in the literature [8].

We have proposed a method that combines the self-consistent Green's functions framework for the treatment of SRC and Fermi liquid theory for the incorporation of LRC to the neutron matter pairing problem with NN interactions. Work on incorporating three-nucleon forces is on-going [26]. The extension of the approach to asymmetric nuclear systems, with the small admixture of protons that is relevant for neutron star matter, will provide the full set of pairing gaps needed for realistic cooling calculations. More importantly, the triplet pairing gaps that we have generated fall below the parametrizations used so far in astrophysical situations [8], and will substantially affect the corresponding cooling scenarios.

Acknowledgments

This material is based upon work supported by the U.S. National Science Foundation under grant PHY-1304242, Consolider Ingenio 2010 Programme CPAN CSD2007-00042, Grant No. FIS2011-24154 from MICINN (Spain), and Grant No. 2014SGR-401 from Generalitat de Catalunya (Spain); by STFC, through Grants No. ST/I005528/1, ST/L005816/1 and ST/J000051; and by NewCompstar, COST Action MP1304.

References

[1] Dean D J and Hjorth-Jensen M 2003 *Rev. Mod. Phys.* **75** 607–656
[2] Dickhoff W H and Van Neck D 2008 *Many-Body Theory Exposed!, 2nd edition* (New Jersey: World Scientific)
[3] Shapiro S and Teukolsky S 2004 *Black Holes, White Dwarfs and Neutron Stars: The Physics of Compact Objects* (Wiley)
[4] Haensel P, Potekhin A and Yakovlev D 2007 *Neutron Stars 1: Equation of State and Structure* Astrophysics and Space Science Library (Springer New York)
[5] Haskell B and Melatos A 2015 *Int. J. Mod. Phys. D* **24** 1530008
[6] Yakovlev D G and Pethick C J 2004 *Annu. Rev. Astron. Astrophys.* **42** 169
[7] Elshamouty K G, Heinke C O, Sivakoff G R, Ho W C G, Shternin P S, Yakovlev D G, Patnaude D J and David L 2013 *Astrophys. J.* **777** 22
[8] Ho W C G, Elshamouty K G, Heinke C O and Potekhin A Y 2015 *Phys. Rev. C* **91** 015806
[9] Page D, Prakash M, Lattimer J and Steiner A W 2011 *Phys. Rev. Lett.* **106** 081101
[10] Baldo M, Bombaci I and Lombardo U 1992 *Phys. Lett. B* **283** 8–12
[11] Müther H and Dickhoff W H 2005 *Phys. Rev. C* **72** 054313
[12] Maurizio S, Holt J W and Finelli P 2014 *Phys. Rev. C* **90** 044003
[13] Rohe D *et al.* 2004 *Phys. Rev. Lett.* **93** 182501
[14] Arrington J, Higinbotham D, Rosner G and Sargsian M 2012 *Prog. Part. Nucl. Phys.* **67** 898
[15] Rios A, , Dickhoff W H and Polls A 2009 *Phys. Rev. C* **79** 064308
[16] Rios A, Polls A and Dickhoff W H 2014 *Phys. Rev. C* **89** 044303
[17] Hen O *et al.* 2014 *Science* **346** 614–7
[18] Dewulf Y, Dickhoff W H, Van Neck D, Stoddard E and Waroquier M 2003 *Phys. Rev. Lett.* **90** 152501
[19] Shen C, Lombardo U and Schuck P 2005 *Phys. Rev. C* **71** 054301 ISSN 0556-2813 URL `http://link.aps.org/doi/10.1103/PhysRevC.71.054301`
[20] Dong J M, Lombardo U and Zuo W 2013 *Phys. Rev. C* **87** 062801 ISSN 0556-2813
[21] Migdal A B 1967 *Theory of Finite Fermi Systems* (Interscience, New York)
[22] Ding D, Rios A, Dickhoff W H, Dussan H, Polls A and Witte S J 2015 Pairing in high-density neutron matter including short- and long-range correlations arXiv:1502.05673 URL `http://arxiv.org/abs/1502.05673`
[23] Machleidt R, Sammarruca F and Song Y 1996 *Phys. Rev. C* **53** R1483
[24] Wiringa R B, Stoks V G J and Schiavilla R 1995 *Phys. Rev. C* **51** 38
[25] Entem D R and Machleidt R 2003 *Phys. Rev. C* **68** 041001
[26] Carbone A, Rios A and Polls A 2014 *Phys. Rev. C* **90** 054322

IOP Conf. Series: Journal of Physics: Conf. Series **940** (2017) 012015 doi:10.1088/1742-6596/940/1/012015

α-induced reaction cross sections in the mass range A $\approx$ 20 – 50: a critical review

Peter Mohr

Diakonie-Klinikum, Diakoniestraße 10, D-74523 Schwäbisch Hall, Germany
Institute for Nuclear Research ATOMKI, P.O. Box 51, H-4001 Debrecen, Hungary

E-mail: WidmaierMohr@t-online.de; mohr@atomki.mta.hu

Abstract. In a recent review it was shown that the cross sections of α-induced reactions in the $A \approx 20-50$ mass range follow a general and smooth trend in most cases. For comparison of cross sections of different targets at various energies the method of reduced cross sections $\sigma_{\rm red}$ and reduced energies $E_{\rm red}$ was used. Four outliers were identified: ^{36}Ar and ^{40}Ar with unusal small cross sections and ^{23}Na and ^{33}S with unusual huge cross sections. New data for ^{23}Na were presented at this NPA-7 conference; contrary to the previous data, these new data fit into the general systematics. In addition, a relation between the most effective energy E_0 for astrophysical reaction rates (the so-called Gamow window) and the reduced energy $E_{\rm red}$ is presented.

1. Introduction

In a recent review [1] the cross sections of α-induced reactions for target nuclei in the mass range $A \approx 20-50$ were studied. For comparison of cross sections of different targets at various energies the method of reduced cross sections $\sigma_{\rm red}$ and reduced energies $E_{\rm red}$ [2] was used:

$$E_{\rm red} = \frac{(A_P^{1/3} + A_T^{1/3}) E_{\rm c.m.}}{Z_P Z_T} \quad (1)$$

$$\sigma_{\rm red} = \frac{\sigma_{\rm reac}}{(A_P^{1/3} + A_T^{1/3})^2} \quad (2)$$

The reduced energy $E_{\rm red}$ takes into account the different heights of the Coulomb barrier in the systems under consideration, whereas the reduced reaction cross section $\sigma_{\rm red}$ scales the measured total reaction cross section $\sigma_{\rm reac}$ according to the geometrical size of the projectile-plus-target system. (All energies are given as $E_{\rm c.m.}$ except explicitly noted.) In the $A \approx 20-50$ mass range under study the total reaction cross section $\sigma_{\rm reac}$ can often be taken from the dominating (α,n) or (α,p) reaction channel. It was found in [1] that most of the experimental cross sections show a general and smooth trend for $\sigma_{\rm red}$ *vs.* $E_{\rm red}$. The two outliers ^{36}Ar and ^{40}Ar show significantly smaller cross sections; however, this finding is based only on one very old experiment [3]. Surprisingly, two recent experiments on ^{23}Na [4] and ^{33}S [5] show unexpected huge cross sections. In addition, the data for ^{23}Na [4] show a steeper energy dependence, and the data for ^{33}S show a flatter energy dependence compared to most of the other nuclei in this mass region.

The present study presents two extensions to the recent review [1]. First, a relation between the reduced energy $E_{\rm red}$ and the most effective energy E_0 for astrophysical reaction rates (the so-called Gamow window) is derived. Second, the review [1] is updated with two data sets for the previous outlier ^{23}Na which were presented at this NPA-7 conference [6, 7] and have been published now [8, 9]. These new data for ^{23}Na show that ^{23}Na should not be considered as outlier any more. Some further information on the other outlier ^{33}S was already given in [10].

2. Gamow window and reduced energy $E_{\rm red}$

The astrophysical reaction rate $N_A\langle\sigma v\rangle$ is defined by the Maxwellian-averaged cross section $\langle\sigma v\rangle$

$$\langle\sigma v\rangle = \int_0^\infty \Phi(v)\,\sigma(v)\,v\,dv = \left(\frac{8}{\pi\mu}\right)^{1/2} \frac{1}{(kT)^{3/2}} \int_0^\infty E\,\sigma(E)\,\exp\left(\frac{-E}{kT}\right) dE \tag{3}$$

For an energy-independent astrophysical S-factor the integrand in Eq. (3) shows a maximum at the energy E_0 which is given by

$$E_0 \approx 0.122\,\mathrm{MeV}\left(Z_P^2 Z_T^2 A_{\rm red} T_9^2\right)^{1/3} \tag{4}$$

with the charge numbers Z_P and Z_T of projectile and target, the reduced mass number $A_{\rm red} = \frac{A_P A_T}{AP+A_T}$, and the plasma temperature T_9 in 10^9 Kelvin. Although it has been shown that the underlying simple approximation $S(E) \approx$ const. does not hold exactly [11], the energy E_0 is still a reasonable estimate for the most effective energy, and the energy window around E_0 is usually called Gamow window.

According to Eq. (1), the corresponding reduced energy $E_{\rm red,0}$ is given by

$$E_{\rm red,0} = \frac{(A_P^{1/3} + A_T^{1/3})}{Z_P Z_T} E_0 \approx 0.122\,\mathrm{MeV} \times T_9^{2/3} \times f(Z_P, Z_T, A_P, A_T) \tag{5}$$

where the function $f(Z_P, Z_T, A_P, A_T)$ is given by

$$f(Z_P, Z_T, A_P, A_T) = \left(\frac{A_P A_T}{Z_P Z_T}\right)^{1/3} \times \frac{1 + (A_P/A_T)^{1/3}}{[1 + (A_P/A_T)]^{1/3}} \tag{6}$$

From the exponents of 1/3 in Eq. (6) already a relatively smooth dependence of $E_{\rm red,0}$ on the mass and charge numbers of projectile and target can be expected. A closer inspection of Eq. (6) shows that there is a further compensation between the first and second factor in Eq. (6) which leads to a reduced Gamow energy $E_{\rm red,0}$ which is almost independent of target charge and mass for α-induced reactions ($Z_P = 2$, $A_P = 4$). The first factor increases from light nuclei (with $N_T \approx Z_T$ and thus $A_T/Z_T \approx 2$) towards heavier nuclei (with $N_T \approx 1.5\,Z_T$ or $A_T/Z_T \approx 2.5$). The second factor decreases with increasing target mass number A_T. Some numerical examples for $f(Z_P, Z_T, A_P, A_T)$ are $f(^{20}\mathrm{Ne}) = 2.367$, $f(^{51}\mathrm{V}) = 2.288$, and $f(^{208}\mathrm{Pb}) = 2.165$. In the mass range under study f varies only by a few per cent, leading to

$$E_{\rm red,0} \approx 0.284\,\mathrm{MeV} \times T_9^{2/3} \tag{7}$$

for the mass range under study. Explicitly, the numbers are $E_{\rm red,0} \approx 0.284\,\mathrm{MeV}$ for $T_9 = 1$, $E_{\rm red,0} \approx 0.451\,\mathrm{MeV}$ for $T_9 = 2$, and $E_{\rm red,0} \approx 0.591\,\mathrm{MeV}$ for $T_9 = 3$.

3. Update for ^{23}Na + α

The total reaction cross section of α + ^{23}Na is dominated by the ^{23}Na(α,p)^{26}Mg reaction at low energies. The ^{23}Na(α,n)^{26}Al reaction has a negative Q-value of about -3 MeV.

Two new experimental data sets for the ^{23}Na(α,p)^{26}Mg reaction have been presented at this NPA-7 conference [6, 8, 7, 9]. The results of both experiments agree with each other, but are in disagreement with the previous result of [4].

At the Aarhus accelerator forward kinematics was applied [6, 8]. The target composition of the NaCl target was carefully monitored using Rutherford scattering on ^{23}Na and $^{\rm nat}$Cl. Because of the relatively low beam current, no significant deterioration of the target was observed. Angular distributions for the p_0 and p_1 protons were measured over a broad angular range. Thus, the derived absolute cross sections of the ^{23}Na(α,p)^{26}Mg reaction should be reliable within the given uncertainties in [8] which are of the order of 15 – 25 per cent.

At the ISAC facility at TRIUMF a helium-filled gas cell was mounted in the TUDA scattering chamber for a measurement in inverse kinematics [7, 9]. The number of target nuclei was controlled via the gas pressure, and the number of beam particles was measured by Rutherford backscattering on the gas cell entrance window and an additionally mounted thin gold foil. Also this procedure should lead to a reliable absolute normalization of these experimental data. Absolute uncertainties of slightly below 20 % at higher energies and about 30 – 40 % at the lowest energies are reported in [9].

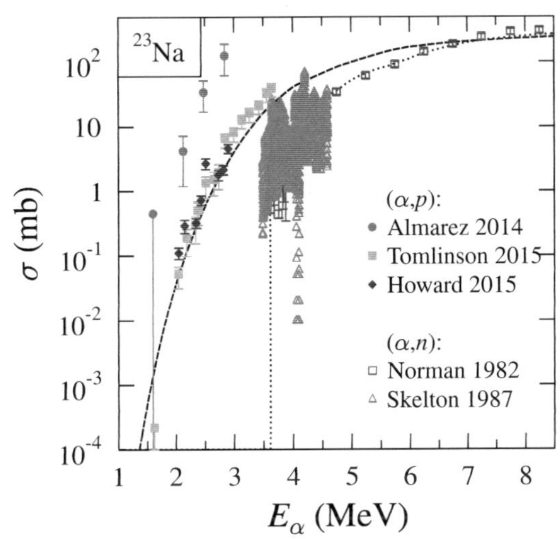

Figure 1. Cross sections σ of the ^{23}Na(α,p)^{26}Mg and ^{23}Na(α,n)^{26}Al reactions *vs.* laboratory energy $E_{\rm lab}$. The experimental data have been taken from [4, 8, 9, 14, 15]. The new data by Howard *et al.* [8] and Tomlinson *et al.* [9] are significantly lower than the earlier data by Almarez-Calderon *et al.* [4]. The calculations are taken from [1]. Further discussion see text.

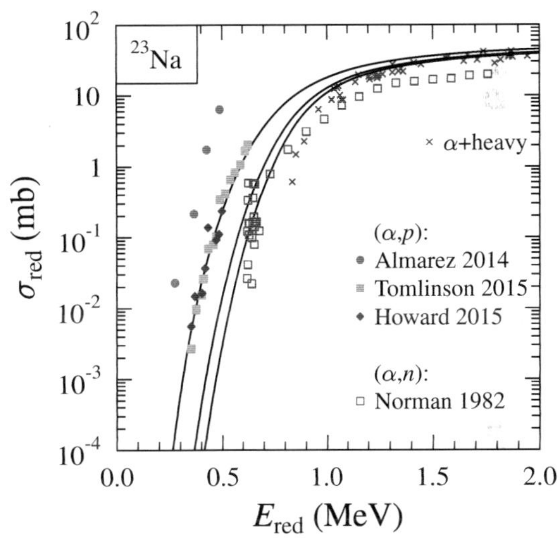

Figure 2. Reduced cross section $\sigma_{\rm red}$ *vs.* reduced energy $E_{\rm red}$ for the ^{23}Na(α,p)^{26}Mg reaction. The reduced cross sections $\sigma_{\rm red}$ slightly increase towards lower target masses; the expected range is indicated by three calculations for ^{21}Ne, ^{36}Ar, and ^{51}V (taken from [1]; for details see there). The new data [8, 9] fit into the general systematics whereas the earlier data [4] are much higher.

In both experiments the target can be considered as relatively thick ($\approx 80 - 100$ keV in the Aarhus experiment and ≈ 140 keV in the TRIUMF experiment in the c.m. system). Thus, the data show a relatively smooth energy dependence because of the averaging of the cross section in a broad energy interval. The high data point of the Aarhus experiment at $E_{\rm lab} \approx 2.5$ MeV corresponds to a strong resonance which has been seen also in earlier experiments [12, 13]. The new data are shown together with previous data for the ^{23}Na(α,p)^{26}Mg [4] and ^{23}Na(α,n)^{26}Al reactions [14, 15] in Fig. 1 as cross sections *vs.* $E_{\rm lab}$ and in Fig. 2 as $\sigma_{\rm red}$ *vs.* $E_{\rm red}$. These figures are updates of the corresponding Figs. 56 and 57 of the review [1].

It is obvious from Fig. 1 that the new data from Howard *et al.* [8] and Tomlinson *et al.* [9] are in good agreement with each other and with the prediction from the statistical model (taken from [1]). Contrary, the previous data by Almarez-Calderon *et al.* [4] are more than one order of magnitude higher. Fig. 2 shows the same data as reduced cross sections $\sigma_{\rm red}$ vs. reduced energies $E_{\rm red}$. Here it has been found in [1] that there is a smooth trend for increased $\sigma_{\rm red}$ towards lower target masses. The typical range of $\sigma_{\rm red}$ for the $A \approx 20 - 50$ mass range is indicated by three calculations for the total reaction cross sections of α + ^{21}Ne, α + ^{36}Ar, and α + ^{51}V (full lines from left to right in Fig. 2; for details see [1]). The new data of [8, 9] are located very close to the expectation from systematics.

4. Summary and Conclusions

It has been shown in [1] that α-induced reaction cross sections for targets in the $A \approx 20 - 50$ mass range show a relatively smooth and systematic behavior. This behavior can be nicely visualized using so-called reduced cross sections $\sigma_{\rm red}$ and reduced energies $E_{\rm red}$. The present work provides a simple relation between the reduced energy $E_{\rm red}$ and the most effective energy E_0 for astrophysical reaction rates (the so-called Gamow window).

Four exceptions from the general systematics of α-induced reaction cross sections have been identified in [1]: ^{36}Ar and ^{40}Ar with smaller cross sections [3], ^{23}Na [4] and ^{33}S [5] with larger cross sections. New data [8, 9] for ^{23}Na which have been presented first at this NPA-7 conference [6, 7] supersede the previous data [4]. Contrary to the previous data by [4], these new data [8, 9] indicate a regular behavior of ^{23}Na. New data for the remaining outliers ^{36}Ar, ^{40}Ar (and ^{38}Ar where no data are available) and ^{33}S would be helpful to confirm or reject the irregular behavior of these nuclei.

Acknowledgments

I thank J. Tomlinson and A. M. Howard for providing their data before final publication. This work was supported by OTKA (K101328 and K108459).

References

[1] Mohr P 2015 *Europ. Phys. J.* A **51** 56
[2] Gomes P R S, Lubian J, Padron I, Anjos R M 2005 *Phys. Rev.* C **71**, 017601
[3] Schwartz R B, Corbett J W, Watson W W 1956 *Phys. Rev.* **101** 1370
[4] Almarez-Calderon S *et al.* 2014 *Phys. Rev. Lett.* **112** 152701
[5] Bowers M, Kashiv Y, Bauder W, Beard M, Collon P, Lu W, Ostdiek K, Robertson D 2013 *Phys. Rev.* C **88** 065802
[6] Hubbard N J *et al.* 2015 these proceedings.
[7] Tomlinson J R *et al.* 2015 these proceedings.
[8] Howard A M, Munch M, Fynbo H O U, Kirsebom O S, Laursen K L, Diget C A, Hubbard N J 2015 *Phys. Rev. Lett.* **115** 052701
[9] Tomlinson J R *et al.* 2015 *Phys. Rev. Lett.* **115** 052702
[10] Mohr P 2014 *Phys. Rev.* C **89** 058801
[11] Rauscher T 2010 *Phys. Rev.* C **81** 045807
[12] Kuperus J 1964 *Physica* **30**, 2253
[13] Whitmire D P and Davids C N 1974 *Phys. Rev.* C **9** 996
[14] Norman E B, Chupp T E, Lesko K T, Schwalbach P, Grant P J 1982 *Nucl. Phys.* **A390** 561
[15] Skelton R T, Kavanagh R W, Sargood D G 1987 *Phys. Rev.* C **35** 45

Search for resonant states in ^{10}C and ^{11}C and their impact on the primordial ^{7}Li abundance

F Hammache[1], A Coc[2], N de Séréville[1], I Stefan[1], P Roussel[1], M Assié[1], L Audouin[1], D Beaumel[1], S Franchoo[1], B Fernandez-Dominguez[3], S Fox[4], C Hamadache[2], J Kiener[2], A Laird[4], B Le Crom[1], A. Lefebvre-Schuhl[2], L Lefebvre[1], I Matea[1], A Matta[1], G Mavilla[1], J Mrazek[5], P Morfouace[1], F de Oliveira Santos[6], A Parikh[7], L Perrot[1], A M Sanchez-Benitez[8], D Suzuki[1], V Tatischeff[2], P Ujic[6] and Marine Vandebrouck[1]

[1] Institut de Physique Nucléaire d'Orsay, UMR8608, IN2P3-CNRS, Université Paris sud 11, 91406 Orsay, France
[2] CSNSM, IN2P3-CNRS, Université Paris Sud, 91405 Orsay, France
[3] Universidade de Santiago de Compostela, E-15786 Santiago, Spain
[4] Department of Physics, University of York, Heslington, York YO10 5DD, United Kingdom
[5] Nuclear Physics Institute ASCR, 250 68 Rez, Czech Republic
[6]GANIL, CEA/DSM-CNRS/IN2P3, Caen, France
[7] Departament de Física i Enginyeria Nuclear, Universitat Politècnica de Catalunya, E-08036 Barcelona, Spain
[8] Departamento de Física Aplicada, Universidad de Huelva, E-21071 Huelva, Spain

E-mail: hammache@ipno.in2p3.fr

Abstract. The cosmological ^{7}Li problem arises from the significant discrepancy of about a factor 3 between the predicted primordial ^{7}Li abundance and the observed one. The main process for the production of ^{7}Li during Big-Bang nucleosynthesis is the decay of ^{7}Be. Many key nuclear reactions involved in the production and destruction of ^{7}Be were investigated in attempt to explain the ^{7}Li deficit but none of them led to successful conclusions. However, some authors suggested recently the possibility that the destruction of ^{7}Be by ^{3}He and ^{4}He may reconcile the predictions and observations if missing resonant states in the compound nuclei ^{10}C and ^{11}C exist. Hence, a search of these missing resonant states in ^{10}C and ^{11}C was investigated at the Orsay Tandem-Alto facility through ^{10}B(^{3}He,t)^{10}C and ^{11}B(^{3}He,t)^{11}C charge-exchange reactions respectively. After a short overview of the cosmological ^{7}Li problem from a nuclear physics point of view, a description of the Orsay experiment will be given as well as the obtained results and their impact on the ^{7}Li problem.

1. Introduction

The Big-Bang model of the Universe is mainly supported by three observational evidences: the expansion of the Universe, the Cosmic Microwave Background (CMB), and the primordial or Big-Bang Nucleosynthesis (BBN) of light nuclei like ^{2}H, ^{3,4}He, and ^{7}Li. The observed primordial abundances for D and ^{3,4}He agree well with the predictions of the BBN theory together with the precise WMAP cosmic baryon density while ^{7}Li observations lie below the BBN+WMAP expectations by a factor of $\sim$ 3 [1, 2]. This observed deficit constitutes the so-called lithium

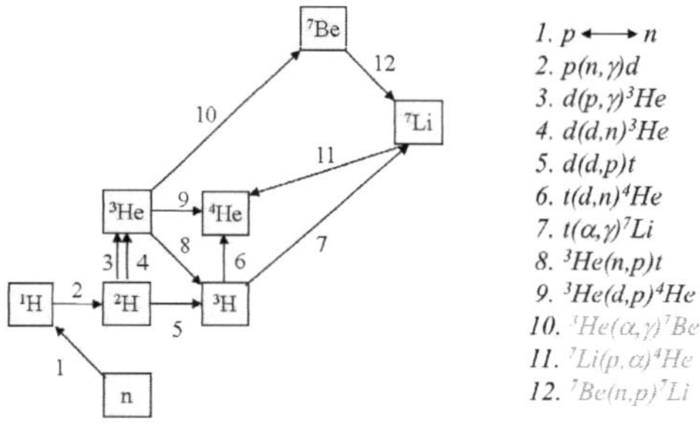

Figure 1. BBN nuclear reactions network.

problem. Many scenarios were proposed to try to explain this anomaly. The first one is related to the possibility of systematic errors in the extraction of Li abundances from the observed atomic spectra due to the used atmosphere models. But this is very unlikely since the various models give nearly the same results [3]. The second explanation is the possibility that the atmospheric ^{7}Li may have partially depleted from the atmosphere of the metal poor halo stars through rotationally induced mixing and/or diffusion [4]. But then a uniform destruction mechanism of ^{7}Li is needed all over this Spite plateau which exhibits a very little dispersion of the observed ^{7}Li abundance as a function of the metallicity of the stars. Some astrophysicists claim that the ^{7}Li problem point towards physics beyond standard model such as decay of super-symmetric particles, variation of the fundamental constants...[5, 6]. But before investigating these far going possibilities, one has to be sure about the nuclear data involved.

The main process for the production of the BBN ^{7}Li is the decay of ^{7}Be. Hence any reaction which produces or destroys ^{7}Be has an impact on ^{7}Li. Among the 12 BBN standard reaction network displayed in Fig. 1, the most important reaction which produces ^{7}Be is ^{3}He+^{4}He reaction and the main reaction which destroys it is ^{7}Be(n,p)^{7}Li followed by ^{7}Li(p,α)α. The cross section of ^{3}He(α,γ)^{7}Be reaction was measured by several groups using several methods and its knowledge is nowadays better than 8% according to the latest evaluation of solar fusion cross sections of Adelberger et al. [7]. Concerning ^{7}Be(n,p) reaction, its cross section is very well known according to the latest analysis of BBN rates performed by Descouvemont et al. [8] as well as the following reaction ^{7}Li(p,α)α.

Recently some authors [9, 10, 11, 12] investigated the possibility that missed resonances in some secondary destruction channels of ^{7}Li and ^{7}Be with p,n, d,t,^{3}He may be responsible of the ^{7}Li deficit. From their investigations, three promising candidates came out: ^{7}Be+d, ^{7}Be+^{3}He and ^{7}Be+^{4}He reactions. According to Cyburt et al. [12], the $5/2^+$ state close to 16.8 MeV excitation energy in the compound nucleus ^{9}B may enhance significantly the ^{7}Be+d reaction rate, thus leading to an appreciable depletion of ^{7}Li, if the deuteron width of the state is between 10-40 keV. For ^{7}Be+^{3}He channel, the presence of 1^- or 2^- state close to 15 MeV excitation energy in ^{10}C with a resonance energy between 50 to 100 keV and having a narrow width may reconcile the predicted primordial ^{7}Li with the observed one [9]. For ^{7}Be+^{4}He reaction, the solution may come from an existence of a hypothetical state close to 7.8 MeV in the compound nucleus ^{11}C and having a total width between 30-160 keV [11].

^{7}Be+d reaction was investigated by various works and was dismissed as possible solution to the ^{7}Li problem. Indeed, the ^{7}Be(d,p)2α cross section measurement [13] showed no enhancement of the cross section in the BBN region. The ^{7}Be(d,d) measurement [14] didn't observe any resonance in the region of interest and the extracted upper limit for the deuteron width of about 1 keV was found much less than the needed width to solve the ^{7}Li problem. And finally the recent study [15] of ^{7}Be(d,p)2α reaction rate using the spectroscopic properties of the state of interest at 16.8 MeV of ^{9}B (Ex=16.80 MeV±10 keV, Γ=81(5) keV, J^π=5/2$^+$) obtained from ^{9}Be(^{3}He,t)^{9}B measurement, showed that the state at 16.8 MeV can not enhance significantly ^{7}Be(d,p)2α reaction rate and its impact on ^{7}Li abundance is less than 4%.

So what about ^{7}Be+^{3}He and ^{7}Be+^{4}He? the case of ^{10}C is appealing because no states between 10 and 16.5 MeV are known and the state of interest lies close to 15 MeV. For ^{11}C, the excited states up to 9 MeV were studied through various indirect reactions and no state close to 7.8 MeV is reported. However, no dedicated measurements in this narrow energy region were carried out so one can not exclude that a weakly state in this energy region has been missed.

2. Experiment and results

A search for the missing levels has been performed at the ALTO facility [16]. ^{10}C and ^{11}C nuclei were populated with the (^{3}He,t) charge exchange reaction on ^{10}B and natural Boron targets. (^{3}He,t) measurements were also performed on C and Si_2O_4 targets because of the contamination of the boron targets by ^{12}C and ^{16}O nuclei. The emitted tritons were detected in the focal plane of Split-Pole spectrometer first by a position-sensitive gas detector and then by ΔE proportional gas counter. For ^{10}B(^{3}He,t) the tritons were detected at four different angles, 7°,10°,13° and 15° and for ^{11}B(^{3}He,t), they were detected at two angles, 7° and 10°. A good identification of the tritons and the deuterons was achieved by only using the position versus ΔE measured spectrum, see Fig. 2.

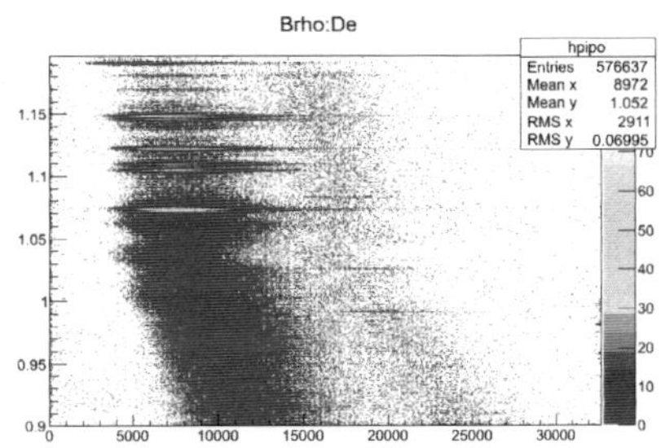

Figure 2. Bρ versus energy loss identification spectrum

The measured bρ position spectra for ^{11}C and ^{10}C at the angle of 10° are displayed in Fig. 3 and Fig. 4 respectively. The well populated peaks in ^{11}C spectrum are the already well known ^{11}C levels. No new peaks are observed at 10° in the energy region of interest between 7.79 and 7.9 MeV, neither at 7°. The very small "peaks" observed in this energy region are only statistical fluctuations. Thanks to the very large signal to background ratio measured in this spectrum, we can assert that it is very unlikely that a new state in ^{11}C exists in this energy region. Moreover, all the known states in the mirror nucleus ^{11}B below 9 MeV excitation energy have their counterpart in ^{11}C nucleus. For the ^{10}C case, no additionnal states in ^{10}C are observed around 15 MeV (see Fig. 4, blue histogram), the only states we observe are those coming from the ^{16}O contamination of the target as one can see in the spectrum (red histogram) obtained from the measurement on Si_2O_4 target. The same results were obtained at 7°, 13° and 15° measured angles. But since the background in this case is very important and the signal to background ratio is 10 times less than in the case of ^{11}C, we can not exclude that a state may be hidden in the background. However, from a chi2 study [16] of a simulated assumed state close to 15 MeV with various widths and various populated cross sections on the top of measured background, we could draw an exclusion zone in the plane of the charge-exchange cross section versus the width of the assumed state (see figure 4 in [16]) and conclude that any 1^- or 2^- state of ^{10}C in the excitation energy region around 15 MeV should have very likely, if present, a total width larger than 590 keV with a 95% CL to escape our detection.

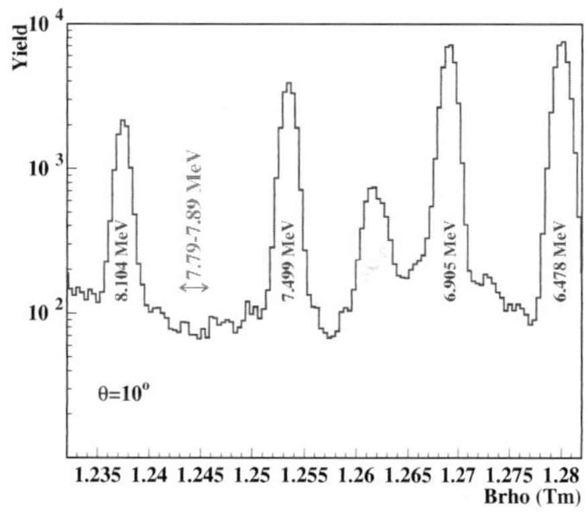

Figure 3. ^{11}C Bρ position spectrum measured at 10°.

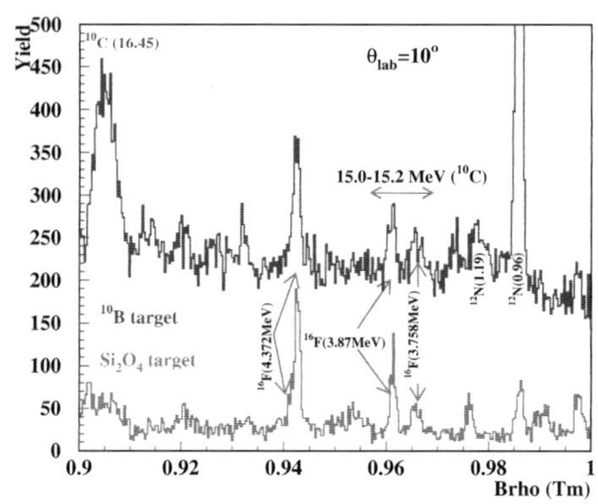

Figure 4. ^{10}C Bρ position spectrum measured at 10°.

Reaction rate calculations for the two only possible open channels, ^{7}Be(^{3}He,p)^{9}B and ^{7}Be(^{3}He,α)^{6}Be were performed assuming a 1^- state in the compound nucleus ^{10}C having a total width equal to our deduced lower limit, 590 keV, and even a three times lower value, 200 keV in case the differential charge-exchange cross section is three times smaller than the expected minimum one. The calculated ^{7}Be+^{3}He reaction rates [16] were included in a BBN nucleosynthesis calculation and were found to have no impact on the primordial ^{7}Li/H abundance. In conclusion, our results exclude ^{7}Be+^{3}He and ^{7}Be+^{4}He reactions as possible solution to ^{7}Li.

Finally With Orsay results [16] and those of previous works [13, 14, 15] concerning other possible resonant reaction channels, we may even say that the solution to the ^{7}Li problem has very likely to be found outside of nuclear physics.

References

[1] R.H. Cyburt, B.O. Fulop and K. A. Olive, J. Cosmology and Astroparticle Phys. 11, 12 (2008)
[2] A. Coc, E. Vangioni, *J. of Physics Conference Series* **202**, 012001 (2010).
[3] M. Asplund and K. Lind, Proceedings of the International Astronomical Unio, 5, (S268) 191 (2009)
[4] O. Richard, G. Michaud and J. Richer, Astrophys. J. 619, 538 (2005)
[5] M. Kusakabe et al., Phys. Rev D 76, 121302 (R) (2007)
[6] A. Coc, N. J. Nunes, K. A. Olive and E. Vangioni, Phys. Rev. D 76, 023511 (2007)
[7] E. G. Adelberger et al., Rev. Mod. Phys. 83, 195 (2011)
[8] P. Descouvemont, A. Adahchour, C. Angulo, A. Coc and E. Vangioni-Flam, At. Data Nucl. Data Tables **88**, 203 (2004).
[9] N. Chakraborty, B. D. Fields, K. Olive, Phys. Rev. D **83**, 063006 (2011).
[10] C. Broggini, L. Canton, G. Fiorentini, F.L. Villante, J. Cosmology Astropart. Phys. **06**, 030 (2012).
[11] O. Civitarese, M.E. Mosquera, Nucl. Phys. A **898**, 1 (2013).
[12] R.H. Cyburt, M. Pospelov, Int. J. Mod. Phys. E **21**, 1250004 (2012).
[13] C. Angulo, E. Casarejos, M. Coude, P. Demaret, P. Leleux et al., Astrophys. J. **630**, L105 (2005)
[14] P. D. O'Malley, D. W. Bardayan, A. S. Adekola, S. Ahn, K. Y. Chae et al., Phys. Rev. C **84**, 042801 (2011).
[15] O. S. Kirsebom, B. Davids, Phys. Rev. C **84**, 058801 (2011) and referrences therein
[16] F. Hammache, A. Coc, N. de Séréville, I. Stefan, P. Roussel et al., Phys. Rev. C **88**, 062802(R) (2013), `arXiv:1312.0894 [nucl-ex]`.

IOP Conf. Series: Journal of Physics: Conf. Series **940** (2017) 012017 doi:10.1088/1742-6596/940/1/012017

Trojan Horse cross section measurements and their impact on primordial nucleosynthesis

R.G. Pizzone[1], R. Spartá[1,2], C. Bertulani[3], C. Spitaleri[1,2], M. La Cognata[1], L. Lamia[2], A. Mukhamedzhanov[4], A. Tumino[1,5]
[1]INFN - Laboratori Nazionali del Sud, Catania, Italy
[2]Dipartimento di Fisica e Astronomia, Università degli Studi di Catania, Catania, Italy
[3]Department of Physics and Astronomy, Texas A&M University, Commerce, TX 75025, USA
[4] Cyclotron Institute, Texas A& M University, College Station, TX, USA
[5] Università degli Studi di Enna "Kore", Enna, Italy

E-mail: rgpizzone@lns.infn.it

Abstract. Big Bang Nucleosynthesis (BBN) nucleosynthesis requires several nuclear physics inputs and, among them, an important role is played by nuclear reaction rates. They are among the most important input for a quantitative description of the early Universe. An up-to-date compilation of direct cross sections of d(d,p)t, d(d,n)^{3}He and ^{3}He(d,p)^{4}He reactions is given, being these ones among the most uncertain bare-nucleus cross sections.

An intense experimental effort has been carried on in the last decade to apply the Trojan Horse Method (THM) to study reactions of relevance for the BBN and measure their astrophysical S(E)-factor. The result of these recent measurements is reviewed and compared with the available direct data. The reaction rates and the relative error for the four reactions of interest are then numerically calculated in the temperature ranges of relevance for BBN ($0.01<T_9<10$) and compared with up-to-date reaction rate compilations. Their value were therefore used as input physics for primordial nucleosynthesis calculations in order to evaluate their impact on the calculated primordial abundances of D, ^{3,4}He and ^{7}Li. These ones were then compared with the observational primordial abundance estimates in different astrophysical sites. A comparison was also performed with calculations using other reaction rates compilations available in literature.

1. Introduction

Over the last decades Big Bang Nucleosynthesis (BBN) has emerged as one of the pillars of the Big Bang model, together with the Hubble expansion and the Cosmic Microwave Background (CMB) radiation [1]. BBN probes the Universe to the earliest times, from a fraction of second to few minutes. It involves events that occurred at temperatures below 1 MeV, and naturally plays a key role in forging the connection between cosmology and nuclear physics [2]. Focusing only on the products of the BBN, according to the Standard Big Bang Nucleosynthesis model (SBBN), only the formation of light nuclei (^{2}H,^{3,4}He,^{7}Li) is predicted in observable quantities, starting from protons and neutrons. Today, with the only exception of ^{3}He and lithium, the abundances of these isotopes in the appropriate astrophysical environments are rather consistent with SBBN predictions [3]. A comparison between the primordial abundances from WMAP observations and the calculated ones constrains the baryon-to-photon ratio, η, which is the only free parameter in

IOP Conf. Series: Journal of Physics: Conf. Series **940** (2017) 012017 doi:10.1088/1742-6596/940/1/012017

the presently accepted model of the SBBN. A recent observation yields $\eta = 6.16 \pm 0.15 \times 10^{-10}$ [4], which is the value that we adopt in our calculations.

BBN nucleosynthesis requires several nuclear physics inputs and, among them, an important role is played by nuclear reaction rates. Due to the relatively small amount of key nuclear species involved in the BBN nuclear reaction network, only 12 reactions play a major role [5]. The reaction rates are calculated from the available low-energy cross sections for reactions which are also a fundamental information for a number of other still unsolved astrophysical problems, e.g. the so called "lithium depletion" either in the Sun or in other galactic stars [6, 7]. Cross sections should be measured in the astrophysically relevant Gamow window [8], of the order of few hundreds keV. In the last decades these reactions have been widely studied and, in particular, great efforts have been devoted to their study by means of direct measurements at the relevant astrophysical energies, sometimes in underground laboratories [9, 10]. However, for many of the relevant reactions, no direct experiments exist at astrophysical energies (mostly because of difficulties connected with the presence of the Coulomb barrier in charged particle induced reactions) and the cross section within the Gamow window has to be extrapolated from higher energy measurements. Alternative and challenging ways to obtain σ_b for charged-particles at sub-Coulomb energies have been provided by indirect methods such as the Coulomb dissociation method [11, 12] and the ANC (Asymptotic Normalization Coefficient) [13]. Among them, the Trojan-horse Method (THM) [14] is particularly suited to investigate binary reactions induced at astrophysical energies by neutrons or charged particles by using appropriate three-body reactions. It allows one to avoid both Coulomb barrier suppression and electron screening effects, thus preventing the use of unreliable extrapolations. In the next sections we will show the calculations of the reaction rates based also on the THM measurements of the cross sections σ_b. For recent reviews on the THM see [15, 16]). Thus, the method can be regarded as a powerful indirect technique to get information about bare nucleus cross section for reactions of astrophysical interest, which leads to new reaction rates determination. Some of the reactions of interest for the SBBN , i.e. ^{7}Li(p,α)^{4}He, ^{2}H(d,p)^{3}H, ^{2}H(d,n)^{3}He, ^{3}He(d,p)^{4}He, were studied by means of the THM in the energy range of interest and their measurements were performed in an experimental campaign which took place in the last decade [17, 18, 19, 20, 21, 22, 23, 24, 25, 26].

1.1. Reaction rates with TH data

The reaction rates for the the four reactions mentioned above (from a compilation of direct and THM data, as reported in [27]), have been calculated numerically. Then, we fitted the rates with the parametrization displayed in Equation 1. This is the common procedure adopted in previous works (see, e.g., [28, 29, 30]). For the 4 reactions of interest, we have fully included the experimental errors from measurements, allowing us to evaluate the respective errors in the reaction rates. The numerical results are then fitted with the expression

$$N_A\left<\sigma v\right> = \exp\left[a_1 + a_2 \ln T_9 + \frac{a_3}{T_9} + a_4 T_9^{-1/3} + a_5 T_9^{1/3} + a_6 T_9^{2/3} + a_7 T_9 + a_8 T_9^{4/3} + a_9 T_9^{5/3}\right], \quad (1)$$

which incorporates the relevant temperature dependence of the reaction rates during the BBN. The a_i coefficients for the ^{2}H(d,p)^{3}H and the ^{2}H(d,n)^{3}He reactions are given for both THM and direct measurements as well as for the direct ones in Table 1, while the coefficients for the ^{3}He(d,p)^{4}He and ^{7}Li(p,α)^{4}He reaction rate expression are given in Table 2. The direct data were considered from the compilation described in [27] for energies above 100 keV for ^{3}He(d,p)^{4}He and ^{7}Li(p,α)^{4}He and for energies above 10 keV for ^{2}H(d,p)^{3}H and ^{2}H(d,n)^{3}He, in order to avoid the enhancement due to the electron screening in the direct data.

For all the cases we noticed that deviations of up to 20% are obtained from previous compilations.

a_i	^{2}H(d,p)^{3}H (present)	^{2}H(d,p)^{3}H (direct)	^{2}H(d,n)^{3}He (present)	^{2}H(d,n)^{3}He (direct)
a_1	14.996	20.255	16.1787	13.3209
a_2	−2.4127	−0.63670	−1.9372	−2.9254
a_3	2.8261×10^{-3}	7.7756×10^{-5}	2.0671×10^{-3}	4.0072×10^{-3}
a_4	−5.3256	−4.2722	−5.0226	−5.6687
a_5	6.6125	−1.0758	5.7866	10.1787
a_6	2.4656	2.3211	-2.039×10^{-2}	0.1550
a_7	−3.8702	−1.3062	−0.7935	−2.5764
a_8	1.6700	0.38274	0.2678	1.1967
a_9	−0.25851	-5.0848×10^{-2}	-3.1586×10^{-2}	−0.1807

Table 1. Table with reaction rate parameters (appearing in Eq. 1) for ^{2}H(d,p)^{3}H and ^{2}H(d,n)^{3}He evaluated from the present work and S-factors from direct measurements.

a_i	^{3}He(d,p)^{4}He (present)	^{3}He(d,p)^{4}He (direct)	^{7}Li(p,α)^{4}He (present)	^{7}Li(p,α)^{4}He (direct)
a_1	20.4005	38.9078	17.6686	17.5315
a_2	1.3850	5.9512	−1.1549	−1.397
a_3	-1.2982×10^{-2}	-1.6061×10^{-2}	-4.4059×10^{-4}	6.9425×10^{-4}
a_4	−4.1193	−2.1962	−8.5485	−8.7921
a_5	12.2954	−20.5983	4.6683	5.7430
a_6	−15.2114	1.5636	−0.7858	−2.4092
a_7	5.4147	0.7040	−2.3208	0.6434
a_8	−0.5048	−0.1877	2.0628	1.290
a_9	-4.3372×10^{-2}	2.9419×10^{-2}	−0.4747	−0.3467

Table 2. Table with reaction rate parameters (appearing in Eq. 1) for ^{3}He(d,p)^{4}He and ^{7}Li(p,α)^{4}He evaluated from present work and S-factors from direct measurements.

2. Discussion and Conclusion

The reaction rates of 4 of the main reactions of the BBN network in the temperature range ($0.001 < T_9 < 10$), namely, ^{2}H(d,p)^{3}H, d(d,n)^{3}He, ^{3}He(d,p)^4, ^{7}Li(p,α)^{4}He, have been calculated numerically including the recent THM measurements [31, 32, 33, 34, 35, 36, 37, 38]. The uncertainties of experimental data for direct and THM data have been fully included for the above reactions. The extension of the same methodology to the other reactions forming the BBN reaction network will be examined in a forthcoming paper. The parameters of each reaction rates as given in Eq. 1 are reported in Tables 1 and 2. The obtained reaction rates are compared with some of the most commonly used compilations found in the literature. The reaction rates calculated in the present work are used to calculate the BBN abundance for ^{3,4}He, D and ^{7}Li. The obtained abundances are in agreement, within the experimental errors, with those obtained using the compilation of direct reaction rates. Moreover, a comparison of our predictions with the observations for primordial abundance of ^{3,4}He, D and ^{7}Li show an agreement for ^{3,4}He and D, while showing a relevant discrepancy for ^{7}Li [27]. The present results show the power of THM as a tool for exploring charged particle induced reactions at the energies typical of BBN.

Acknowledgments

This work was supported by the Italian Ministry of the University under Grant No. RBFR082838 and by the Italian Ministry of University MIUR under the grant "LNS Astrofisica Nucleare (fondi premiali)". This work was partially supported by the US-DOE grants DE-FG02-08ER41533 and DE-FG02-10ER41706. A.M. acknowledges the support by US Department of Energy under

Grant Nos. DE-FG02-93ER40773, DE-FG52- 09NA29467, and DE-SC0004958.

References

[1] G. Steigman 2007, Ann. Rev. Nucl. Part. Sci. 57, 463
[2] B. D. Fields and S. Sarkar 2006, J. Phys. G33, 220
[3] G. Israelian 2012, Nature 489, 37
[4] E. Komatsu et al. 2011, Ap. J. Sup. 192, 18
[5] E.W. Kolb and M.S. Turner 1990, "The Early Universe", Addison-Wesley
[6] R. Weymann and E. Moore 1963, Ap. J. 137, 552
[7] D. Ezer and A.G.W. Cameron 1963, Icarus 1, 422
[8] C. Iliadis 2007, "Nuclear Physics of Stars", Wiley
[9] R. Bonetti et al. 1999, Phys. Rev. Lett. 82, 5205
[10] C. Casella et al. 2002, Nucl. Phys. A706, 203
[11] G. Baur, C.A. Bertulani & H Rebel 1986, Nucl. Phys. A458, 188
[12] C.A. Bertulani & A. Gade 2010, Phys. Rep. 485, 195
[13] A. Mukhamezhanov et al. 2008, Phys. Rev. C 78, 0158042008
[14] C. Spitaleri et al. 2003, Nucl. Phys. A 719, 99c
[15] C. Spitaleri et al. 2011, Physics of At. Nucleus 74, 1725
[16] R.E. Tribble et al. 2014, Rep. Progr. Phys. 77, 10,106901
[17] A. Tumino et al. 2008, Phys. Rev. C 78, 064001
[18] R.G. Pizzone et al. 2005, A. & A. 438, 779
[19] L. Lamia et al., 2013, Ap. J. 768, 65
[20] M. La Cognata et al. 2011, Ap. J. L. 739, L54
[21] L. Lamia, et al. 2012, J. Phys. G 39, 015106
[22] L. Lamia, M. La Cognata, C. Spitaleri, B. Irgaziev, R.G. Pizzone 2012, Phys. Rev. C 85, 025805
[23] S. Romano et al. 2006, Eur. Phys. J. A 27, 221
[24] Q. Wen et al. 2008, Phys. Rev. C 78, 035805
[25] R.G. Pizzone et al. 2011, Phys. Rev. C, 83, 045801
[26] S. Cherubini et al., 2015, Phys. Rev. C, 92, 015805
[27] R.G. Pizzone et al. 2014, Ap. J.,786, 112
[28] M.S. Smith, L.H. Kawano and R.A. Malaney 1993, Ap. J. 85, 219
[29] R.H. Cyburt 2004, Phys. Rev. D 70, 023505
[30] A. Coc, S. Goriely, Y. Xu, M. Saimpert, and E. Vangioni 2012, Astrophys. J. 744, 18
[31] L. Lamia et al. 2012, A. & A. 541, 158
[32] M. Lattuada et al. 2001, Ap. J. 562, 1076
[33] A. Tumino et al. 2011, Phys. Lett. B 700 (2), 111
[34] A. Tumino et al. 2014, Ap. J., 785, 96
[35] M. La Cognata et al. 2005, Phys. Rev. C 72, 065802
[36] R.G. Pizzone et al. 2003, A. & A. 398, 423
[37] R.G. Pizzone et al. 2013, Phys. Rev. C 87, 025805
[38] Li C. et al., 2015, Phys. Rev. C, 92, 025805

IOP Conf. Series: Journal of Physics: Conf. Series **940** (2017) 012018 doi:10.1088/1742-6596/940/1/012018

ν-driven winds from the remnant of binary neutron star mergers

A Perego

Institut für Kernphysik, Technische Universität Darmstadt, Schlossgartenstraße 2, D-64289 Darmstadt, Germany

E-mail: albino.perego@physik.tu-darmstadt.de

Abstract. We present a 3D hydrodynamic study of the neutrino-driven winds that emerge from the remnant of a neutron star merger, represented by a thick accretion disc orbiting around a massive neutron star. This strong baryonic wind is blown out by neutrino absorption on free baryons inside the disc. It expands within a few tens of ms along the original binary rotation axis. If the central object survives for at least 200ms, the mass ejected in the wind can reach 5% of the initial mass of the accretion disc. Due to the intense neutrino irradiation, matter ejected in the wind increases its electron fraction between 0.3 and 0.4, producing weak r-process nucleosynthesis yields. We predict a distinct UV/optical transient associated with the wind ejecta that peaks from a few hours to a few days after the merger.

1. Introduction

Binary compact objects coalesce due to the emission of gravitational waves (GWs), carrying away orbital angular momentum and energy [1]. In addition to be primary targets for terrestrial GW detectors, such as aLIGO and aVIRGO, compact binary mergers involving at least one neutron star (NS) are expected to copiously emit neutrinos of all flavours (ν's), with luminosities comparable to the ones of a core collapse supernova event [2]. After the discovery of the first double NS binary, its has been soon recognized that the decompression of cold NS matter in mergers can represent a favourable site for the production of the heaviest elements in the Universe, in particular of rapid neutron capture process (r-process) elements [3]. In these systems, the mutual tidal interaction and the actual collision between the compact objects expel a fraction of the original mass in a dense accretion disc around a black hole or a (possibly gravitational unstable) massive neutron star (MNS). A fraction of this matter can even become unbound within a few ms (*dynamic ejecta*). Interactions between ν's and matter in the remnant, as well as the secular evolution of the accretion disc, are also expected to unbind matter in the so-called *ν-driven wind* and in the *viscous ejecta*, respectively. The decay of radioactive elements in the different ejection channels can power electromagnetic (EM) transients [4], whose detection in association with GW signals can directly prove the link between compact binary merger and r-process nucleosynthesis. The possible detection of such an EM signal in the afterglow of two short gamma-ray-bursts has recently corroborated the idea that binary compact mergers are also the progenitors of these powerful burst [5, 6, 7].

In this proceeding, we explore the origin and the properties of the ν-driven wind that emerges from the remnant of a double binary neutron star merger [8].

2. Method

To study the aftermath of a binary neutron star merger and the development of a ν-driven wind, we set up a radiation-hydrodynamics (HD) simulation. We take our initial conditions from the final stages of an equal mass binary NS merger simulation. The coalescence has been studied using a smoothed-particle hydrodynamics (SPH) code [9], complemented by a multi-flavour neutrino leakage scheme. We map the SPH data 10 ms from the beginning of the merger into the equally-spaced Cartesian grid (resolution: 1 km) of our 3D Eulerian hydrodynamics code `FISH` [10]. Neutrino-matter interaction is implemented via a multi-energy, multi-flavor leakage scheme. It provides spectral neutrino cooling rates as smooth interpolations between diffusion and free-streaming emission rates. The obtained neutrino luminosities are used to compute neutrino particle densities in optically thin regions, via a ray-tracing algorithm. Neutrino absorption rates are finally computed as a product of the local neutrino absorptivity and particle density. We use the HS tabulated nuclear equation of state ([11]) with the TM1 parametrization. We place 10^5 tracer particles, i.e. passively advected Lagrangian particles, in the disc around the MNS to record the local properties of matter. After an initial relaxation phase of $\sim$ 10 ms, when only neutrino cooling processes are considered, the system is evolved for $\sim$ 200 ms under the influence of both neutrino cooling and heating. The adopted uniform grid ensures a high spatial resolution in the disc and in the dilute wind, but it is not enough for the dense MNS, characterized by large density gradients. Therefore, the innermost part of the MNS ($\rho > 10^{12}\,\mathrm{g\,cm^{-3}}$) is treated as a stationary, axisymmetric rotating object.

3. Results

3.1. Neutrino emission

Neutrinos diffusing from the MNS and from the innermost part of the disc power an intense neutrino luminosity. This luminosity is only slightly decreased during the first hundreds of ms, due to the large energy reservoir in the central compact object and to the intense accretion rate onto it ($\dot{M} \sim 0.5\,M_\odot\,\mathrm{s^{-1}}$). The high neutron richness of the disc favours positron capture and ν_e absorption on free neutrons. As a consequence, $\bar{\nu}_e$ luminosities and mean energies are larger than the ones for ν_e's ($E_{\nu_e} \approx 11\,\mathrm{MeV}$ and $E_{\bar{\nu}_e} \approx 15\,\mathrm{MeV}$, while $L_{\bar{\nu}_e} \approx 3 \times 10^{52}\,\mathrm{erg\,s^{-1}}$ and $L_{\nu_e} \approx 1.7 \times 10^{52}\,\mathrm{erg\,s^{-1}}$). $\nu_{\mu,\tau}$'s and $\bar{\nu}_{\mu,\tau}$'s, which decouple from matter at the edge of the MNS, i.e. much deeper than electron flavour neutrinos, have larger mean energies ($E_{\nu_{\mu,\tau}} \approx 18\,\mathrm{MeV}$) but lower luminosities ($L_{\nu_{\mu,\tau}} \approx 9 \times 10^{51}\,\mathrm{erg\,s^{-1}}$). Due to the larger opacity provided by the disc along the equatorial direction, the luminosity is highly anisotropic and the isotropised luminosities in the polar direction are up to 3 times larger than the ones in the equatorial plane.

3.2. Wind dynamics and properties

A few tens of milliseconds after the beginning of the simulations, a low density wind develops from the disc (figure 1, see also [12]). The energy and momentum deposition operated by ν_e's and $\bar{\nu}_e$'s in the low density part of the disc ($\rho \lesssim 10^{10}\,\mathrm{g\,cm^{-3}}$) expand the merger debris and drives matter orbiting around the MNS on wider trajectories on the time scale of a few orbital periods. The wind originates mainly from the region of the disc located 50-100 km from the MNS and expands far from it, initially parallel to the rotational axis. As the distance from the central compact object increases, the vertical motion turns into a radial expansion. The ν irradiation of the initially extreme neutron rich matter in the disc changes the electron fraction in the wind. In particular, the dominant ν_e absorption on free neutrons rises Y_e towards larger values, ($Y_e \approx 0.42$). The entropy in the wind is initially increased by the ν energy deposition, before the expansion proceeds almost adiabatically. The evolution of the density and Y_e inside the wind can be seen in figure 2. Around 100 ms after the beginning of the simulation, the disc and the wind have reached a quasi-stationary state. While matter in the wind expands and cools, the free protons form first α-particles and, later at NSE freeze-out, neutron-rich heavy

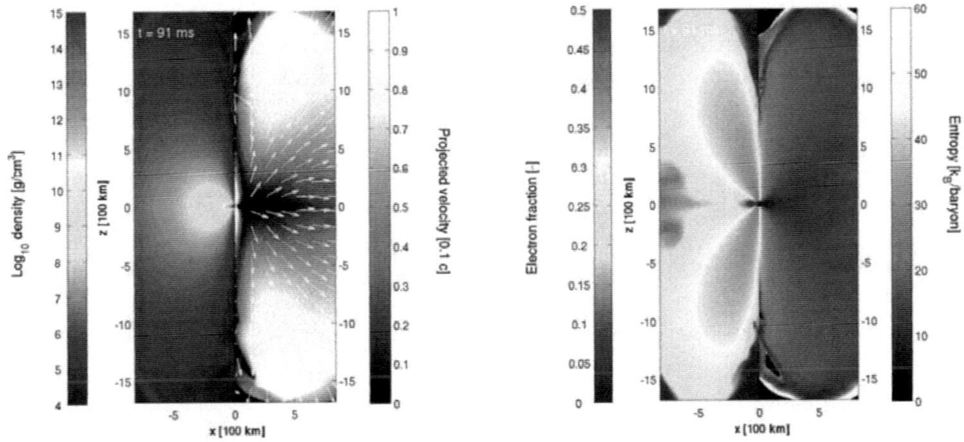

Figure 1. 2D slices ($y = 0$ plane) of the 3D simulation box of relevant quantities taken 90 ms after the beginning of our simulations. In the left figure, we represent the density (left) and the projected velocity (right). In the right figure, the electron fraction (left) and the entropy (right).

nuclei surrounded by free neutrons (with an initial density of $n_n \sim 10^{30}\,\mathrm{cm}^{-3}$). The associated release of nuclear binding energy is crucial to ultimately unbind matter in the wind.

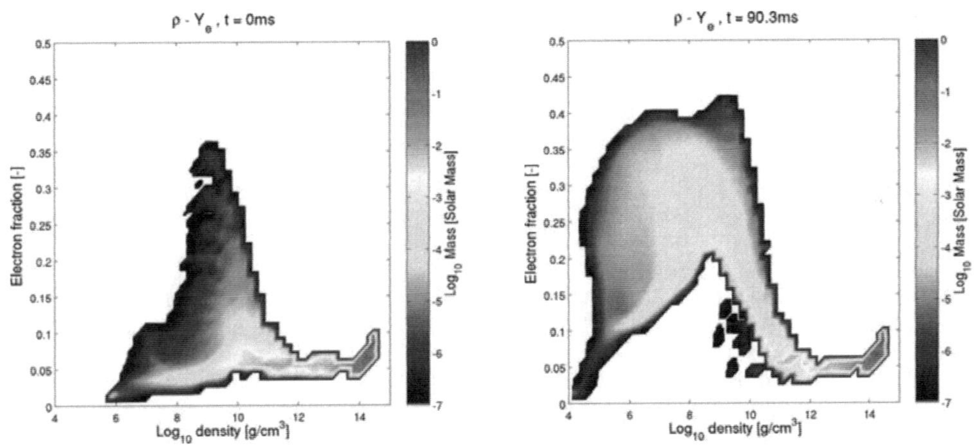

Figure 2. 2D histograms of density-Y_e conditions occurring inside our computational domain at $t = 0\,\mathrm{ms}$ (left) and at $t = 90\,\mathrm{ms}$ (right). Color coded is the amount of mass that experiences the thermodynamical conditions indicated by the location on the plane, at a given time. The development of the wind is visible in the right panel, where an increasing amount of matter is expanding at lower density ($\rho \lesssim 10^{10}\mathrm{g\,cm}^{-3}$) with higher Y_e, due to ν-matter interactions.

3.3. Wind ejecta, nucleosynthesis and EM signal

After 100 ms, $\sim 2 \times 10^{-3} M_\odot$ of matter in the wind has become unbound, while at 200 ms the mass of the ejecta has increased up to $\sim 9 \times 10^{-3} M_\odot$. This corresponds to 1% and 5% of the initial mass of the disc, respectively. Due to the stronger ν irradiation, matter expanding closer to the rotational axis has a larger electron fraction, closer to the equilibrium value. At early times ($t \lesssim 100\,\mathrm{ms}$), the competition between the expansion and the weak equilibrium time scales produces a wide spread of Y_e in the ejecta, ranging from 0.2 to 0.4. At later times, the prolonged

ν irradiation increases the electron fraction towards larger mean values ($Y_e \sim 0.33$) and reduces the relevant Y_e interval between 0.3 and 0.4. The velocities of the ejecta in the wind are non-relativistic and span a broad range between $0.06c$ and $0.1c$, while the entropy is usually between 15 and 20 $k_{\rm b}$ baryon^{-1}. The larger values for both the expansion velocity and the entropy are reached close to the rotational axis, where the neutrino influence is larger. We analyse the ejected tracer particles and we post-process them using a full nuclear reaction network. We obtain a weak r-process, characterized by the production of elements in the mass range $80 < A < 130$, and a negligible fraction of lanthanides and actinides [13]. The combination of these ejecta with the yields expected from the dynamic ejecta [14, 15, 16, 19], and from the viscous evaporation of the disc [17, 18], indicates that neutron star mergers can be (one of) the astrophysical site(s) for the full r-process nucleosynthesis. We also calculate the EM signal expected from the wind ejecta [13]. We predict luminosity peaks in the range between 4 hours and 4 days after the merger, bluer than the IR peak expected from the dynamic ejecta around 10 days after the merger [20, 21].

4. Conclusions

We have shown that the ν's emitted from the remnant of a binary NS merger can trigger the formation of a baryonic wind within a few tens of ms after a binary NS merger. This wind has potentially relevant implications for the nucleosynthesis and for the EM transient expected from these compact binary mergers. Further investigations, including, for example, GR or more accurate neutrino treatments, are required to sharpen our understanding of this scenario.

Acknowledgments

AP gratefully acknowledges support by the Nuclear Astrophysics Virtual Institute to participate in Nuclear Physics in Astrophysics VII Conference, and the use of computational resources provided by the Swiss SuperComputing Center (CSCS), under the allocation grants s414. He also thanks D Martin and A Arcones for useful discussions.

References

[1] Peters P C 1964 *Phys. Rev.* **136** B1224
[2] Eichler D, Livio M, Piran T and Schramm D N 1978 *Nature* **340** 126
[3] Lattimer J M and Schramm D N 1974 *ApJ* **192** L145
[4] Li L and Paczyński B 1998 *ApJL* **507** L59
[5] Berger E, Fong W and Chornock R 2013 *ApJL* **774** L23
[6] Tanvir N R, Levan A J W, Fruchter A S, Hjorth J, Hounsell R A, Wiersema K and Tunnicliffe R L 2013 *Nature* **500** 547
[7] Yang B, Jin Z-P, Li X, Covino S, Zheng X-Z, Hotokezaka K, Fan Y-Z, Piran T and Wei D-M 2015 *Nature Communications* **6** 7323
[8] Perego A, Rosswog S, Cabezón R M, Korobkin O, Käppeli R, Arcones A and Liebendörfer M 2014 *MNRAS* **443** 3134
[9] Rosswog S and Price D 2007 *MNRAS* **379** 915
[10] Käppeli R, Whitehouse S, Scheidegger S, Pen U-L and Liebendörfer M 2011 *ApJS* **195** 20
[11] Hempel M and Schaffner-Bielich J 2011 *Nucl. Phys. A* **837** 210
[12] Dessart L, Ott C D, Burrows A, Rosswog S and Livne E 2009, *ApJ* **690** 1681
[13] Martin D, Perego A, Arcones A, Thielemann F-K, Korobkin O and Rosswog S 2015, accepted by *ApJ* (*Preprint* arXiv:1506.05048)
[14] Korobkin O, Rosswog S, Arcones A and Winteler C 2012, *MNRAS* **426** 1940
[15] Bauswein A, Goriely S and Janka H-T 2013, *ApJ* **773** 78
[16] Hotokezaka K, Kiuchi K, Kyutoku K, Muranuchi T, Sekiguchi Y, Shibata M, Taniguchi K 2013, *Phys. Rev. D* **88** 044026
[17] Fernández R and Metzger B 2013, *MNRAS* **435** 502
[18] Just O, Bauswein A, Pupillo R A, Goriely S and Janka H-T 2015, *MNRAS* **448** 541
[19] Wanajo S, Sekiguchi Y, Nishimura N, Kiuchi K, Kyutoku K, Shibata M 2014, *ApJL* **789** L39
[20] Metzger B and Fernández R 2014, *MNRAS* **441** 3444
[21] Grossman D, Korobkin O and Rosswog S 2014, *MNRAS* **439** 757

IOP Conf. Series: Journal of Physics: Conf. Series **940** (2017) 012019 doi:10.1088/1742-6596/940/1/012019

Ion implant-β-decay correlation half-lives in a pulsed beam for isotopes beyond N=126

R.Caballero-Folch[1]*, C. Domingo-Pardo[2] and GSI - S410 collaboration[3]

[1]Universitat Politècnica de Catalunya, 08028, Barcelona, Spain

[2]Institut de Física Corpuscular, CSIC-Universitat de València, E-46071 València, Spain

[3]GSI Helmholtzzentrum für Schwerionenforschung GmbH, Planckstrasse 1, D-64291, Darmstadt, Germany

E-mail: roger@baeturia.com

Abstract. Half-lives and β-delayed neutron emission probabilities of several isotopes of Au, Hg, Tl, Pb and Bi, in the neutron-rich region around and beyond N=126, were determined in an experiment performed at the RIB facility of GSI [1]. These are decay properties that are directly related to the final abundance distribution of the elements due to their contribution during the freeze-out of the r-process nucleosynthesis. This contribution summarizes the main aspects of the analysis methodology followed for the determination of β-decay half-lives. The main particularities of the present analysis concern the characterization of the β-background, as well as the effect of the pulsed-beam in the ion-β time correlations.

1. Introduction

β-delayed neutron emission plays a fundamental role in astrophysics for the understanding of the rapid neutron capture process in explosive stellar environments. This effect is more important in the heavy mass region around N=126, where practically all nuclei along the r-process path are expected to be β-delayed neutron emitters. In order to provide new information in this mass region we carried out an experiment at GSI with an ^{238}U beam at 1 GeV/u impinging onto a 1.6 mg/cm^2 Be target. The fragments of interest were selected and identified by means of the Fragment Separator (FRS) [2]. The Silicon Implantation Beta Absorber (SIMBA) [3, 4] was used to measure both ion implants and β-decays. Neutrons were detected by means of the BEta deLayed Neutron (BELEN) detector [5]. The methodology implemented for building the implant-β correlation diagrams is described in section 2. An experimental validation of the analysis technique is reported in section 3. Finally, section 4 summarizes the main results and conclusions.

2. Analysis methodology for the determination of β-decay half-lives

In this work we have followed the analytical approach employed in several previous works at both isol- and fragmentation facilities [3, 6, 7] to analyze the implant-β time-correlation diagram. However, in order to do so, we had to quantify and carry out a systematic study of several aspects involved in the experiment, such as the characterization of the β background, the correlation area inside the silicon layers of SIMBA, between the implantation position of the ions and the position of the decays, and the effect of the beam characteristics used in the measurement.

2.1. Analytical model for implant-β correlations

Several approaches of the Bateman equations [8] were studied in order to determine the value of the half-life. Since all granddaughters of the analyzed species are stable, or very long lived isotopes, in all cases all β decays contributing to the activity decay curve were considered emitted by the parent and the direct daughter, an approximation which is given by equation 1:

$$\sum_{i=1}^{N} \lambda_i N_i(t) \approx (\lambda_1 N_1(t) + \lambda_2 N_2(t)) \tag{1}$$

where $N_i(t)$ is the number of nuclei for each isotope i at certain time t, $\lambda_1 = ln(2)/T_{1/2}$ is the decay constant for the implanted nucleus with an unknown half-life $T_{1/2}$ and λ_2 is the decay constant for the daughter nucleus. Following the time-dependency of the correlation distributions presented in [9], there are two strategies to study the implant-β time- correlations in a certain region around the implants. One is using the first β-decay after the implant, which has a strong dependency with the β efficiency, ε_β. In this case, ε_β is not constant along the energy range, it changes according to the Q_β-value of each isotope, the result is biased when the β-background rate is high and also by the time-dependent β-background characteristic of the pulsed beam-structure of the SIS synchrotron used in this experiment at GSI. On the other hand, the correlation of each implant with all-β events, over a sufficiently broad time-window ($\gg T_{1/2}$), gives a better solution for the present experiment. In this case the dependency on ε_β is affecting by the same amount all terms of the expression which describes the time-correlation probability function, $\rho(\lambda, t)$, and it can be left as a free parameter in the analysis [9]. In this case, the accumulated distribution is described by equation (2),

$$\begin{aligned} N_\beta^{All_\beta}(t) = N_1(0)\rho(\lambda,t)\Delta t &= (\varepsilon_\beta\lambda_1 N_1(t) + \varepsilon_\beta\lambda_2 N_2(t) + \varepsilon_\beta b) \cdot \Delta t \\ &= \varepsilon_\beta \cdot (\lambda_1 N_1(t) + \lambda_2 N_2(t) + b) \cdot \Delta t \end{aligned} \tag{2}$$

where $N_\beta^{All_\beta}(t)$ is the total number of detected decays at a time t, N_i the number of nuclei of each isotope in the chain, b is the β-background corrected by ε_β and Δt corresponds to the bin time-width used in the implant-β time-correlations. The factor $N_1(0) \cdot \varepsilon_\beta$ is determined by the content of the first bin in the time correlation curve. For each distribution, this latter factor is included within the Maximum-Likelihood (ML) analysis and thus, the time dependent factor of the adjusted expression, $\lambda_1 N_1(t) + \lambda_2 N_2(t) + b$ does not depend on the β-efficiency.

3. ^{213}Tl analysis as method benchmark

To analyze the elements present in the analysis affecting the sensitivity and the accuracy of the half-life result, a systematic study based on the decay of ^{213}Tl was carried out. This particular isotope was chosen because its implant statistics were large (1015 implants) and there exist previous half-life measurements in the literature obtained independently using other methods [10, 11].

3.1. Background characterization

The main contribution to the β-background was related to the pulsed beam-structure. It induced a different background level during beam extraction from the SIS-synchrotron (spill width of 1 s) and after it (the spill-off time interval, which was of 3 s). The background was characterized with time-reversal (or time backward) implant-β correlations using only the pixels within the assigned correlation area. This can be observed in figure 1 for both methods, all-*beta* (left) and all-β-off-spill (right) events, over a time window of $\pm$360 s for each implant event. The negative time-interval of these diagrams represents the backward (uncorrelated) background contribution. The average background is of $b = 2$ counts/s for the all-β events and the $b = 1.4$ counts/s in the off-spill correlation mode. The higher background rate in the first case indicates the aforementioned effect of enhanced β-like background during the beam-spill interval. The background level thus determined allows one to adjust the parameter b in the equation 2, which has a flat dependency in time. Once determined, the background parameter b was kept fixed in the ML analysis over the whole time-window, and the unknown parameter λ_1 was derived.

3.2. Effect of the beam structure with all-β and all-β-off-spill correlations

Due to the limited statistics for most of the implanted isotopes the binned Maximum-Likelihood (ML) algorithm [12] was used. It was implemented by including the expression presented in equation 2 in a ROOT/C++ code. Figure 1 shows the result of the analysis after the convergence of the ML algorithm for both distributions with all-β and all-β-off-spill methods. It is worth to

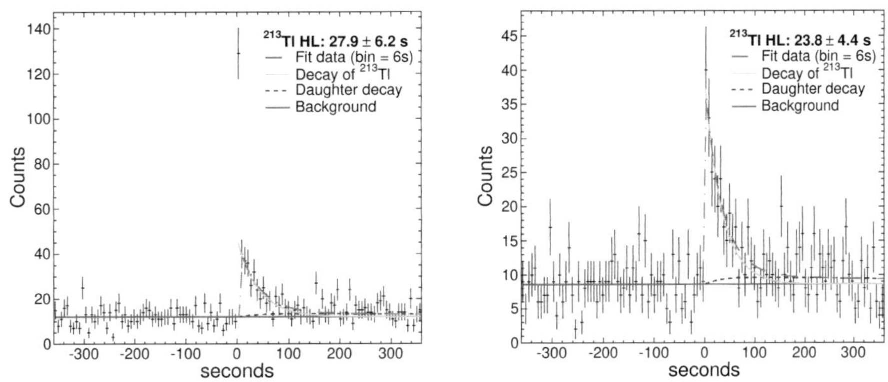

Figure 1. ML analysis of the ^{213}Tl implant-beta correlations, with 9 mm^2 correlation area. Left: All implant-β events. Right: implant-β events off-spill. See text for details.

discuss the large amount of counts in the first bin of the all-β correlation distribution (figure 1-left). This is an effect related to the in-beam mode when the bin time-width is larger or equal to the spill time width. It can be avoided by excluding the first bin from the fit-range, as shown in the analysis presented in figure 1-left. Indeed, the half-life value obtained using both correlation approaches are in perfect agreement within the quoted statistical uncertainty. The goodness of the analytical model in both cases is indicated by the reasonable reduced χ^2-value (χ^2/NDF) obtained, which is of 2.21 and 1.40, respectively.

3.3. Implant-β correlation area

The position of each implanted nucleus is typically determined with an accuracy of ± 1 mm^2, i.e. one pixel in a silicon layer of SIMBA. Selecting β-events detected in SIMBA in a correlation region around the implant position, it was possible to build time correlations like those shown in figure 1. Contributions from decays of other nuclei implanted before or right after the isotope of interest, in the same correlation region can be neglected thanks to the low implantation rate of 0.114 ions/s. Depending on the average range of β-particles in SIMBA and the rate of β-like background events, an optimal correlation area can be determined, which maximizes the peak-to-background P/B ratio. In order to determine it, ^{213}Tl implant-β correlations were performed for several symmetric correlation areas of 9 mm^2, 25 mm^2 and 49 mm^2 around the implant location and for both types of correlation, all-β-events and spill-off β-events (case of 9 mm^2 shown in figure 1-left and right, respectively). These correlation areas correspond to the one-, two- and three-closest pixels around the implant location. Table 1 summarizes the evaluation of each correlation area according to the P/B ratio together with the reduced χ^2-value (χ^2/NDF) obtained for each case. In summary, it can be concluded that a correlation area of 9 mm^2 (3×3

Table 1. Peak-to-Background ratio and χ^2-value (χ^2/NDF) for the implant-β correlation analysis performed.

Correlation	All β-events		Off-spill β-events	
area (mm^2)	P/B	χ^2	P/B	χ^2
9	1.32	2.21	1.34	1.40
25	1.14	1.99	1.11	1.24
49	1.05	1.87	1.04	1.27

pixels, 1 pixel around the implant) seems the most convenient in terms of sensitivity to the decay curve. In this case the P/B-ratio becomes a factor about 20% larger than areas of 25 mm^2 and 49 mm^2. The case of 1 mm^2 was also evaluated but the limited statistics of many isotopes makes this, in general, a worse option.

4. Results and conclusions

To summarize the systematic method for the analysis of the isotopes in this experiment, the half-lives were obtained following these characteristics:

- A constant background value b determined using the backward β-events, with the same time-range and area-correlation as for the forward β-events.
- Implant-β correlation area of 9 mm^2, i.e. one pixel around the implantation pixel.
- Off-spill correlations are better suited for the half-life determination than all β-events approach, mainly due to the lower β-like background rate and generally better χ2-values.

Finally, it is worth emphasizing the consistency of the result obtained for the half-life of ^{213}Tl for all possible analysis conditions within the quoted statistical uncertainties [1].

Acknowledgments

This work was partially supported by the Spanish Ministry of Economy and Competitivity under grants FPA2011-28770-C03-03, CSD200700042, FPA2008-04972-C03-03, AIC-D-2011-0705, FPA2011-24553 and FPA2008-6419, the Spanish Nuclear Security Council (CSN) under a grant of Catedra Argos and by the Helmholtz Association via the Young Investigators Project VH-NG 627 (LISA- Lifetime Spectroscopy for Astrophysics).

References

[1] Caballero-Folch R 2015 *First measurement of beta-decay half-lives and neutron emission probabilities in several isotopes beyond N=126* Ph.D. thesis Universitat Politècnica de Catalunya.
[2] Geissel H *et al.* 1992 *Nucl.Instr.Meth. B* **70** 286–297
[3] Hinke C *et al.* 2012 *Nature* **486** 341–345
[4] Steiger K 2009 *Diploma thesis* Ph.D. thesis Technische Universität München.
[5] Gómez-Hornillos M *et al.* 2014 *Hyp.Int.* **223** 185–194 ISSN 0304-3843
[6] Agramunt J *et al.* 2014 *Nuclear Data Sheets* **120** 74–77
[7] Montes F *et al.* 2006 *Phys.Rev.C* **73** 035801
[8] Bateman H 1910 *Proc. Camb. Philos. Soc.* **15** 423–427
[9] Bernas M *et al.* 1990 *Zeitschrift für Physik A Atomic Nuclei* **336** 41–51
[10] Benzoni G *et al.* 2012 *Phys.Lett.B* **715** 293–297
[11] Chen L *et al.* 2010 *Phys.Lett.B* **691** 234–237
[12] Proakis J G *et al.* 1994 *Communication systems engineering* vol 1 (Prentice-hall Englewood Cliffs)

New theory of stellar convection without the mixing-length parameter: new stellar atmosphere model

Stefano Pasetto[1,2], Cesare Chiosi[3], Mark Cropper[2], Eva K. Grebel[4]

[1] Observatories of the Carnegie Institution of Washington, 813 Santa Barbara, Street, Pasadena, CA 91101, USA

[2] Mullard Space Science Laboratory, University College London, Holmbury , St. Mary, Dorking, Surrey RH5 6NT, United Kingdom

[3] Dept. of Physics & Astronomy "Galileo Galilei", University of Padua, Vicolo dell'Osservatorio, 5, 35141 Padova PD, Italy

[4]Astronomisches Rechen-Institut, Zentrum für Astronomie der Universität Heidelberg, Mönchhofstr 12-14, 69120 Heidelberg, Germany

E-mail: spasetto@carnegiescience.edu

Abstract. Stellar convection is usually described by the mixing-length theory, which makes use of the mixing-length scale factor to express the convective flux, velocity, and temperature gradients of the convective elements and stellar medium. The mixing-length scale is proportional to the local pressure scale height of the star, and the proportionality factor (i.e. the mixing-length parameter) is determined by comparing the stellar models to some calibrator, i.e. the Sun. No strong arguments exist to suggest that the mixing-length parameter is the same in all stars and all evolutionary phases and because of this, all stellar models in the literature are hampered by this basic uncertainty.

In a recent paper [1] we presented a new theory that does not require the mixing length parameter. Our self-consistent analytical formulation of stellar convection determines all the properties of stellar convection as a function of the physical behavior of the convective elements themselves and the surrounding medium. The new theory of stellar convection is formulated starting from a conventional solution of the Navier-Stokes/Euler equations expressed in a non-inertial reference frame co-moving with the convective elements. The motion of stellar convective cells inside convective-unstable layers is fully determined by a new system of equations for convection in a non-local and time-dependent formalism.

The predictions of the new theory are compared with those from the standard mixing-length paradigm with positive results for atmosphere models of the Sun and all the stars in the Hertzsprung-Russell diagram.

1. Introduction

The transfer of energy by convection is of paramount importance in all the stars. High-mass stars, roughly for masses $M > 1.3M_{\odot}$ contain fully convective cores, all stars $M \in [0.1, 100]\,M_{\odot}$ have outer convective envelopes, and finally stars smaller in mass than $M < 0.3M_{\odot}$ are fully convective. Despite its great importance, a satisfactory treatment of stellar convection in stars is still open to debate and a self-consistent treatment of the physics of convective energy transfer is still missing. The ideal goal would be to obtain a set of self-consistent equations, i.e. a set of equations resulting from physical assumptions without ad-hoc parameters to be determined by means of suitable calibrators (usually the Sun).

The most successful theory dealing with the external convection is the mixing-length theory (MLT) developed long ago by [2] and [3]. The MLT is the paradigm reference to which every new theory has to be compared with, because of its success over decades in which it has been used. In the MLT, the convective elements are supposed to travel a mean-free-path l_m [4]. l_m which is assumed to be proportional to the natural distance scale h_P, given by the pressure stratification of the star, the proportionality constant being the mixing-length (ML) parameter Λ_m, defined implicitly as $l_m \equiv \Lambda_m h_P$. The parameter Λ_m is derived from comparing the theoretical luminosity, radius and effective temperature of a stellar model for the Sun to its observational values.

In a recent paper [1], we developed the first theory of stellar convection in which the solar properties are reproduced without making use of free parameters. In the following we will refer to this theory as the scale-free convection (SFC) theory. In this approach the authors obtained a solution for the equations governing stellar atmospheres that self-consistently predict the energy transport, luminosities, radii and effective temperatures all along the evolutionary sequence of a star.

2. A mixing-length free set of equation for stellar atmospheres

The ideas at the base of the SFC theory are in principle simple. Let us think for example of the upward motion of a convective element. The evolution of a single convective cell can be considered as the sum of the upward motion and the expansion. In the MLT only the upward motion is considered. The free-parameter of the MLT stems indeed from the assumptions made to describe the upward motion of the convective elements. Therefore, the only logical alternative in developing a new theory is to consider the expansion of the convective as the main driver of the whole process. To make the upward motion ineffective it is enough to write the equations describing the motion of a convective element in a reference frame co-moving with it. In such a case, all equations are referred to convective element and the latter is at rest. [1] name S_1 this comoving reference, to distinguish it from the inertial reference frame centered on the star and named S_0. In [1] the hydrodynamic equations have been integrated accounting for the non-inertial apparent forces that arise in the treatment of any physical system evolving in S_1. Under the assumption that viscous terms are much smaller than the inertial ones and the magnetic field is negligible, the potential flow approximation can be adopted and suitably formulated in S_1 (mathematical formulation in S_1 is slightly more complicated). In order to keep the equations analytically treatable, [1] limit the analysis to the linear regime. If we limit ourselves to the subsonic regime of the stellar convection, the velocity of a convective element, $\mathbf{v}$, will be much smaller that its expansion rate, $\left\|\frac{d\xi}{dt}\right\| \equiv \left\|\dot{\xi}\right\|$, where ξ is the size of the convective elements. Therefore, a linear theory on the small parameter $\varepsilon \equiv \frac{\|\mathbf{v}\|}{\|\dot{\xi}\|} \ll 1$ can be developed. In particular, within the framework of this linear approximation, in the equations governing the evolution of the expansion rate of a convective element, the role of the inertia of the fluid displaced by the motion of the convective element turns out to be important. In contrast, this term has always been neglected in the literature and the evolution of the convective elements was always studied only in relation to its vertical motion (this led indeed to the problem of the mixing-length scale).

As a result of this approach, we obtain a new system of equations for the energy transfer as a function of the radiative plus conductive flux $\varphi_{\mathrm{rad|cnd}}$, the convective flux φ_{cnv}, the average temperature over pressure gradient $\nabla_e \equiv \left|\frac{d\ln T}{d\ln P}\right|_e$ of the element, and the stellar gradient ∇. Moreover, two extra variables, the mean velocity $\bar{v}$ and the mean size $\bar{\xi}_e$ of the convective elements are obtained as a result of the solution of the system. All these physical quantities

IOP Conf. Series: Journal of Physics: Conf. Series **940** (2017) 012020 doi:10.1088/1742-6596/940/1/012020

are a function of the pressure P, temperature T, density ρ, specific heat at constant pressure c_p, adiabatic gradient of temperature over pressure $\nabla_{\rm ad} \equiv \left|\frac{d\ln T}{d\ln P}\right|_{\rm ad}$, radiative gradient $\nabla_{\rm rad}$, molecular weight gradient $\nabla_\mu \equiv \frac{d\ln\mu}{d\ln P}$, the gravity g, the opacity κ. Finally, all these quantities are a function of the position inside the star and time.

The general form of the system of equations obtained in [1] is:

$$\begin{cases} \varphi_{\rm rad/cnd} &= \frac{4ac}{3}\frac{T^4}{\kappa h_P \rho}\nabla \\ \varphi_{\rm rad/cnd} + \varphi_{\rm cnv} &= \frac{4ac}{3}\frac{T^4}{\kappa h_P \rho}\nabla_{\rm rad} \\ \bar{v}^2 &= \frac{\nabla - \nabla_e - \frac{\varphi}{\delta}\nabla_\mu}{\frac{3h_P}{2\delta\bar{v}\tau} + \left(\nabla_e + 2\nabla - \frac{\varphi}{2\delta}\nabla_\mu\right)}\bar{\xi}_e g \\ \varphi_{\rm cnv} &= \rho c_P T\left(\nabla - \nabla_e\right)\frac{\bar{v}^2\tau}{h_P} \\ \frac{\nabla_e - \nabla_{\rm ad}}{\nabla - \nabla_e} &= \frac{4acT^3}{\kappa\rho^2 c_P}\frac{\tau}{\bar{\xi}_e^2} \\ \bar{\xi}_e &= \frac{g}{4}\frac{\nabla - \nabla_e - \frac{\varphi}{\delta}\nabla_\mu}{\frac{3h_P}{2\delta\bar{v}\tau} + \left(\nabla_e + 2\nabla - \frac{\varphi}{2\delta}\nabla_\mu\right)}\bar{\chi}, \end{cases} \tag{1}$$

where a is the radiation-density constant, and c the speed of light and for the purposes of this paper $\bar{\chi}$ is a function of time linking size to velocity, i.e. a monotonic linear map (a bijection) between time, velocity and size of the convective elements (see Appendix A of [1] for more details). In addition to this we need a numerical procedure to solve all the above equations together with their boundary conditions. We adopt the code for stellar models written by [5] and largely modified and updated by the Padua group: [6] with semi-convection, [7] with ballistic convective overshoot from the core, [8] with envelope overshoot, [9], [10] and [11] with turbulent diffusion, finally the many revision and improvements described in [12], [13], [14]. In the future we will implement the new theory of convection also in the twin-code developed independently by [15]. The SFC theory of [1] and the classical MLT are run in parallel so that comparison is possible.

3. Results: the model matching the Sun

We present here a comparison between the standard MLT and the SFC theory. The results are obtained from solving the system of Eq.1 for each layer of a stellar atmosphere governed by the equations considered in the previous section. We consider the stellar track of [14] best fitting the present position of the Sun on the HRD e.g., $\log_{10}\{L/L_\odot, T_{\rm eff}\} \cong \{0.000, 3.762\}$ with standard chemical composition $\{X, Y\} = \{0.71, 0.27\}$. The results are shown in Fig.1. In the same plot we show also the predictions of the MLT with $\Lambda_m = 1.65$ (the MLT is according to the version presented in [4], so that comparison between SFC theory and MLT is possible. Both the temperature gradients ∇ and ∇_e and fluxes $\varphi_{\rm rad|cnd}$ and $\varphi_{\rm cnv}$ predicted by SFC theory and MLT are in mutual agreement over an impressive range in pressure of almost ten orders of magnitude.

4. Conclusions

We have presented here the first results of the integration of stellar atmospheres with SFC theory developed in [1]. We have set up a numerical code to systematically integrate as function of time and position the equations presented in [1] and have run it in parallel with the standard MLT. All the results achieved by MLT are successfully recovered by the SFC theory without making use of any adjustable free-parameter. We argue that the new theory despite its linear formulation is able to capture the essence of the convection in the stellar atmospheric layers in a simple manner. Furthermore, the SFC theory has a predictive potential that descriptive analysis of numerical simulations still miss. To be able to generate numerical simulations with millions of degrees of freedoms does not automatically mean that we fully understand them. An

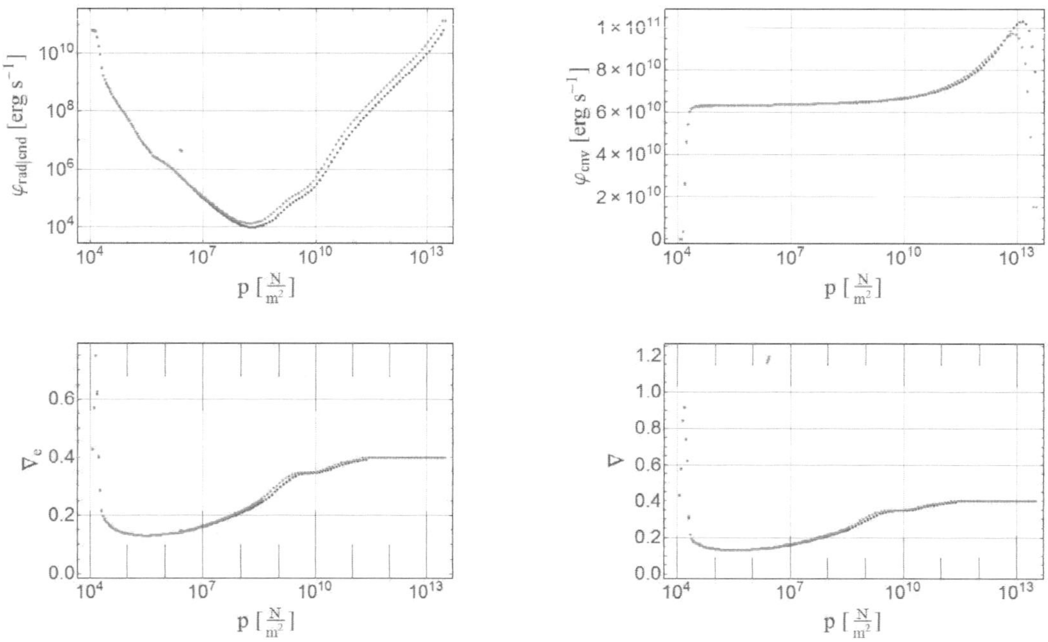

Figure 1. Solar fluxes and temperature gradient profiles for the internal pressure stratification of the star. The upper panels show the expectation for $\varphi_{\rm rad|cnd}$ on the left and $\varphi_{\rm cnv}$ on the right. Yellow refers to our theory, blue to the MLT.

emblematic example of these problems has recently been discussed by [17] where the authors fail to close the equations suggested by their hydrodynamic simulations. Their ultimate goal is to search 1D theory based on 3D simulations passing through a 2D intermediate stage. We dare to claim here that a 1D, parameter-free theory already exists [1] and that this is the right trail to follow.

References

[1] Pasetto S., Chiosi C., Cropper M., Grebel E. K., 2014, MNRAS, 445, 3592
[2] Biermann L., 1951, Zeitschrift fur Astrophysik, 28, 304
[3] Bohm-Vitense E., 1958, Zeitschrift fur Astrophysik, 46, 108
[4] Kippenhahn R., Weigert A., Weiss A., 2012, Stellar Structure and Evolution, Cambridge Press.
[5] Hofmeister E., Kippenhahn R., Weigert A., 1964, Zeitschrift fur Astrophysik, 59, 215
[6] Chiosi C., Summa C., 1970, Ap&SS, 8, 478
[7] Bressan A. G., Chiosi C., Bertelli G., 1981, A&A, 102, 25
[8] Alongi M., Bertelli G., Bressan A., Chiosi C., 1991, A&A, 244, 95
[9] Deng L., Bressan A., Chiosi C., 1996a, A&A, 313, 145
[10] Deng L., Bressan A., Chiosi C., 1996b, A&A, 313, 159
[11] Salasnich B., Bressan A., Chiosi C., 1999, A&A, 342, 131
[12] Bertelli G., Bressan A., Chiosi C., Ng Y. K., Ortolani S., 1994, Memorie della Societ Astronomia Italiana, 65, 689
[13] Bertelli G., Bressan A., Chiosi C., Ng Y. K., Ortolani S., 1995, A&A, 301, 381
[14] Bertelli G., Nasi E., 2001, AJ, 121, 1013
[15] Bertelli G., Nasi E., Girardi L., Chiosi C., Zoccali M., Gallart C., 2003, AJ, 125, 770
[16] Weiss, A., Schlattl, H., 2008, ApSS, 316, 99
[17] Tuteja G. S., Khattar D., Chakraborty B. B., Bansal S., 2010, Int. J. Contemp. Math. Sciences, 5, 1065
[18] Arnett W. D., Meakin C., Viallet M., Campbell S. W., Lattanzio J., Mocak M., 2015, ApJ, 809, 30

IOP Conf. Series: Journal of Physics: Conf. Series **940** (2017) 012021 doi:10.1088/1742-6596/940/1/012021

Nucleosynthesis in the first massive stars

Arthur Choplin[1], Georges Meynet[1], André Maeder[1], Raphael Hirschi[2], Cristina Chiappini[3]

[1]Geneva Observatory, University of Geneva, Maillettes 51, CH-1290 Sauverny, Switzerland
[2]Astrophysics group, Lennard-Jones Laboratories, EPSAM, Keele University, ST5 5BG, Staffordshire, UK
[3]Leibniz-Institut fur Astrophysik Potsdam (AIP), An der Sternwarte 16, 14482 Potsdam, Germany

E-mail: `georges.meynet@unige.ch`

Abstract. The nucleosynthesis in the first massive stars may be constrained by observing the surface composition of long-lived very iron-poor stars born around 10 billion years ago from material enriched by their ejecta. Many interesting clues on physical processes having occurred in the first stars can be obtained based on nuclear aspects. First, in these first massive stars, mixing must have occurred between the H-burning and the He-burning zone during their nuclear lifetimes; Second, only the outer layers of these massive stars have enriched the material from which the very iron-poor stars, observed today in the halo of the Milky Way, have formed. These two basic requirements can be obtained by rotating stellar models at very low metallicity. In the present paper, we discuss the arguments supporting this view and illustrate the sensitivity of the results concerning the [Mg/Al] ratio on the rate of the reaction ^{23}Na(p,γ)^{24}Mg.

1. Some puzzling abundance patterns

If a particularly imaginative theoretician would have imagined the composition of very metal-poor stars formed from the ejecta of the first stellar generations, she/he would likely never have succeeded to explain the very strange abundance patterns shown by these stars. The fact that they present very low iron is per se not a surprise, but the fact that they present very strong overabundances of carbon with respect to iron is at first sight astonishing. Moreover, other light elements like nitrogen, oxygen, sodium, magnesium and aluminum also show excesses with respect to iron[1]. So the question is which kind of sources can predict such abundance patterns? This is the main topic of the present paper. However, before to address that specific question, it is worthwhile to be aware of a few facts.

The stars we are speaking about are named the Carbon-Enhanced Metal Poor stars. These stars have clearly much larger [C/Fe] than the bulk of the normal halo stars (Aoki et al. 2007). Some of them present significant excesses in r- and s-process elements. In that case they are classified as CEMP-r, CEMP-s or CEMP-r/s stars. Some of them present no or only very small excesses in neutron-capture elements and are named CEMP-no stars. Presently 46 stars of this type are known (see table 2 in Maeder & Meynet 2015). The sample increases slowly especially at the very low metallicity end. A way to illustrate this is to note that the number of stars

[1] Actually the abundance ratios of these elements relative to iron are increased with respect to the abundance ratios in the Sun.

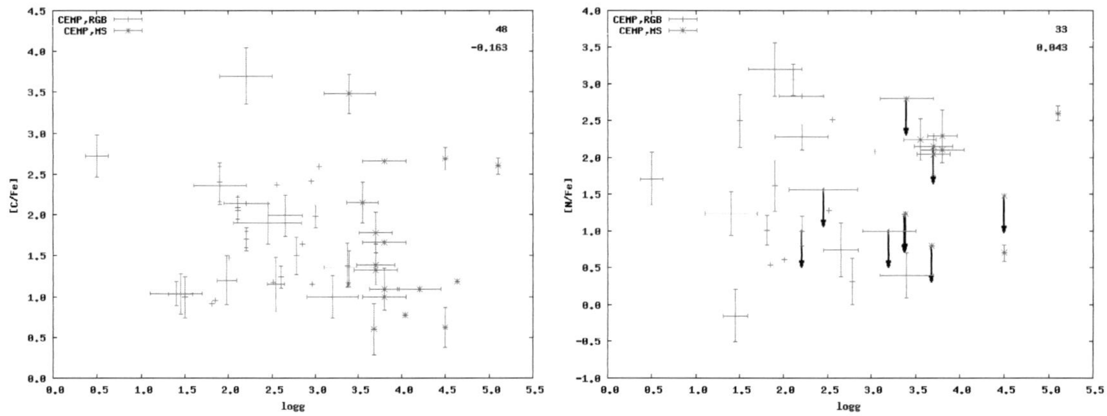

Figure 1. *Left panel:* CEMP stars with [Fe/H] < -2.0, [C/Fe] > 0.9 and [Ba/Fe] < 0.9 plotted in the plane [C/Fe] versus the surface gravity of the stars. Different colors are for main-sequence and red giant stars. *Right panel:* Same stars as those plotted in the left panel that have measured values for the [N/Fe] ratio. The database SAGA has been used to make this plot (see Suda et al. 2008).

mentioned in the review by Beers & Christlieb (2005) with a metallicity [Fe/H] below -4 was 6. Ten years later, in 2015, about 14 stars have been added in that metallicity range (see e.g. Frebel et al. 2015), which rises the number to 20. The 20 stars are all CEMP stars except one SDSS J102915+172927 (Caffau et al 2012).

How did these very metal poor stars form? Why are most of them carbon-rich? Could these abundances be built by the star itself? If not, what was the nature of the nucleosynthetic sources that build up their abundance pattern? These questions are the main topics of the next section.

2. Some clues from nuclear considerations

CEMP-no stars are found among main-sequence and evolved red giant stars. In the left panel of Fig. 1, we can see that the scatter in carbon overabundance is more or less the same for non-evolved main-sequence (MS) stars and evolved red giant (RG) stars. This supports the idea that the carbon enrichment is not just an enrichment of only the outer layers of the stars acquired by the accretion of a thin enriched layer. Would it be the case, then how to explain this similarity between MS and RG stars? At the red giant stage, a deep convective envelope produces a rapid mixing and any superficial enrichment would be rapidly diluted and made either unobservable or at least much smaller than during the MS phase. So what we see at the surface has great chance to be the bulk composition of the whole star, not just of its surface.

The stars plotted in Fig. 1 are very small mass stars (around 0.6-0.8 $M_{\odot}$). These stars cannot produce any carbon in the stellar evolutionary phases in which they are observed. Thus they must have inherited these abundances from the protostellar cloud, from which they formed. The iron content of this protostellar cloud is so low that it must have been enriched by a unique or maybe two events able to enrich their surrounding in short enough timescales for avoiding the cloud to be enriched by many other stars that would otherwise increase the iron content rapidly above the observed level. Thus, the source stars, that produced the enrichment of this protostellar cloud are very likely short-lived massive stars.

The CEMP-no stars clearly show the signature of H- and He-burning. Concerning H-burning, one (among many others) signature comes from the fact that most CEMP-no stars present nitrogen overabundances. This is shown in the right panel of Fig. 1. As for the [C/Fe] ratio, there is no very relevant differences between MS and RG stars. During the RG phase, some nitrogen surface enrichment may occur due to the concomitant effects of the CNO cycle operating in the H-burning shell and the dredge-up due to convection.

There are also very clear signatures of the operation of He-burning reactions. The presence, at the surface of CEMP-no stars, of high abundances of carbon, element produced by the triple α reaction, is one of them (see the left panel of Fig. 1). Are CEMP-no stars made of a mixture of H- and He-burning processed material? Actually, while such a mixture is indeed required, it is still not sufficient. To see that, let us make a very simple thought experiment. Let us consider for instance some material having initially a very low metal content, let us say to fix the idea having a mass fraction of heavy elements (Z) equals to 10^{-4}. If we assume that the heavy elements are distributed like in the Sun, then the initial abundance of CNO elements in mass fraction would correspond to about 3/4 of Z, thus of the order of 10^{-4}. Since the CNO cycle mainly transforms the carbon and oxygen into nitrogen, at the end of the process, the nitrogen abundance would also be around 10^{-4}. The mass fraction of iron would not change and will stay around 1/20 of the initial Z, that means $5{\cdot}10^{-6}$, so that in CNO-processed material one expects a [N/Fe] at most around 1.3 - 0.3 = 1 dex (since the brackets mean relative to the Sun and $(\mathrm{N/Fe})_\odot \simeq 0.3$ dex). Actually this is an upper limit, since we assumed a complete transformation of carbon and oxygen and we did not account for any dilution effect when this material is mixed with the material processed by He-burning.

Now let us consider the star G77-61 which has a surface gravity of log g= 5.1 and an iron abundance of [Fe/H] = −4 (Plez & Cohen 2005). That star is the MS star shown at the right in both panels of Fig. 1. Let us assume that this star was made from the ejecta of a source star having also a [Fe/H] equal to -4 and a solar distribution of the heavy elements. As explained above, one would then obtain an upper limit for the [N/Fe] equals to 1, but this is much too small with respect to the observed value of 2.6. It means that the above scenario is unable to account for the observed N overabundance. Something is missing. We could argue that the source star began with a non solar initial composition, for instance which much larger carbon initial abundance but in that case, it would just shift the problem: how to explain the high carbon abundance in that previous generation of stars? This simple example shows that just mixing material processed by H- and He-burning is not sufficient. Actually, in order to produce the large amounts of nitrogen, it is necessary to assume that some mixing operates between these two nuclear burning regions. If some carbon produced in the He-core diffuses into the H-burning region then this would produce much larger amounts of nitrogen. Thus we can conclude that the abundance patterns of CEMP-no stars provide some clues for the existence of a mixing mechanism in the source stars having occurred during their nuclear lifetimes.

There is still one more conclusion that we can deduce just from nuclear physics arguments. We just saw that the abundance patterns of CEMP-no stars present the signature of He-burning, but this signature does not overwhelm the one of the H-burning process. Too much He-burning material would produce very high [C/N] ratios due to the fact that carbon is so much more abundant in the He-burning region than in the H-burning one. This would make impossible to understand a [C/N] ratio of about 0 in the case of G77-61. It would also be in contradiction with the very low $^{12}\mathrm{C}/^{13}\mathrm{C}$ ratio measured at the surface of this star which is around 5. This value is equal to the value expected from CNO processing alone (with some carbon produced by He-burning injected into the H-burning shell to account for the high carbon abundances). If too much material from the He-burning region would contribute in making the material from which this star formed, then one would obtain high [C/N] and $^{12}\mathrm{C}/^{13}\mathrm{C}$ ratios that are not observed. This supports the view that no or only a very small amount of the He-burning zone has enriched that material.

3. From nuclear considerations to a stellar model

Let us now see how the above conditions can be realized in a star. Having material processed by H- and He-burning nuclear reactions is per se not constraining since these two nuclear networks are active in any massive stars. Now we need mixing to occur between these two zones. Meynet & Maeder (2002a), Hirschi (2007) have proposed that rotation might trigger this mixing, producing large amounts of nitrogen at low metallicity. Chiappini (2013) in a recent review discusses many other indications supporting rotational mixing from nucleosynthetic arguments. In brief, the arguments favoring rotational mixing, and a more efficient mixing at low metallicity are the following: first the fraction of rapidly rotating stars is larger when the metallicity is decreasing (Maeder et al. 1999, Martayan et al. 2007). Second the rotational mixing is more efficient, all other parameters being kept the same, when the metallicity is lower (Maeder & Meynet 2001). This is due to the fact that stars at low metallicity are more compact. Also, in metal-poor massive stars, the distance between the H- and He-burning zone decreases since the lack of CNO elements implies that the H-burning shell occurs at higher temperature and thus in a zone nearer from the He-burning core. All these considerations produce the interesting situation that the mixing which is indeed needed at very low metallicity to explain the CEMP-no stars, occurs only at low or very low metallicity. It is absent at solar metallicity for instance. This is quite consistent with the fact that at solar metallicity there are no evidences for processes requiring such mixing like the production of primary nitrogen[2].

Another interesting aspect of this kind of models is that they might also produce conditions for triggering stellar winds even if the source massive star had a very low initial metallicity. This comes from the fact that the rotational mixing can enrich the surface in CNO elements that might enhance the opacity of the outer layers during the red supergiant phase and thus triggers important mass losses through radiatively driven stellar winds (Meynet & Maeder 2002b). The fact that the matter is ejected with a low velocity may help star formation to occur rapidly from a mixture of such stellar winds and ISM. The matter ejected at high velocity during the supernova ejecta might on the other hand escape from the potential well. So we see that rotation might produce both interesting abundance patterns and at the same time favorable conditions for star formation to occur rapidly from a mixture of the ejected material with some ISM.

Such a model requires, on the other hand, that the metallicity of the source stars be equal or higher (in case of dilution) than the metallicity of the CEMP-no stars, because the wind is not enriched in iron. Thus, in that case, the source star would not be a Pop III star but a very metal poor massive star of likely second generation. Another possibility would be that the source star would be primordial. In that case, it is needed that some ejection of iron occurs at the moment of the faint supernova event as proposed by Umeda & Nomoto (2005). This last scenario could also occur in case of a non-zero metallicity stars.

At the moment it is difficult to disentangle between these two possibilities. One can wonder whether in case of Pop III star, there would still be possible to account for the small amounts of strontium and barium that are observed at the surface of some CEMP-no stars. If these elements are produced by the r-process then this would indicate some contribution from the supernova explosion and it could explain their presence even starting from a Pop III star. If they are produced by the s-process then likely the source star was not a Pop III stars, since in that case the production of s-process elements might be nearly zero due to the lack of iron seeds (it might be not completely zero due to the fact that other elements could serve as seeds for neutron-captures at low metallicity). Further investigations are needed to explore this point.

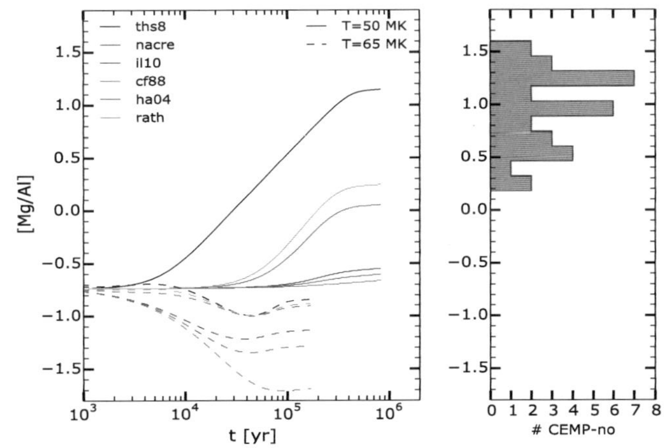

Figure 2. *Left panel:* [Mg/Al] ratio as a function of time in a hydrogen burning box where we inject ^{12}C, ^{16}O and ^{22}Ne. 6 rates are tested for the reaction ^{23}Na(p,γ)^{24}Mg at $T_6 = 50$ and 65. Tested rates are from Cyburt et al. (2010) (ths8), Angulo et al. (1999) (nacre), Iliadis et al. (2010) (il10), Caughlan & Fowler (1988) (cf88), Hale et al. (2004) (ha04) and Rauscher & Thielemann (2000) (rath). *Right panel:* observed [Mg/Al] distribution for CEMP-no stars.

4. Nuclear reaction rate uncertainties

The results of models for the source stars depends on many physical ingredients such as the way of treating the rotational mixing for instance. Among the physical ingredients, there is also the impact of some uncertain nuclear reaction rates. In order to see more clearly what could be the impact of a given nuclear reaction rate, it is common to study the nucleosynthesis in a box at a fixed temperature and density and see how the abundances obtained change when different nuclear reaction rates are used. In Fig. 2, we show the result of such an experiment. The density is set to 1 g cm^{-3}. The initial abundances in the box are taken from the Hydrogen burning shell of a 20 $M_\odot$ model at the beginning of the core Helium burning phase. The computation is done at two temperatures $T_6 = 50$ (plain lines) and $T_6 = 65$ (dashed lines) which are typical temperature of H-shell burning. To model the mixing occurring between the He- and H-burning region, ^{12}C, ^{16}O and ^{22}Ne are injected into the box at constant rates $R_{^{12}C} = 10^{-8}$ yr^{-1}, $R_{^{16}O} = 10^{-8}$ yr^{-1} and $R_{^{22}Ne} = 10^{-10}$ yr^{-1}. Those rates represents the mass fraction we add per year of a given element. We chose those species because they correspond in general to the most abundant species diffusing by rotational mixing from the He-burning core into the H-burning shell of a low metallicity massive stellar model. The nitrogen produced in the H-shell may diffuse at its turn into the He-core. The chain ^{14}N(α,γ)^{18}F(,$e^+\nu_e$)^{18}O(α,γ)^{22}Ne, occurring in the He-core, allows the synthesis of a large amount of ^{22}Ne, which also diffuses into the H-burning shell. We note that the amount of ^{22}Ne entering in the H-burning shell is lower (at least 2 dex lower than the amount of ^{12}C and ^{16}O) so that the rate $R_{^{22}Ne}$ is taken 100 times smaller than the rates $R_{^{12}C}$ and $R_{^{16}O}$. Six rates for the reaction ^{23}Na(p,γ)^{24}Mg are tested, taken from the database JINA REACLIB database (Cyburt et al. 2010).

The duration of the simulation depends on the temperature : the Hydrogen is consumed more rapidly if higher temperatures are chosen. The injection of ^{22}Ne ignites the Ne-Na chain

[2] Primary nitrogen is produced from carbon and oxygen synthesized by the star in its He-burning zone, while secondary nitrogen is produced from carbon and oxygen already present in the star at its formation and thus resulting from one (or a few) previous generation(s) of star.

and also the Mg-Al chain, if the temperature is high enough. $T_6 = 50$ (plain lines) is a too low temperature for the magnesium to burn into aluminum. Magnesium is then only synthesized, essentially through the reaction ^{23}Na(p,γ)^{24}Mg. At $T_6 = 65$, the reaction ^{26}Mg(p,γ)^{27}Al becomes significant so that magnesium is also destroyed, overall reducing the [Mg/Al] ratios compared to the $T_6 = 50$ case.

These numerical experiments can be used to indicate which rates are still not known with a sufficient accuracy. Here we see that still large uncertainties pertain the rate of the reaction ^{23}Na(p,γ)^{24}Mg in the low temperature range ($T_6 = 50 - 65$), producing therefore very different outcomes for the [Mg/Al] ratios. On the right panel of Fig. 1, the distribution of the observed [Mg/Al] ratios in CEMP-no stars is shown. We see that the observed scatter is smaller than the scatter due to the uncertainties in the nuclear physics. Such test cannot be used to constrain the nuclear reaction rates, since the final result depends on some arbitrary chosen ingredients of the numerical experiment, but at least it might show the needs from improvements on the side of the nuclear physics before some relevant conclusions can be drawn from comparisons between models and observations.

5. Conclusions

We discussed the possibility of constraining the first massive stars by observing the surface composition of CEMP-no stars. We suggested that a mixing between the hydrogen and helium burning region of the source star is necessary to reproduce the abundance pattern of CEMP-no stars and that only the outer layers should be expelled. These outer layers were likely expelled through strong winds or by a faint supernova. We also pointed out the need of a better determination of the ^{23}Na(p,γ)^{24}Mg reaction at relevant temperatures for hydrogen burning.

References

[1] Angulo C, Arnould M, Rayet M, Descouvemont P, Baye D, Leclercq-Willain C, Coc A, Barhoumi S, Aguer P, Rolfs C, Kunz R, Hammer J W, Mayer A, Paradellis T, Kossionides S, Chronidou C, Spyrou K, degl'Innocenti S, Fiorentini G, Ricci B, Zavatarelli S, Providencia C, Wolters H, Soares J, Grama C, Rahighi J, Shotter A and Lamehi Rachti M 1999 *Nuclear Physics A* **656** 3–183
[2] Aoki W, Beers T C, Christlieb N, Norris J E, Ryan S G and Tsangarides S 2007 *ApJ* **655** 492–521
[3] Beers T C and Christlieb N 2005 *ARA&A* **43** 531–580
[4] Caffau E, Bonifacio P, François P, Spite M, Spite F, Zaggia S, Ludwig H G, Steffen M, Mashonkina L, Monaco L, Sbordone L, Molaro P, Cayrel R, Plez B, Hill V, Hammer F and Randich S 2012 *A&A* **542** A51
[5] Caughlan G R and Fowler W A 1988 *Atomic Data and Nuclear Data Tables* **40** 283
[6] Chiappini C 2013 *Astronomische Nachrichten* **334** 595–604
[7] Cyburt R H, Amthor A M, Ferguson R, Meisel Z, Smith K, Warren S, Heger A, Hoffman R D, Rauscher T, Sakharuk A, Schatz H, Thielemann F K and Wiescher M 2010 *ApJS* **189** 240–252
[8] Frebel A, Chiti A, Ji A P, Jacobson H R and Placco V M 2015 *ApJ* **810** L27
[9] Hale S E, Champagne A E, Iliadis C, Hansper V Y, Powell D C and Blackmon J C 2004 *Phys. Rev. C* **70** 045802
[10] Hirschi R 2007 *A&A* **461** 571–583
[11] Iliadis C, Longland R, Champagne A E and Coc A 2010 *Nuclear Physics A* **841** 251–322
[12] Maeder A, Grebel E K and Mermilliod J C 1999 *A&A* **346** 459–464
[13] Maeder A and Meynet G 2001 *A&A* **373** 555–571
[14] Maeder A and Meynet G 2015 *A&A* **580** A32
[15] Martayan C, Frémat Y, Hubert A M, Floquet M, Zorec J and Neiner C 2007 *A&A* **462** 683–694
[16] Meynet G and Maeder A 2002 *A&A* **381** L25–L28
[17] Meynet G and Maeder A 2002 *A&A* **390** 561–583
[18] Plez B and Cohen J G 2005 *A&A* **434** 1117–1124
[19] Rauscher T and Thielemann F K 2000 *Atomic Data and Nuclear Data Tables* **75** 1–351
[20] Suda T, Katsuta Y, Yamada S, Suwa T, Ishizuka C, Komiya Y, Sorai K, Aikawa M and Fujimoto M Y 2008 *PASJ* **60** 1159–
[21] Umeda H and Nomoto K 2005 *ApJ* **619** 427–445

IOP Conf. Series: Journal of Physics: Conf. Series **940** (2017) 012022 doi:10.1088/1742-6596/940/1/012022

γ-process reaction studies via in-beam γ-ray spectroscopy at HORUS

Philipp Scholz, Jan Mayer, Lars Netterdon, and Andreas Zilges

Institute for Nuclear Physics, University of Cologne, Zülpicher Strae 77, 50937 Cologne, Germany

E-mail: pscholz@ikp.uni-koeln.de

Abstract. The predictions for reaction rates from the Hauser-Feshbach statistical model as they are needed for γ-process reaction-network calculations crucially depend on the accuracy of the nuclear physics input-parameters like nuclear-level densities, γ-ray strength functions, and particle+nucleus optical-model potentials. At the Institute for Nuclear Physics of the University of Cologne, several charged-particle induced reactions were investigated via the in-beam method with HPGe detectors at the high-efficiency γ-ray spectrometer HORUS. One outstanding feature of this experimental method is the access to partial cross-sections, thus, cross-sections for the population of excited states in the reaction product via a radiative capture of an charged particle. Here, recent results for the reactions ^{89}Y(p,γ)^{90}Zr and ^{112}Sn(α,γ)^{116}Te will be presented.

1. Introduction

Despite the enormous effort and progress in the field during the last years, the nucleosynthesis processes for the p nuclei are still not fully understood [1]. However, although several other processes in various astrophysical scenarios were proposed, the γ-process [2] is still believed to be responsible for the largest contributions to the abundance of the p nuclei for masses $A > 100$. Speaking about network calculations for the γ-process means considering thousands of reactions on many mainly unstable nuclei [3] and, therefore, relying on theoretically calculated reaction rates within the Hauser-Feshbach statistical model [4]. The accuracy of these calculated reactions rates, however, strongly depends on the adopted theoretical models for nuclear physics input-parameters as nuclear-level densities (NLD), γ-ray strength functions and particle + nucleus optical-model potentials (OMP). In order to constrain or invalidate different models for these input-parameters, precise cross-section measurements of nuclear reactions inside or close to the astrophysical relevant energy region are needed. In the last years, many different charged particle induced reactions at sub-Coulomb energies were investigated in Cologne and a selection should be shown here. After the experimental setup for nuclear astrophysics experiments in Cologne will be presented in Sec. 2, a selection of recent results on studied charged-particle induced reactions will be discussed in Sec. 3.

2. In-beam γ-ray spectroscopy at HORUS

All in-beam experiments presented in the following were carried out at the Institute for Nuclear Physics of the University of Cologne. The charged-particle beam was delivered by the 10 MV FN-Tandem accelerator to the experimental setup at HORUS. This high-efficiency γ-ray

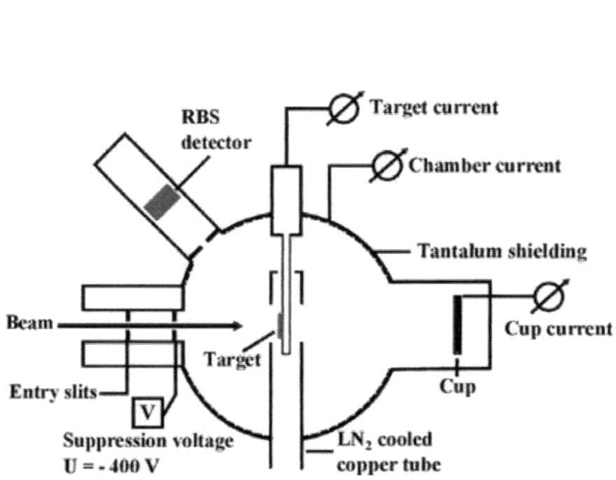

Figure 1. Schematical illustration of the target chamber which was used for the experiments at HORUS. The pictures was taken from Ref. [5].

spectrometer consists of 14 high-purity germanium detectors (HPGe) which are partly equipped with BGO-shields for an active Compton background suppression. The target chamber, see Fig. 1, which was specifically designed for nuclear astrophysics experiments is coated from the inside with tantalum. This high-Z shielding hampers reactions of the charged particle beam with the target chamber material. For a reliable cross section determination, recording the beam current, thus, the number of impinging particles, is of utmost importance. For this, the deposited charge by the beam particles is read out at three independent positions at the same time: on the target itself, the chamber, and at the build-in Faraday Cup behind the target. In addition, a suppression voltage of $U = -400V$ at the entrance of the target chamber prevents secondary electrons to leave the chamber. A particle detector for Rutherford Backscattering Spectrometry measurements assures the monitoring of the target stability throughout the whole time of irradiation. Around the target position, a cooling trap at temperatures of liquid nitrogen prevents condensation of residual molecules on the target surface.

3. In-beam reaction studies for the γ-process nucleosynthesis

3.1. The ^{89}Y*(p,γ)*^{90}Zr *reaction*

The ^{89}Y(p,γ)^{90}Zr reaction is located in a region of the chart of nuclei where the p isotopes are normally underproduced within γ-process reaction-network calculations [6]. Since the total cross section of this reaction was measured twice before [7,8], this reaction was an excellent candidate for a commissioning experiment of the nuclear astrophysics setup at HORUS in Cologne [5]. Moreover, since ^{90}Zr was investigated using nuclear-resonance fluorescence before [9], it was possible to extract valuable input for the adopted γ-ray strength function in the compound nucleus of this reaction. For five different proton energies between 3.65 MeV and 4.7 MeV total cross-section values could be obtained. In addition, via the detection of the prompt γ-rays seven partial cross-section values could be measured which are particularly sensitive to the input of the γ-ray strength function entering the statistical model calculations between γ-ray energies of 7.71 MeV and 12.98 MeV. Figure 2 shows some of the obtained values in comparison with theoretical predictions calculated using the TALYS code in version 1.6 [11]. The dashed line in Fig. 2 corresponds to calculations using the γ-ray strength from Ref. [9]. As it can be seen in Fig. 2, the values from the statistical model calculation cannot reproduce the experimental ones properly. However, a rather good reproduction for the total and all partial cross sections could be obtained by adjusting the γ-strength in ^{90}Zr simultaneously for all the different channels. The results of the statistical model calculations using this adjusted γ-strength is shown as a blue band in Fig. 2, labeled as "Fit" The band represents the uncertainty in the statistical

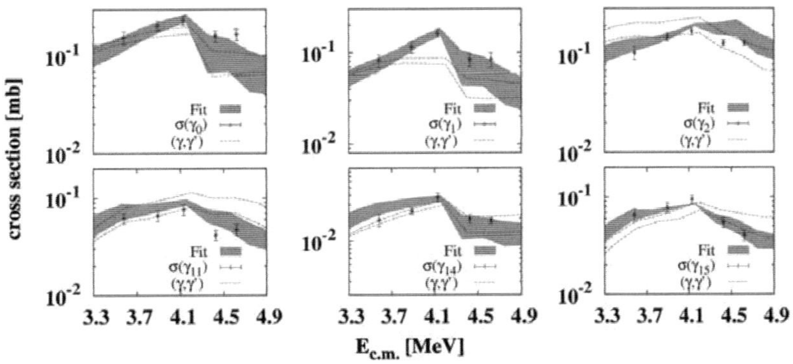

Figure 2. Some of the measured partial cross-sections compared to statistical model calculations using the (γ,γ') data from Ref. [9] as input for the γ-ray strength function in ^{90}Zr [10]. The yellow shaded area corresponds to the results using the adjusted γ-ray strength functions.

model calculations one obtains by varying all the other input-parameters. It turned out that an adequate description of the experimental values needs additional γ-strength between $E_\gamma = 10$ MeV and $E_\gamma = 12$ MeV as well as between $E_\gamma = 12$ MeV and $E_\gamma = 13.5$ which was not seen in Ref. [9]. For more information, see Ref. [10].

3.2. The ^{112}Sn$(\alpha,\gamma)^{116}$Te reaction

The radiative α-capture reaction on the p nucleus ^{112}Sn was also investigated by means of the in-beam technique with HPGe detectors [12]. At four center-of-mass energies between 10.1 MeV and 11.5 MeV, i.e. below the neutron threshold, total and partial cross sections could be measured. In this energy region, the total cross section of the reaction as calculated in the scope of the statistical model is almost exclusively sensitive to the γ-ray strength functions and the α-OMP [13]. Since partial cross sections for this reactions could be extracted, comparing them to statistical model calculations based on different models for the γ-ray strength function made it possible to reduce this uncertainty. A very good agreement was found by using the microscopic HFB + QRPA model [14]. However, constraining the input for the γ-ray strength function entering the statistical model calculations enabled tracing back the remaining deviations between the measured total cross sections and the calculated ones to the input for the α-OMP. Different TALYS calculations based on various models for the α-OMP were not able to reproduce the experimental values (see Fig. 3). In detail, each of the calculated value sets tend to underestimate the cross-section values at the highest energies which already indicates that the calculations are penalizing the absorption in comparison to the scattering of the α-particles due to an incorrect value for the depth of the real part of the α-OMP. After adjusting this value as well as the proton-width, a very good agreement was found for the disperive α-OMP of Demetriou *et al.* [15] ("TALYS Fit" in Fig. 3). For more information, see Ref. [12].

4. Outlook

The experimental program measuring radiative α-capture reactions via the in-beam method with HPGe detectors was continued by the end of 2014. At α-particle energies between 12.0 MeV and 13.5 MeV, the total cross section of the ^{108}Cd$(\alpha,\gamma)^{112}$Sn was measured at HORUS. After the irradiation, the targets were placed inside the Cologne Clover Counting Setup [18-20] in order to additionally measure the ^{108}Cd$(\alpha$,n$)^{111}$Sn reaction cross section. The latter experiment was continued at the beginning of 2015 where cross-section values for ^{108}Cd$(\alpha$,n$)^{111}$Sn

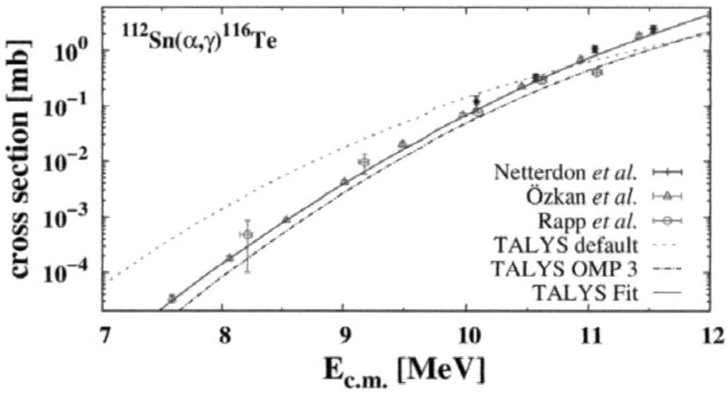

Figure 3. Measured total reaction cross-section values for the $^{112}\mathrm{Sn}(\alpha,\gamma)^{116}\mathrm{Te}$ reaction and the values from references [16] and [17] compared to different TALYS calculations [12]. "TALYS OMP3" and "TALYS Fit" denote the calculations using the α-OMP of Ref. [15] before and after adjusting the depth of the real-part of the OMP as well as the proton- and γ-width. See text for details.

reaction could be measured by means of the activation method down to 10.2 MeV. The analysis of these experiments is still ongoing but the data is expected to be published in 2015.

5. Acknowledgments

The authors thank A. Dewald and the whole accelerator staff at the Institute for Nuclear Physics of the University of Cologne for providing excellent beams. In addition, the authors acknowledge K.O. Zell and A. Blazhev for the target preparation. This work has been supported by the ULDETIS project within the UoC Excellence Initiative institutional strategy. P.S. is supported by the Bonn-Cologne Graduate School of Physics and Astronomy.

References

[1] M Wiescher, Europhysics Letters **109** (2015) 62001.
[2] S. E. Woosley and W. M. Howard, Astrophys. J. Suppl. Ser. **36** (1978) 285.
[3] M. Arnould and S. Goriely, Phys. Rep. **384** (2003) 1.
[4] W. Hauser and H. Feshbach, Phys. Rev. **87** (1952) 366.
[5] L. Netterdon *et al.*, NIM A **754** (2014) 94.
[6] M. Rayet *et al.*, Astron. Astrophys. **298** (1995) 517.
[7] P. Tsagari *et al.*, Phys. Rev. C **70** (2004) 015802.
[8] S. Harissopulos *et al.*, Phys. Rev. C **87** (2013) 025806.
[9] R. Schwengener *et al.*, Phys. Rev. C **78** (2008) 064314.
[10] L. Netterdon *et al.*, Phys. Lett. B **744** (2015) 358.
[11] A. J. Koning, S. Hilaire, and M. C. Duijvestijn, TALYS-1.0, Proceedings of the International Conference on Nuclear Data for Science and Technology, April 22-27, 2007, Nice, France, editors O.Bersillon, F.Gunsing, E.Bauge, R.Jacqmin, and S.Leray, EDP Sciences, 2008, p. 211-214.
[12] L. Netterdon *et al.*, Phys. Rev. C **91** (2015) 035801.
[13] T. Rauscher, Astrophys. J. Supple. Ser. **201** (2012) 26.
[14] S. Goriely, E. Khan, and M. Samyn, Nucl. Phys. A **739** (2004) 331
[15] P. Demetriou, C. Grama, and S. Goriely, Nucl. Phys. A **707** (2002) 253.
[16] N. Ozkan *et al.*, Phys. Rev. C **75** (2007) 025801.
[17] W. Rapp *et all.*, Phys. Rev. C **78** (2008) 025804.
[18] A. Sauerwein *et al.*, Phys. Rev. C **84** (2011) 045808.
[19] L. Netterdon *et al.*, Nucl. Phys. A **916** (2013) 149.
[20] P. Scholz *et al.*, Phys. Rev. C **90** (2014) 065807.

IOP Conf. Series: Journal of Physics: Conf. Series **940** (2017) 012023 doi:10.1088/1742-6596/940/1/012023

Detection of MeV scale neutrinos and the solar energy paradigm

Aldo Ianni

Canfranc Underground Laboratory, Paseo de los Ayerbe, s/n, E-22880 Canfranc Estación, Spain

E-mail: aianni@lsc-canfranc.es

Abstract. The fundamental solar energy paradigm establishes that the energy in the Sun is due to a series of nuclear reactions which turn hydrogen into helium. In particular, for the Sun the fundamental reaction corresponds to $p + p \to d + e^+ + \nu_e + 0.42$ MeV. This is a very slow process which drives the evolution of the Sun over a timescale of 10^9 years. Electron neutrinos produced in the core interact only weakly and travel almost undisturbed from the core to the surface. They are a unique probe to explore the interior of stars. Observations of solar neutrinos are discussed in the paper. The solar energy paradigm proposed in 1938 by H. Bethe has been measured in realtime by Borexino at 10% level in 2014. This observation allows to probe the solar stability over a 10^5 years timescale. At present, solar neutrinos offer the opportunity to understand the new Solar Abundance Problem, that is our lack of knowledge of the chemical composition of the Sun. Therefore, improving solar neutrino measurements is of great interest for astrophysics. At the same time, a better determination of some astrophysical factors will reduce uncertainties on predictions to better identify a possible inadequate assumption in the solar model.

1. Introduction

The source of energy in our Sun and in H-burning stars makes electron neutrinos through the following process:

$$4p \to^4 He + 2e^+ + 2\nu_e \quad (1)$$

Eq. (1) produces about 26.73 MeV of energy, including positrons annihilation. Neutrinos from Eq. (1) are known as solar neutrinos. These neutrinos have an average energy of 0.53 MeV and carry about 2% of the solar energy. The total energy per unit of time (luminosity) released by the Sun is $(2.40 \pm 0.01) \times 10^{39}$ MeV/s [1]. Hydrogen burning works by means of two processes known as pp-chain and CNO-cycle. These processes consist of a number of interactions which turn hydrogen into helium as from Eq. (1) [2]. This solar energy paradigm was proposed in 1938 by H. Bethe in a comprehensive and systematic work [3]. However, the first proto idea behind this solar paradigm goes back to 1868 when Joseph N. Lockyer discovered helium on the solar surface. Another fundamental information in this puzzle came in 1956 when by means of radiometric methods applied on meteorites the age of the Earth was determined to be $(4.55 \pm 0.07) \times 10^9$ years. This observation can only be supported with a Sun driven by nuclear fusion as proposed by H. Bethe. Experimental evidence of this theory came in 1968 with the first measurement of solar neutrinos [4]. As a matter of fact, solar neutrinos can stream from the center to the surface of the Sun with an extremely low probability of interaction and carry information on the nuclear processes taking place in the core of our star. Therefore, solar neutrinos are a unique probe to understand how stars work. In figure 1 we show the burning rate in stars as a function of the central temperature. The black dot shows the position of our Sun. In this case the size of the star is such that the dominant energy contribution comes from the pp-chain.

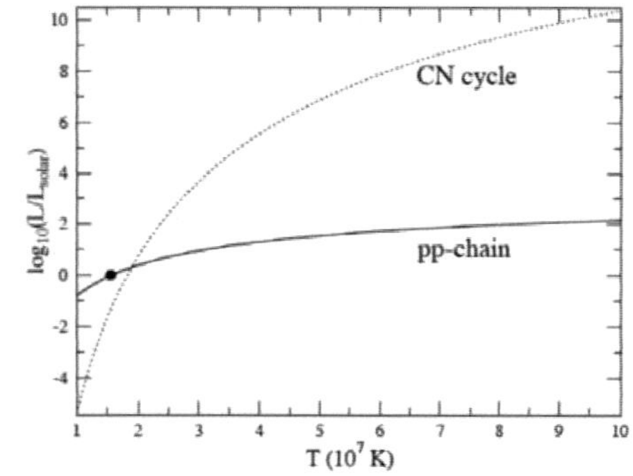

Figure 1. Burning rate in stars from the pp-chain and the CNO-cycle. For our Sun (black dot) the dominant contribution comes from the pp-chain.

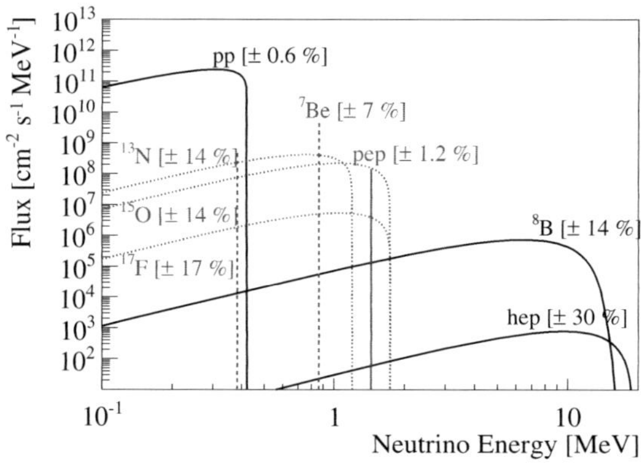

Figure 2. Energy spectrum of solar neutrinos flux at Earth.

The Solar Standard Model (SSM) is a general framework to study the characteristics of stars [2]. In particular, observation of solar neutrinos offer an opportunity to test the SSM and the use of some fundamental assumptions. The SSM is a reference model to understand how stars work. In the SSM it is assumed energy generation through hydrogen burning; hydrostatic equilibrium between the gravitational inward force and the one due to the pressure from the heat produced by nuclear fusion; the primordial core metal abundance equal to the today's surface metal abundance (zero-age homogeneous Sun). In the SSM boundary conditions include the value of the present metal abundance on the surface, $(Z/X)_{surf}$, and the solar luminosity, $L_\odot$ (X is the abundance of hydrogen and Z that of elements heavier than helium). The age, $(4.57 \pm 0.02) \times 10^9$ years, and the mass, $(1.9886 \pm 0.0003) \times 10^{30}$ kg, are fixed. The model which, after iterative evolution, matches the boundary conditions produces as output the present value for: the solar neutrino fluxes, the depth of the convective zone, the surface helium abundance, Y_{surf}, the density and sound speed profiles. In figure 2 we show the solar neutrino fluxes at Earth and predicted uncertainties obtained by the SSM. It turns out that the pp-chain and CNO-cycle make neutrinos from a number of reactions. The most abundant neutrinos (91%) are named pp-neutrinos and are sub-MeV particles. The CNO neutrinos from β decays of ^{13}N, ^{15}O and ^{17}F account for only 0.8% of the total flux. The ^{8}B neutrinos measured by Homestake

IOP Conf. Series: Journal of Physics: Conf. Series **940** (2017) 012023 doi:10.1088/1742-6596/940/1/012023

in 1968 accounts for only 0.009%. In the SSM neutrino fluxes are given as a power function of fundamental parameters. As an example, the pp and ^{7}Be neutrino fluxes are written:

$$\phi_{pp} \propto S_{11}^{+0.14} \cdot S_{33}^{+0.03} \cdot S_{34}^{-0.06} \cdot S_{1,14}^{-0.02} \cdot L_{\odot}^{+0.73} \cdot \tau_{\odot}^{-0.07} \cdot Op_{\odot}^{+0.14} \cdot (Z/X)^{-0.08} \tag{2}$$

$$\phi_{^7Be} \propto S_{11}^{-0.97} \cdot S_{33}^{-0.44} \cdot S_{34}^{+0.88} \cdot L_{\odot}^{+3.56} \cdot \tau_{\odot}^{+0.69} \cdot Op_{\odot}^{-1.49} \cdot (Z/X)^{+0.59} \tag{3}$$

where $L_{\odot}$, $\tau_{\odot}$ and $Op_{\odot}$ are the solar luminosity, age and opacity, S_{ij} astrophysical factors for reactions taking place in the core of the Sun [2]. S_{33} for $^3He +^3 He \rightarrow^4 He + 2p$. S_{34} for $^3He +^4 He \rightarrow^7 Be + \gamma$. S_{11} for $p + p \rightarrow^2 H + e^+ + \nu_e$. $S_{1,14}$ for $p +^{14} N \rightarrow^{15} O + \gamma$. The astrophysical factors are known at the level of 5 - 7% and give an important contribution to the predicted uncertainty of neutrino fluxes [1, 2]. For ^{7}Be the 5.2% (5.4%) experimental uncertainty on $S_{33}(S_{34})$ accounts for 2.2%(4.6%) in the uncertainty of the predicted value for the flux. Therefore, improving the error on astrophysical factors, it is important considering that ^{7}Be and ^{8}B fluxes are measured at 5% and 3% level, respectively.

Helioseismology, the study of propagation of acoustic waves in the solar interior, is another important method to probe the SSM in addition to solar neutrinos. In fact from helioseismology we could determine the depth of the convective zone and Y_{surf}. Before 2005 the SSM used as input for chemical abundances the so-called GS98 composition [5]. We point out that the CNO elements abundances can only be inferred from the solar photosphere. In 2005 a new model for the solar photosphere was proposed [6] and later revised in 2009 [7]. The SSM which makes use of this latter composition is referred to as SSM-AGSS09. The new model has led to a downward revision of the heavy elements abundances up to 30-40% with respect to SSM-GS98. The SSM developed with GS98 is in agreement with helioseismology measurements. On the contrary, the SSM developed with AGSS09 shows large disagreement. This is shown in figure 3. The AGSS09 is thought to be based on a much more robust understanding of the solar photosphere. This conflict between SSM predictions and heliosesimology is known as the Solar Abundance Problem. At present this disagreement means that we do not know exactly the solar chemical composition. The zero-age homogeneous Sun assumption could be wrong.

The Solar Abundance Problem replaces the previous Solar Neutrino Problem. This latter has been known since 1968 following the first observation of solar neutrinos. The measured flux of solar neutrinos was only 1/3 of the predicted flux by the SSM. This deficit was later confirmed by other experiments using gallium [8, 9, 10], water [11, 12] and liquid scintillator [13]. This anomaly was understood in terms of neutrino oscillations [14]. In 2002 the SNO experiment proved that neutrino oscillations can solve the observed deficit [15]. The flux of ^{8}B was measured to be consistent with the SSM. Figure 4 shows the Solar Neutrino Problem with respect to GS98 predictions by the SSM.

2. Solar neutrinos observations and the solar energy paradigm

In this Section we review solar neutrino observations and discuss implications for the solar energy paradigm. In table 1 a summary of the solar neutrino experiments is provided [16]. At present two experiments, Super-Kamiokande and Borexino, are taking data in realtime on solar neutrinos. The aim of new measurements is to improve the understanding of the MSW [14] effect in the framework of solar neutrino oscillations and solve the Solar Abundance Problem. This latter target can only be solved by detecting CNO neutrinos. This might be possible in Borexino [16]. In table 2 solar neutrino measurements are reported together with SSM predictions. We notice that ^{8}B and ^{7}Be fluxes are measured with an uncertainty smaller than the one for predictions. The luminosity constraint corresponds to the assumption that $L_{\odot} = L_{\nu}$, with L_{ν} being the solar neutrino luminosity. In table 2 the pp flux is given with and without the luminosity constraint.

IOP Conf. Series: Journal of Physics: Conf. Series **940** (2017) 012023 doi:10.1088/1742-6596/940/1/012023

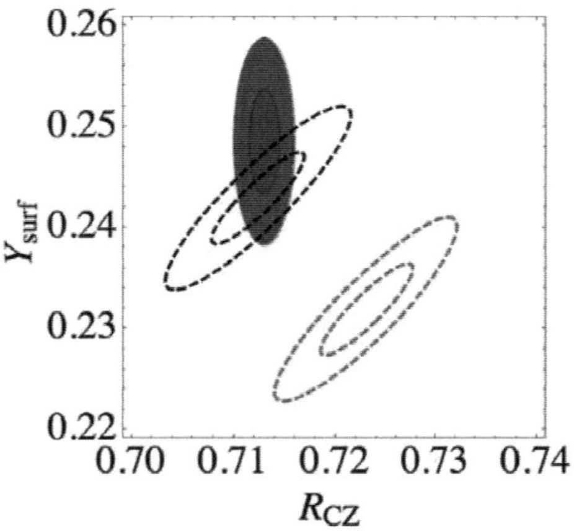

Figure 3. Comparison between the SSM predictions and helioseismology observations. Filled ellipse show 1 and 3σ contours for helioseismology measurements. The upper dashed ellipses correspond to the high-metallicity SSM with GS98. The lower dashed ellipses correspond to the low-metallicity SSM with AGSS09.

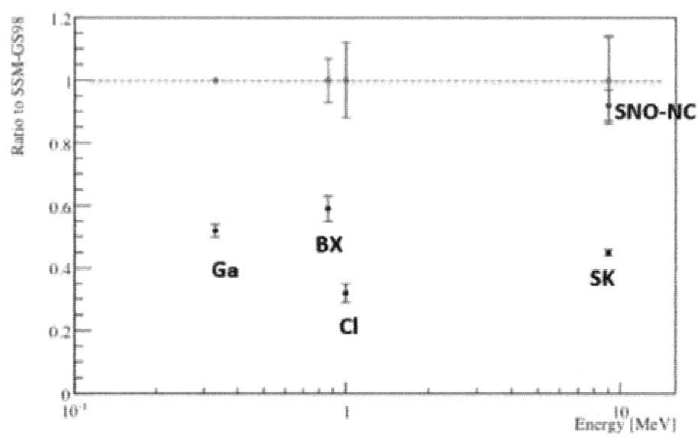

Figure 4. The Solar Neutrino Problem. The dashed horizontal line corresponds to the SSM predictions with GS98 abundances. Measurements from solar neutrino experiments are shown.

Detection of pp solar neutrinos from $p+p \rightarrow d+e^{+}+\nu_{e}+0.42$ MeV provides the opportunity to probe the slowest process which drives the evolution of our Sun on a 10^9 years timescale and compare the neutrino luminosity with the solar luminosity. Moreover, due to the fact that photons produced in the core take some 10^5 years to stream to the surface, this observation explores the solar variability over this time window. pp solar neutrinos have been detected in realtime by Borexino [17]. Because of the low energy carried by these neutrinos, the major challenges faced for this measurement are the content of ^{14}C contamination in the liquid scintillator and the rate of pile-up events due to ^{14}C. The ^{14}C β decay has an endpoint at 0.156 MeV, thus a low detection energy threshold is fundamental for such a measurement. The extreme radio purity of Borexino allows to perform this measurement with a threshold at 0.05 MeV. The flux measured in Borexino is equal to $\phi_{pp} = (6.6 \pm 0.7) \times 10^{10}$ cm^{-2}s^{-1}. From this measurement we could determine the neutrino luminosity: $L_{\nu}(pp - chain)/L_{\odot} = 1.04 \pm 0.07$ and $L_{CNO}/L_{pp} < 0.08(2\sigma)$. SSM predicts that $L_{CNO}/L_{pp} \sim 0.7\%$. A direct measurement of the $L_{\nu}/L_{\odot}$ ratio is important to disregard any exotic energy production in the Sun. The pp solar neutrino measurement performed in Borexino in 2014 is a direct observation of the solar energy paradigm proposed in 1938.

Table 1. Solar neutrino experiments.

Detector	Active mass	Threshold [MeV]	Data taking
Homestake	615tons C_2Cl_4	0.814	1968-1994
Kamiokande II/III	3kton H_2O	9/7.5/7.0	1986-1995
SAGE	50tons molten metal Ga	0.233	1990-present
GALLEX GNO	30.3tons $GaCl_3 - HCl$	0.233	1991-1997 1998-2003
Super-Kamiokande	50ktons H_2O	5 7 4.5 3.5	1996-2001 2003-2005 2006-2008 2008-present
SNO	1kton D_2O	3.5	1999-2006
Borexino	300tons C_9H_{12}	0.2	2007-present

Table 2. Solar neutrino fluxes: best-fit results against predictions. Best-fit for pp are determined with and without the luminosity constraint.

Source	SSM-GS98 (high-Z)	SSM-AGSS09 (low-Z)	data
pp [$\times 10^{10}$ cm^{-2}s^{-1}]	$5.98(1 \pm 0.006)$	$6.03(1 \pm 0.006)$	$6.03(1^{+0.005}_{-0.01})$ $6.3(1 \pm 0.06)$
pep [$\times 10^{8}$ cm^{-2}s^{-1}]	$1.44(1 \pm 0.012)$	$1.47(1 \pm 0.012)$	$1.63(1 \pm 0.21)$
^{7}Be[$\times 10^{9}$ cm^{-2}s^{-1}]	$5.00(1 \pm 0.07)$	$4.56(1 \pm 0.07)$	$4.99(1 \pm 0.05)$
^{8}B [$\times 10^{6}$ cm^{-2}s^{-1}]	$5.58(1 \pm 0.13)$	$4.59(1 \pm 0.13)$	$5.33(1 \pm 0.03)$
CN-cycle [$\times 10^{8}$ cm^{-2}s^{-1}]	$5.25(1 \pm 0.14)$	$3.76(1 \pm 0.14)$	< 6.3 (95%)
p-value pp, Be,B, pep	0.88	0.58	

3. Conclusions

More than 40 years of solar neutrino observations has led to determine the most of the neutrino fluxes, some at a better accuracy than predictions (see table 2). The energy production of the Sun has been proved at 10% level and solar variability studied over a timescale of 10^5 years. A fundamental challenge to face in the coming years is the detection of CNO neutrinos. This measurement could help solving the Solar Abundance Problem providing important information on the chemical composition of the Sun which is yet not well known. Solving this controversy could prove a basic assumption of the SSM is inadequate. At the same time, a better

IOP Conf. Series: Journal of Physics: Conf. Series **940** (2017) 012023 doi:10.1088/1742-6596/940/1/012023

experimental determination of astrophysical factors improves the opportunity to probe the SSM. In fact, S_{34}, S_{17} and $S_{1,14}$ are known at 5.4%, 7.7% and 7.5%, respectively [1]. These uncertainties give a dominant contribution to the overall uncertainty in the predictions of ^{7}Be, ^{8}B and CNO solar neutrino fluxes at the level of 5-8%. In conclusion, being the Sun a laboratory to understand how stars work, improve our understanding of the Sun is of great interest. This challenge needs an effort from experiments on solar neutrinos and from experiments which can make a better measurement of some astrophysical factors.

References

[1] A. Serenelli, C. Pena-Garay and W.C. Haxton, Phys.Rev. D87 (2013) 4, 04300.
[2] John N Bahcall, *Neutrino Astrophysics*, 1989, Cambridge University Press.
[3] H.A. Bethe, Phys. Rev. 55 (1939) 434.
[4] R.J. Davis, D.S. Harmer and K.C. Hoffman, Phys. Rev. Lett. 20 (1968) 1205-1209.
[5] N. Grevesse and A. J. Sauval, Space Sci. Rev. 85 (1998) 161.
[6] M. Ashland, N. Grevesse and A.J. Sauval, Astronomical Society of the Pacific Conference Series, 336 (2005) 25.
[7] M. Asplund, N. Grevesse, A.J. Sauval and P. Scott, ARA&A, 47 (2009) 481.
[8] A.I. Abazov et al., SAGE collaboration, Phys. Rev. Lett. 67 (1991) 3332-3335.
[9] J.N. Abdurashitov et al., SAGE collaboration, Phys. Rev. Lett. 83 (1999) 4686.
[10] W. Hampel et al. Phys. Lett. B 447 (1999) 127.
[11] K.S. Hirata et al., Kamiokande-II collaboration, Phys. Rev. Lett. 63 (1989) 16.
[12] Y. Fukuda et al., Super-Kamiokande collaboration, Phys. Rev. Lett. 82 (1999) 2430.
[13] G. Alimonti et al., Borexino collaboration, Nucl. Instrum. Meth. A 600 (2009) 568.
[14] C. Gonzalez-Garcia and M. Maltoni, JHEP 1309:152, 2013. C. Gonzalez-Garcia, Phys.Dark Univ. 4 (2014) 1-5.
[15] Q.R. Ahmad et al., SNO collaboration, Phys. Rev. Lett. 87 (2001) 071301.
[16] A. Ianni, Phys.Dark Univ. 4 (2014) 44-49.
[17] G. Bellini et al., Borexino collaboration, Nature 512 (2014) 383.

IOP Conf. Series: Journal of Physics: Conf. Series **940** (2017) 012024 doi:10.1088/1742-6596/940/1/012024

Nuclear astrophysics at FRANZ

R Reifarth[1], S Dababneh[1,2], S Fiebiger[1], J Glorius[1], K Göbel[1], M Heil[3], P Hillmann[1], T Heftrich[1], C Langer[1], O Meusel[1], R Plag[3], S Schmidt[1], Z Slavkovská[1], D Veltum[1], M Weigand[1], C Wiesner[1], C Wolf[1], A Zadeh[1]

[1] Goethe University Frankfurt, Frankfurt, Germany
[2] Al-Balqa Applied University, Salt, Jordan
[3] GSI Helmholtzzentrum f. Schwerionenforschung GmbH, Darmstadt, Germany

E-mail: `reifarth@physik.uni-frankurt.de`

Abstract. The neutron capture cross section of radioactive isotopes for neutron energies in the keV region will be measured by a time-of-flight (TOF) experiment. NAUTILUS will provide a unique facility realizing the TOF technique with an ultra-short flight path at the FRANZ setup at Goethe-University Frankfurt am Main, Germany. A highly optimized spherical photon calorimeter will be built and installed at an ultra-short flight path. This new method allows the measurement of neutron capture cross sections on extremely small sample as needed in the case of ^{85}Kr, which will be produced as an isotopically pure radioactive sample. The successful measurement will provide insights into the dynamics of the late stages of stars, an important independent check of the evolution of the Universe and the proof of principle.

1. Introduction

The general features of the FRANZ facility, which is currently under construction, are already reported [1]. The FRANZ facility was developed to measure (n,γ) cross sections of astrophysically interesting isotopes with unprecedented sensitivity [2]. Recently also a program for (p,γ) experiments at FRANZ has been developed [3, 4]. This article concentrates on a new technique, which will be developed to determine neutron capture cross sections using the time-of-flight technique in combination with extremely small samples. This is beyond the standard program of FRANZ.

2. The origin of the elements

The synthesis of the naturally occurring chemical elements of mass equal or heavier than carbon started about 500 million years after the Big Bang - the beginning of our Universe. Apart from the primordial H, He and Li abundances the elements were and still are synthesized in stars. A clear and quantitative picture of nucleosynthesis in the different stages of stellar evolution constitutes the basis for our understanding of the chemical history of the Universe. Furthermore, understanding the chemical history leads to strong constraints on the age of the Universe.

The solar abundances distribution of the chemical elements can be explained by different processes, which contribute to the overall nucleosynthesis and occur during different stages of a star's life. The lightest elements - hydrogen, helium and traces of lithium - are produced in the Big Bang while the elements up to iron are synthesized during stellar burning phases by charged-particle induced fusion reactions. During the extremely hot last stellar burning phase - the silicon burning - the isotopes around iron, which are most tightly bound, are produced in the nuclear statistical equilibrium, forming the famous iron peak around mass number 56. Almost all the elements with higher proton numbers than iron are produced through neutron-induced processes [5]. There are two major processes, the rapid neutron-capture process (r process) and the slow neutron capture process (s process). The s and r processes are reflected in the solar abundance distribution of the elements as a double-peak structure occurring at mass numbers corresponding to closed neutron shells. Only a few isotopes on the proton-rich side of the valley of stability get significant contributions from other processes.

3. Beyond iron - Neutron capture processes

The r process, producing about half of the elemental abundances, takes place in explosive scenarios such as supernovae. The neutron flux is so high and the neutron capture times are so short that a nucleus will almost always capture several neutrons before it undergoes β-decay. Thus, the r process follows a path that is shifted far towards the neutron-rich isotopes. Here, the β-decay half-lives are much shorter for isotopes very rich in neutrons, so that shortly after the neutron source terminates, the products of the r process will β-decay back to the valley of stability.

In contrast, the main component of the s process takes place in thermally pulsing Red Giant stars (T-AGB). These stars have left the main sequence after finishing hydrogen core burning. They are about 50 to 100 times bigger and cooler than our sun while their masses are comparable. The s process [6, 7] starts with an iron seed exposed to free neutrons. Since the neutron densities are low, an unstable isotope produced by a series of neutron capture reactions will almost always β-decay back to its stable isobar before capturing another neutron. Thus, the s process builds up the elements following the neutron-rich side of the nuclear valley of stability until it terminates in the lead-bismuth region.

4. The s-process branching at ^{85}Kr

The exact pathway of the s process depends on the conditions in the star. Starting at the very abundant iron group, all elements up to bismuth could, in principle, be produced by a sequence of reactions in which a series of neutron captures produces an unstable isotope that quickly decays to the next higher element through β-decay and then waits for the next neutron capture. If, however, conditions in the star make the rates for neutron capture comparable to the rate of β-decay for a particular isotope, the s-process path branches at that isotope: a fraction of the isotope transforms via neutron capture, while the other fraction β-decays.

The branching ratio, or relative likelihood, for the different reactions depends on the physical conditions in the interior of the star - temperature, neutron density and electron density. At higher neutron densities with all other conditions equal, more nuclei of a given isotope capture a neutron before having the chance to β-decay. Thus, the branching ratios deduced from the isotopic ratios observed in stellar material could provide the tools to effectively constrain modern models of the stars where the nucleosynthesis occurs. Therefore the fundamental rates, or cross sections, for neutron capture and β-decay are essential.

A particularly interesting branching of the s process occurs in the region around ^{85}Kr (Figure 1). The s-process path indicated by gray arrows has branch points at the unstable isotopes (blue squares). ^{85}Kr has an isomeric state, ^{85m}Kr, which always decays before neutron capture - either through β-decay or through internal conversion to the ground state. If the ^{85}Kr(n,γ) cross section is increased, the neutron capture occurs with enhanced probability and ^{86}Kr is produced more abundantly. This would therefore affect the abundance ratio of ^{84}Kr to ^{86}Kr. During r-process nucleosynthesis very neutron-rich isotopes are produced that decay back to the valley of stability. Stable isotopes (black squares) end these β-decay chains so that the

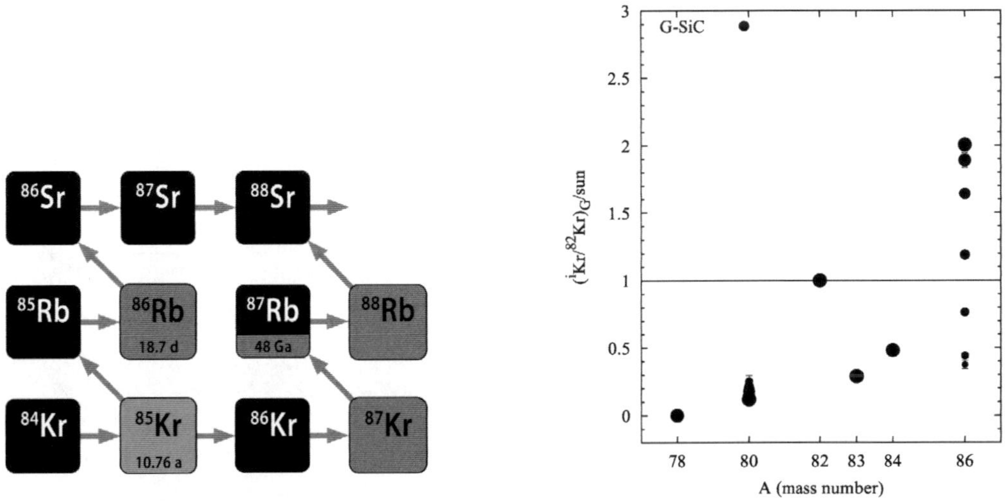

Figure 1. Left: Reaction network during *s*-process nucleosynthesis around the important branch point ^{85}Kr (10.76 yr). Right: Distribution of Kr isotopes in presolar grains [8].

stable isotope ^{86}Sr is produced only in the *s* process (*s*-only isotope). If the very long-lived ^{87}Rb were stable, ^{87}Sr would also be a *s*-only isotope.

5. ^{85}Kr(n,γ) cross section - Constraining stellar parameters

An additional exciting aspect of this region is the availability of observational data of the stellar abundances of those isotopes from stardust. These extremely small diamond-like grains are known to be formed in the outer regions of Red Giant stars. Under certain circumstances, such grains survive their long journey through interstellar space and participate in the formation of new stars and their planets. Therefore we can find them nowadays on earth. Since the grains are extremely small - only a few micrometers in diameter - it was necessary to develop greatly improved experimental equipment to reach the sensitivity needed for isotopic-abundance measurements of several isotopes from a single grain. It is possible to extract information about all stable isotopes of krypton from a single grain - hence, from a single star. Therefore the interesting ^{84}Kr/^{86}Kr ratio can be observed in presolar SiC grains originating from T-AGB stars [8] as shown in the right part of Figure 1. The competition between β-decay and neutron capture at ^{85}Kr is reflected in the observed ^{84}Kr/^{86}Kr ratio. The interpretation of the observed ratio as a measure of stellar parameters is currently hampered by the only very poorly known ^{85}Kr(n,γ) cross section [9, 10, 11].

6. ^{85}Kr(n,γ) cross section - calibrating a nuclear cosmochronometer

The decay of the long-lived ^{87}Rb (48 Gyr) affects the solar abundance ratio of the otherwise *s*-only isotopes ^{86,87}Sr. While the small r-process contribution to the solar abundance of ^{87}Rb can be determined by a comparison with the stable, close-by r-only nuclei ^{96}Zr and ^{100}Mo, the by far more important contribution from the *s* process crucially depends on the ^{85}Kr(n,γ) cross section. If this cross section is smaller than assumed in the models, ^{87}Rb is less produced, and, hence, the ratio of ^{86}Sr to ^{87}Sr is less influenced - the Universe appears to be younger. If the ^{85}Kr(n,γ) cross section is bigger than assumed, the ^{86}Sr to ^{87}Sr ratio is strongly modified by the decays of ^{87}Rb - the Universe appears older. A detailed understanding of the *s*- and r-contribution to the ^{87}Rb abundance would therefore allow the determination of the onset of

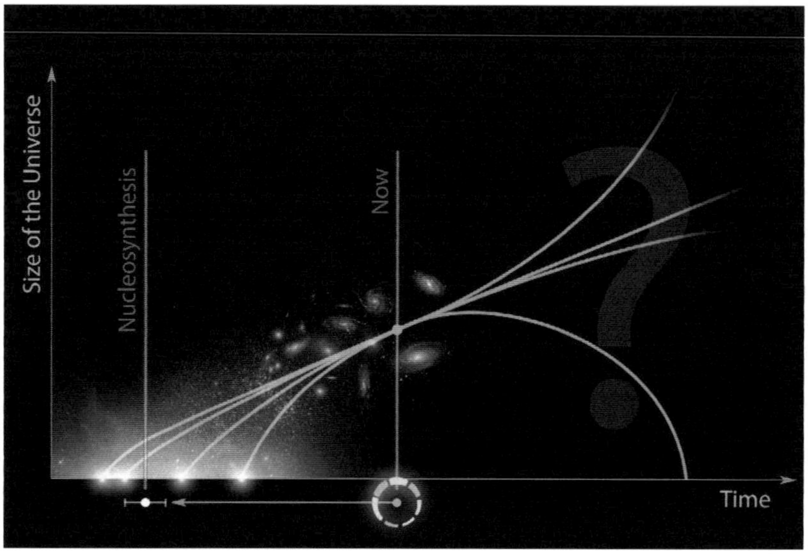

Figure 2. Starting with the big bang, the Universe is expanding ever since. Depending on the parameters of current models, the future could be accelerated expansion, the asymptotic stand still or even a collapse. Cosmological models help to couple the past (age of the Universe) with the future destiny. The exact knowledge of the time since the onset of the synthesis of the heavy elements allows to constrain the age of the Universe.

the *s*- and *r* process in the history of the Universe [12]. Similarly to the determination of the age of the solar system based on isotopic ratios, nuclear cosmochronometers offer a constraint of the age of the Universe (14 Gyr), which is completely independent of astronomical observations. Also here, the biggest remaining uncertainty of this Rb/Sr clock is the ^{85}Kr(n,γ) cross section.

The age of the Universe has been determined by the Re/Os clock [13] with the method described. The result has an overall uncertainty of about 1 Gyr. The unsatisfying outcome stems from major uncertainties of theoretical estimates. On the one hand, the neutron capture cross section of a low-lying nuclear state in ^{187}Os cannot be measured directly, but is extremely important at stellar temperatures. On the other hand, the nuclear clock itself induces problems: the β-decay half-life of ^{187}Re is about 109 times shorter under stellar conditions than under terrestrial or interstellar conditions. The reason for this dramatic change of the β-decay half-life is the bound-state β-decay in stellar interiors. In the hot plasma ^{187}Re may be fully ionized so that the decay electron remains in a bound state instead of being emitted into the continuum, drastically increasing the decay probability [14].

Both nuclear effects are not relevant in the case of the Rb/Sr clock [9, 15]. If the uncertainty of the value for the neutron capture cross section of ^{85}Kr is reduced to around 10% or less, the age determined by the Rb/Sr clock should result in a final uncertainty of less than 0.5 Gyr. This would put strong constraints on the different scenarios for the expansion of the Universe (Figure 2), possibly deciding on the models for a closed, flat or open Universe.

7. Measuring (n,γ) reactions with very short flight path and radioactive samples

The present layout of the FRANZ facility barely provides the high neutron fluxes needed to perform measurements on radioactive isotopes with comparably hard γ-ray emission like ^{85}Kr [2, 16]. Since the neutron production is already at the limits of the current technology, one option is to get closer to the neutron production target to increase the solid angle covered by

IOP Conf. Series: Journal of Physics: Conf. Series **940** (2017) 012024 doi:10.1088/1742-6596/940/1/012024

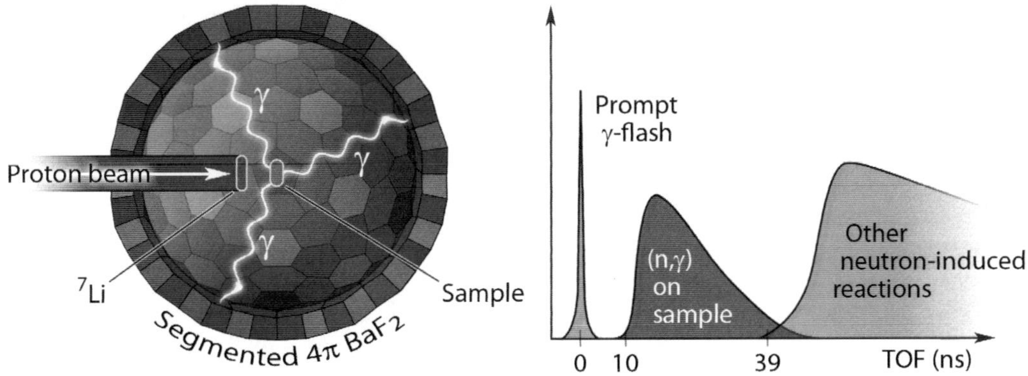

Figure 3. Left: Schematic setup of the planned neutron capture experiment with an ultra-short flight path of only 4 cm. Right: Anticipated time-of-flight spectrum [17].

the sample material. It is possible to perform a TOF measurement with sufficient accuracy even with a flight path as short as a few centimeters (Figure 3, left).

The only feasible solution is to produce the neutrons inside a spherical γ-detector and distinguish between background from interactions with the detector material and the signal from neutron captures on the sample based on the time after neutron production as illustrated in Figure 3, right. First, an initial γ-flash, occurring when the protons hit the neutron production target, is detected. Then the prompt γ-rays produced in the (n,γ) reaction at the sample induce a signal in the detector. Later, the neutrons from other reactions, such as scattering in the detector material, arrive in the detector with large time of flights, travelling at much slower speed than the γ-rays, and produce background.

A detailed investigation of the geometry of the setup at an ultra-short flight path has been performed. In contrast to the calorimeters used in such TOF experiments so far [5, 18, 19], the calorimeter shell has to be much thinner in order to allow the neutrons to escape quick enough. The geometry is based on the DANCE array [18, 20], which was designed as a high efficiency, highly segmented 4π BaF_2 array. It consists of up to 162 crystals of 4 different shapes, each covering the same solid angle. The high segmentation distributes the envisaged high count rate over many channels, leading to a significant increase of the maximal tolerable total count rate that can be processed by the DAQ. The DANCE array has an inner radius of 20 cm and a thickness of 12 cm. The combination of the fastest decay component of any scintillator known so far, very low neutron sensitivity and high γ-ray detection efficiency makes BaF_2 by far the best suited scintillator material for this purpose.

The advantage of this setup is the greatly enhanced neutron flux. Because of the reduction of the flight path from 1 m to 4 cm, the neutron flux will be increased by almost 3 orders of magnitude. The reduced time-of-flight resolution resulting in a reduced neutron-energy resolution is still sufficient for astrophysical and applied purposes. A comparison to the DANCE setup at the Los Alamos Natl. Laboratory shows that, despite the much shorter flight path (4 cm vs. 20 m), a much better time resolution (1 ns vs. 125 ns) will be achieved at the proposed setup. Because of the different time structure of the proton beam at the FRANZ facility the energy resolution will almost be the same for both setups.

8. Summary and Outlook

Radioactive isotopes become more and more in reach of current experimental research. Proton induced reaction studies on unstable nuclei are important and are being addressed also with new techniques [21]. Neutron induced reaction studies are difficult on stable, very difficult on unstable nuclei [5]. FRANZ is designed to address many of those requirements. The NAUTILUS project aims at determining the neutron capture cross section of ^{85}Kr in the astrophysically important energy regime. Current and future ring designs promise breakthroughs in experimental techniques even for the determination of neutron-induced reaction cross sections [22].

Acknowledgments

This project was supported by the European Research Council under the European Union's Seventh Framework Programme (FP/2007-2013)/ERC Grant Agreement n. 615126, HIC for FAIR and NAVI.

References

[1] Wiesner C, Chau L P, Dinter H, Droba M, Heilmann M, Joshi N, Mader D, Metz A, Meusel O, Müller I, Noll D, Podlech H, Ratzinger U, Reichau H, Reifarth R, Schempp A, Schmidt S, Schweizer W, Volk K and Wagner C 2010 *AIP Conference Proceedings* **1265** 487–492
[2] Reifarth R, Chau L P, Heil M, Käppeler F, Meusel O, Plag R, Ratzinger U, Schempp A and Volk K 2009 *PASA* **26** 255
[3] Göbel K, Glorius J, Koloczek A, Pignatari M, Reifarth R, Schach R and Sonnabend K 2015 *EPJ Web of Conferences* **93** 03006
[4] Endres A, Arda C, Erbacher P, Glorius J, Göbel K, Hinrichs O, Mevius E, Reich M, Sonnabend K, Thomas B and Thomas T 2015 *EPJ Web of Conferences* **93** 03007
[5] Reifarth R, Lederer C and Käppeler F 2014 *Journal of Physics G Nuclear Physics* **41** 053101
[6] Pignatari M, Gallino R, Heil M, Wiescher M, Käppeler F, Herwig F and Bisterzo S 2010 *Ap. J.* **710** 1557–1577
[7] Käppeler F, Gallino R, Bisterzo S and Aoki W 2011 *Reviews of Modern Physics* **83** 157–194
[8] Pignatari M, Gallino R, Amari S and Davis A M 2006 *Memorie della Societa Astronomica Italiana* **77** 897
[9] Bao Z Y, Beer H, Käppeler F, Voss F, Wisshak K and Rauscher T 2000 *Atomic Data Nucl. Data Tables* **76** 70
[10] Abia C, Busso M, Gallino R, Dominguez I and Straniero Oand Isern J 2001 *Ap. J.* **559** 1117
[11] Raut R, Tonchev A P, Rusev G, Tornow W, Iliadis C, Lugaro M, Buntain J, Goriely S, Kelley J H, Schwengner R, Banu A and Tsoneva N 2013 *Physical Review Letters* **111** 112501
[12] Beer H and Walter G 1984 *Astrophys. Space Sci.* **100** 243–253
[13] Fujii K, Mosconi M, Mengoni A, Domingo-Pardo C, Käppeler F et al 2010 *Phys. Rev. C* **82** 015804
[14] Bosch F, Faestermann T, Friese J, Heine F, Kienle P, Wefers E, Zeitelhack K, Beckert K, Franzke B, Klepper O, Kozhuharov C, Menzel G, Moshammer R, Nolden F, Reich H, Schlitt B, Steck M, Stöhlker T, Winkler T and Takahashi K 1996 *Phys. Rev. Lett.* **77** 5190 – 5193
[15] Takahashi K and Yokoi K 1987 *Atomic Data Nucl. Data Tables* **36** 375
[16] Couture A and Reifarth R 2007 *Atomic Data and Nuclear Data Tables* **93** 807
[17] Reifarth R, Haight R C, Heil M, Käppeler F and Vieira D J 2004 *Nucl. Instr. Meth. A* **524** 215
[18] Reifarth R, Bredeweg T A, Alpizar-Vicente A, Browne J C, Esch E I, Greife U, Haight R C, Hatarik R, Kronenberg A, O'Donnell J M, Rundberg R S, Ullmann J L, Vieira D J, Wilhelmy J B and Wouters J M 2004 *Nucl. Instr. Meth. A* **531** 530
[19] Wisshak K, Guber K, Käppeler F, Krisch J, Müller H, Rupp G and Voss F 1990 *Nucl. Instr. Meth. A* **292** 595 – 618
[20] Heil M, Reifarth R, Fowler M M, Haight R C, Käppeler F, Rundberg R S, Seabury E H, Ullmann J L, Wilhelmy J B, Wisshak K and Voss F 2001 *Nucl. Instr. Meth. A* **459** 229
[21] Mei B, Aumann T, Bishop S, Blaum K, Boretzky K, Bosch F, Brandau C, Bräuning H, Davinson T, Dillmann I, Dimopoulou C, Ershova O, Fülöp Z, Geissel H, Glorius J, Gyürky G, Heil M, Käppeler F, Kelic-Heil A, Kozhuharov C, Langer C, Le Bleis T, Litvinov Y, Lotay G, Marganiec J, Münzenberg G, Nolden F, Petridis N, Plag R, Popp U, Rastrepina G, Reifarth R, Riese B, Rigollet C, Scheidenberger C, Simon H, Sonnabend K, Steck M, Stöhlker T, Szücs T, Sümmerer K, Weber G, Weick H, Winters D, Winters N, Woods P and Zhong Q 2015 *Phys. Rev. C* **92** 035803
[22] Reifarth R and Litvinov Y A 2014 *Phys. Rev. ST Accel. Beams* **17**(1) 014701

IOP Conf. Series: Journal of Physics: Conf. Series **940** (2017) 012025 doi:10.1088/1742-6596/940/1/012025

Photodisintegration reactions for nuclear astrophysics studies at ELI-NP

C Matei[1], D Balabanski[1], D M Filipescu[1] and O Tesileanu[1]

[1] Extreme Light Infrastructure - Nuclear Physics / Horia Hulubei National Institute for R&D in Physics and Nuclear Engineering, Bucharest-Magurele, 077125, Romania

E-mail: catalin.matei@eli-np.ro

Abstract. Extreme Light Infrastructure - Nuclear Physics facility will come online in Bucharest-Magurele, Romania, in 2018 and will deliver high intensity laser and brilliant gamma beams. We present the physics cases and instruments proposed at ELI-NP to measure capture reactions by means of the inverse photodisintegration reaction. We propose to study the $^{16}O(\gamma,\alpha)^{12}C$ reaction using a Time Projection Chamber detector with electronic readout. Several other reactions, such as $^{24}Mg(\gamma,\alpha)^{20}Ne$ and reactions on heavy nuclei relevant in the p-process, are central to stellar evolution and will be investigated with a proposed Silicon Strip Detector array and a 4π neutron detector. The status of the experimental facilities and first-day experiments will be presented in detail.

1. The ELI-NP facility

The Extreme Light Infrastructure - Nuclear Physics (ELI-NP) consists of two major components: the High Power Laser System and the Gamma Beam System (GBS) [1]. ELI-NP will allow either combined or stand-alone experiments using the high-power laser and the gamma beam.

The GBS is an advanced source of gamma-ray photons able to produce beams of monochromatic and high spectral density gamma-ray photons. The gamma-ray beam is produced through Compton backscattering of laser light off an accelerated electron beam. The main specifications of the system are: photon energy tunable in the range 0.2-19.5 MeV, high degree of linear polarization ($\geq$ 99%), small bandwidth ($\leq$ 0.5%), spectral density larger than 10^4 photons/s/eV, and source spot sizes of about 10-30 microns.

The GBS will consist of two stages: a low-energy stage in which gamma rays are produced with energies up to 3.5 MeV and a high-energy stage in which gamma rays are produced with energies up to 19.5 MeV. Each stage is equipped with a dedicated diagnostics system for tuning and beam optimization. Additional monitoring instruments will be associated with the experimental stations for measuring the intensity and energy of the gamma beam.

2. Proposed measurements

Many nuclear reactions relevant in stellar environments proceed through very small cross sections which are difficult to determine experimentally. Photodissociation reactions are in principle easier to study as the phase space factor enhances the cross section with respect to the inverse process, provided that a high quality gamma beam is available, such as the one delivered by the ELI-NP facility.

IOP Conf. Series: Journal of Physics: Conf. Series **940** (2017) 012025 doi:10.1088/1742-6596/940/1/012025

2.1. The $^{16}O(\gamma,\alpha)^{12}C$ reaction

The ^{12}C$(\alpha,\gamma)^{16}$O reaction is one of the most important reactions in nuclear astrophysics as its reaction rate strongly influences the present C/O ratio in the Universe. The ^{12}C(α,γ) cross section has been measured experimentally down to energies around 1 MeV but must be extrapolated to helium-burning energies around 300 keV. This extrapolation is difficult as the cross section is a mixture of ground state transitions and cascade transitions with a complicated energy dependence and interference pattern. The ^{12}C$(\alpha,\gamma)^{16}$O cross section is composed of 3 major components: $E1$, $E2$ ground state transitions, and several weaker cascade transitions. Modeling the evolution and explosion of massive stars requires a 10% uncertainty [2] on the ^{12}C$(\alpha,\gamma)^{16}$O reaction cross section but the present uncertainties are of the order of 30%.

The cross section at 300 keV can only be determined through a combination of experimental measurements and theoretical extrapolation. There are two energy ranges which are important for the study the ^{12}C$(\alpha,\gamma)^{16}$O cross section: at energies below 1 MeV to approach the Gamow peak or at higher energies to constrain the R-matrix extrapolation.

Several measurements of the ^{12}C$(\alpha,\gamma)^{16}$O reaction using gamma-ray detectors have been carried out with center of mass energies in the vicinity of 1.0 MeV. However, the astrophysical S-factors were determined with very low accuracies and one cannot rule out a low value of the extrapolated E1 S-factor. The new data also point out to a significant ambiguity in the value of the extrapolated E2 S-factor [3]. The goal of the measurements at ELI-NP will be to measure detailed cross sections and angular distributions, thus obtaining accurate values of the E2/E1 ratio. One of the advantages of measuring the photodissociation of ^{16}O is a gain in cross section due to detailed balance. Such an experiment requires gamma-ray beams in the range between 8 and 10 MeV. For example, a measurement of an angular distribution (1500 counts) with the gas e-TPC detector described in Sec. 3.1, at $E_\gamma = 8.26$ MeV with beam intensity available at ELI-NP will require 21 days of beam time [5].

The majority of ^{12}C$(\alpha,\gamma)^{16}$O experimental and theoretical work over the last 50 years have concentrated in the energy region below 10 MeV. Measuring at higher energies, all the way to 14 MeV, would allow to include more states in the R-matrix analysis and to reduce the uncertainty of the extrapolation [6]. The aim of such measurements at ELI-NP would be to determine angular distributions and cross sections between known resonances with statistical precision of 3% or better. For example, a measurement of the weakest cross section between the 2^+ states at 9.84 and 11.52 MeV will require 12 hours of beam time for each experimental point.

2.2. Other (γ,α) and (γ,p) reactions

The rate of ^{28}Si destruction is established by the photodisintegration of ^{24}Mg, making its reaction rate critically important to stellar models of silicon burning [4]. A direct ^{24}Mg photodissociation measurement using gamma beams of energies between 10.4 and 12 MeV will allow us to determine a much more accurate cross section needed in pre-supernova nuclear reaction network calculations. We expect 0.5 events/s in this energy region using the SSD array (see Sec. 3.2) with a 100 μg ^{24}Mg target and a conservative beam intensity [5].

One of the neutron sources for s-process nucleosynthesis in massive stars is the ^{22}Ne$(\alpha,n)^{25}$Mg reaction. Although the study of low-lying resonances in ^{22}Ne have been investigated before, there is considerable uncertainty in the their strength. For example, the 470 keV resonance, could yield up to 1 event/hour when measured with the e-TPC detector at ELI-NP [5].

The ^{18}O$(p,\gamma)^{19}$F reaction could be the origin of the loss of catalytic materials from the CNO cycle, providing at the same time a link to the NeNa cycle. Large uncertainties dominate the strength of some of the low-lying resonances with the resonance at 90 keV known only with an upper limit. We expect a count rate of 5 events per day with the e-TPC and the beam intensities available at ELI-NP [5].

2.3. p-process reactions

There are 35 nuclides classified as the p-process nuclides ranging from ^{74}Se to ^{196}Hg that are produced by re-processing the pre-existing seed nuclei produced in the s- and r-processes, where photodisintegration plays a primary role [4]. In the p-process nucleosynthesis, we measure photodissociation processes, including (γ,n), (γ,α) and (γ,p) reactions. Nuclei such as ^{74}Se, ^{78}Kr, ^{84}Sr, ^{92}Mo, and ^{96}Ru show a very strong dependence on the (γ,p) cross section. Another critical process is the ^{96}Ru(γ,α)^{92}Mo reaction, whose cross section dramatically changes the final abundance of ^{96}Ru.

Measurements of (γ,n) cross sections for the p-process nuclei, known as the destruction cross section, have never been measured because of their low natural abundance. Among the p-process nuclei, two odd-odd nuclei ^{180}Ta and ^{138}La draw highest attention and will be measured with the highest priority [7].

3. Proposed instruments

We consider a large area Silicon Strip Detector (SSD) and a gas Time Projection Chamber (e-TPC) to address important (γ,α) and (γ,p) reactions in nuclear astrophysics [5]. A 4π neutron detector based on ^{3}He proportional counters is proposed for photoneutron cross section measurements relevant for the p-process [7].

3.1. Time Projection Chamber Detector

The detector which is most suitable to investigate the multi alpha-particle decay of light nuclei such as ^{12}C and ^{16}O and the cross section of astrophysically-relevant (γ,p) or (γ,α) reactions, is a gaseous detector in which the gas acts at the same time as target for the nuclear reaction and detection medium. For this purpose an active target TPC with electronic readout (e-TPC) is proposed [5].

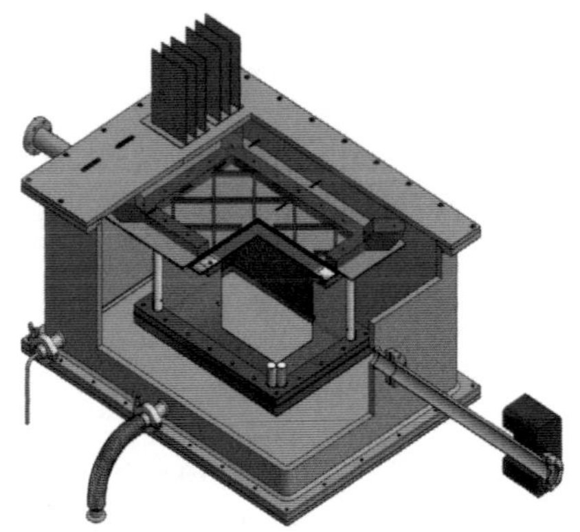

Figure 1. The internal structure of the e-TPC detector.

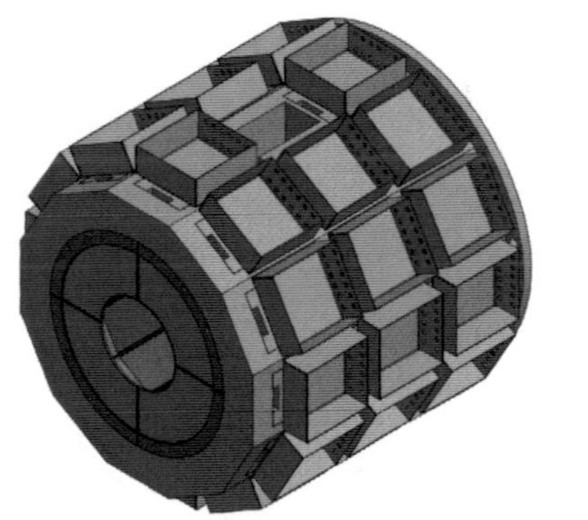

Figure 2. Drawing of the proposed silicon barrel detector array.

The active volume of the chamber, in which the reaction happens and the decay products are detected, will have a length of 35 cm and a square cross-section of 20 cm × 20 cm, centered around the beam axis with a window for the gamma beam and another on the side for an alpha source. The schematic diagram of the e-TPC detector is shown in Fig. 1. The multiplication of the drifting electrons will be achieved by a sequence of 35 cm × 20 cm Gas Electron Multiplier (GEM) foils fabricated at CERN. Combination of the 2D position in the collection plane with time information will allow the 3D reconstruction of the reaction products which is vital for

IOP Conf. Series: Journal of Physics: Conf. Series **940** (2017) 012025 doi:10.1088/1742-6596/940/1/012025

reactions in which more than two particles are involved or when more events happen within a short time-window of the information collection.

3.2. The Large Area Silicon Strip Detector

Silicon detectors represent an excellent solution for detecting charged particles: they guarantee exceptional energy resolution, with almost 100% efficiency, the thresholds can be set very low, they are very little sensitive to neutrons, gamma rays and electrons, which constitute the beam induced background. A silicon array would make it possible to measure reactions on solid targets and all the reactions on heavy nuclei intervening in the p-process. In the past years, SSD arrays have been successfully designed for studies of nuclear astrophysics reactions (e.g. ORRUBA [8]).

The array shown in Fig 2 consists of X3 silicon-strip detectors manufactured by Micron Semiconductor arranged into a barrel configuration [5]. The X3 are 4-strip detectors 4 cm wide, position sensitive along the longitudinal axis (7.5 cm long), leading to a position resolution better than 1 mm. The barrel can be made up of 3 rings of 12 position sensitive detectors, for a total angular coverage of 100° in the laboratory system. This setup results in a very compact design with a good angular resolution for reactions such as ^{24}Mg$(\gamma,\alpha)^{20}$Ne. Since position is determined by charge partition, the number of electronic channels is strongly reduced, as about 300 channels would be necessary for the whole barrel. The angular coverage is extended by using end cap detectors such as the assembly of four QQQ3 segmented detectors by Micron Semiconductor. If necessary, smaller angles can be covered by adding additional smaller rings at the two ends of the barrel.

3.3. High-efficiency 4π neutron detector

We propose to use a high-efficiency 4π triple-ring detector based on ^{3}He proportional counters for measuring (γ, n) reactions [7]. The active detection elements are 20 cylindrical counters (2.5 cm diameter and 49.5 cm length), each containing 12 bars of ^{3}He. The proportional counters are placed equally spaced in three concentric rings of 80, 140 and 200 mm diameter respectively. The first ring contains 4 detectors, while the other two contain 8 detectors each. The neutron moderator is a cube of 36 × 36 × 50 cm^3 made of polyethylene. The moderator is covered by additional 5-cm-thick polyethylene plates with 1-mm-thick cadmium metal for background neutron suppression. GEANT4 and MCNP simulations have been used to calculate an efficiency between 60 and 75 % for neutrons below 1 MeV.

4. Conclusions

The Gamma Beam System at ELI-NP will provide quasi-monochromatic, very intense gamma-ray beam with a high degree of polarizability starting in 2018. An ambitious experimental program to measure photodisintegration reactions relevant to nuclear astrophysics has been proposed. Several instruments are funded and in various stages of prototyping or execution: an active target TPC with electronic readout, a large area silicon detector array, and a 4π high-efficiency neutron detector.

References

[1] EuroGammaS Association, Technical Design Report EuroGammaS proposal for the ELI-NP Gamma beam System *Preprint* arXiv:1407.3669 physics.acc-ph
[2] W A Fowler 1984 *Rev. Mod. Phys.* **56** 149
[3] M. Gai 2013 *Phys. Rev.* C **88** 062801(R)
[4] C Iliadis 2007 *Nuclear Physics of Stars* (Weinheim: Wiley-VCH Verlag)
[5] ELI-NP 2015 *GBS TDR4 - Charged Particle Detection at ELI-NP* to be published
[6] D B Sayre 2012 *Phys. Rev. Lett.* **109** 142501
[7] ELI-NP 2015 *GBS TDR2 - Gamma Above Neutron Threshold* to be published
[8] S D Pain 2008 *AIP Conf. Proc.* **1090** 570

IOP Conf. Series: Journal of Physics: Conf. Series **940** (2017) 012026 doi:10.1088/1742-6596/940/1/012026

Latest results from LUNA

Rosanna Depalo for the LUNA collaboration
Dipartimento di Fisica e Astronomia and INFN Padova, via Marzolo 8, 35131 Padova (Italy)
E-mail: rdepalo@pd.infn.it

Abstract. A precise knowledge of the cross section of nuclear fusion reactions is a crucial ingredient in understanding stellar evolution and nucleosynthesis. At stellar temperatures, fusion cross sections are extremely small and difficult to measure.
Measuring nuclear cross sections at astrophysical energies is a challenge that triggered a huge amount of experimental work. A breakthrough in this direction was the first operation of an underground accelerator at the Laboratory for Underground Nuclear Astrophysics (LUNA) in Gran Sasso, Italy. The 1400 meters of rocks above the laboratory act as a natural shield against cosmic radiation, suppressing the background by orders of magnitude.
The latest results achieved at LUNA are discussed, with special emphasis on the ^{22}Ne(p,γ)^{23}Na reaction. Future perspectives of the LUNA experiment are also illustrated.

1. Introduction

Nuclear fusion reactions provide most of the energy radiated by stars. Indeed, the whole life of a star consists of a sequence of phases in which heavier and heavier elements are burnt inside the stellar core. Moreover, nuclear reactions are responsible for the synthesis of the elements both in the early Universe and in stars.
Theoretical models have been developed to try to reproduce the elemental abundances observed in the stellar surface. An important input for these models is the thermonuclear reaction rate for all the nuclear reactions involved.
At astrophysical temperatures, the kinetic energy of the interacting particles follows a Maxwell-Boltzmann distribution and is usually much lower than the Coulomb repulsion between the nuclei. Therefore, nuclear reactions can only occur by quantum mechanical tunneling and the cross section decreases exponentially with the energy [1].
As a consequence of the interplay between the energy distribution of nuclei and the tunneling probability through the Coulomb barrier, thermonuclear reactions can only occur in a well defined energy range, called the Gamow peak. At Gamow energies, nuclear cross sections become extremely small (of the order of 10^{-9} - 10^{-12} barns), therefore, in typical experimental conditions, the expected counting rate can be much smaller than the background in a detector.
In a gamma ray detector, the background origin is twofold: below 2.6 MeV it is maily due to the decay of environmental radioactive isotopes (uranium and thorium chains and ^{40}K). This background can be substancially reduced by shielding the detector with high Z and high density material (usually lead or copper). Above 2.6 MeV, the main source of environmental background is cosmic radiation. At sea level, most of the cosmic radiation is made of muons. The most efficient way to suppress the muon-induced background is to perform experiments in underground laboratories.

IOP Conf. Series: Journal of Physics: Conf. Series **940** (2017) 012026 doi:10.1088/1742-6596/940/1/012026

2. The LUNA experiment

The Laboratory for Underground Nuclear Astrophysics is located at Gran Sasso National Laboratories (LNGS), Italy [2]. The laboratory is shielded against cosmic radiation by 1400 meters of rock (3800 meters of water equivalent). This guarantees a six orders of magnitude reduction in the cosmic muon flux and a three orders of magnitude reduction in the neutron flux. As a consequence, the background in a gamma ray detector is also suppressed by three to five orders of magnitude.

The Laboratory for Underground Nuclear Astrophysics is equipped with a 400 kV accelerator providing high intensity proton or alpha beam. The beam can be delivered either to a solid target or to a windowless gas target. Different gamma-ray or particle detectors can be used, depending on the characteristics of the nuclear reaction to be studied.

3. Latest results

Since the installation of the LUNA 400 kV accelerator, several nuclear reactions have been studied relevant for stellar evolution and nucleosynthesis [3, 4, 5, 6, 7, 8, 9] as well as Big Bang nucleosynthesis [10]. The last two years were dedicated to the investigation of the ^{17,18}O(p,α)^{14,15}N reactions of the CNO cycle [11] and the ^{22}Ne(p,γ)^{23}Na reaction of the NeNa cycle.

In the following, the latest results on the ^{22}Ne(p,γ)^{23}Na reaction are discussed.

3.1. The ^{22}Ne(p,γ)^{23}Na reaction

The neon-sodium cycle of hydrogen burning contributes to the synthesis of the isotopes between ^{20}Ne and ^{23}Na in Red Giant Branch (RGB) stars, Asymptotic Giant Branch (AGB) stars and classical novae explosions.

The synthesis of sodium in RGB stars is still puzzling. Observations of galactic globular clusters show that the surface abundance of sodium in RGB stars anticorrelates with the oxygen abundance [12]. A possible explanation for this anticorrelation involves the pollution of the interstellar medium with material processed through hydrogen burning reactions at high temperatures in AGB stars. In the hydrogen burning shell of AGB stars, oxygen is efficiently destroyed by the CNO cycle and sodium is mainly produced by the ^{22}Ne(p,γ)^{23}Na reaction of the NeNa cycle [13].

Another scenario where the ^{22}Ne(p,γ)^{23}Na reaction is active are classical novae explosions. A sensitivity study showed that the ^{22}Ne(p,γ)^{23}Na reaction rate uncertainty strongly affects the final abundances of neon, sodium and magnesium isotopes, demonstrating the need for new experimental efforts on the ^{22}Ne(p,γ)^{23}Na cross section [14].

The ^{22}Ne(p,γ)^{23}Na Gamow window for AGB stars and classical novae extends from 50 to 600 keV. In this energy range, the proton capture on ^{22}Ne is dominated by a large number of resonances. None of the resonances below 436 keV has ever been observed in either direct or indirect experiments (tab. 1). Moreover, the spin and parity assignment to the resonances is often uncertain and the mere existence of the three resonances at 71, 105 and 215 keV is still debated since it has been tentatively reported in [16] but has not been observed in subsequent experiments [17, 18].

As a consequence, the ^{22}Ne(p,γ)^{23}Na reaction rates reported in the widely adopted compilations by NACRE [19] and C. Iliadis et al. [20] are three orders of magnitude discrepant in the energy range of interest for hydrogen burning in AGB stars.

As a first step, a feasibility test was performed using the setup of the previous ^{2}H(α,γ)^{6}Li experiment [21]. For this test, neon gas with natural isotopic composition was used (90.48% ^{20}Ne, 0.27% ^{21}Ne and 9.25% ^{22}Ne).

Table 1. Summary of literature resonance strengths for ^{22}Ne(p,γ)^{23}Na resonances below 400 keV proton energy.

$E_{res,LAB}$ [keV]	$\omega\gamma$ [eV] J. Görres (direct) [15]	NACRE [19]	Iliadis et al. [20]
29	-	-	$\leq 2.6 \cdot 10^{-25}$
37	-	$(6.8 \pm 1.0) \cdot 10^{-15}$	$(3.1 \pm 1.2) \cdot 10^{-15}$
71	$\leq 3.2 \cdot 10^{-6}$	$\leq 4.2 \cdot 10^{-9}$	-
105	$\leq 6.0 \cdot 10^{-7}$	$\leq 6.0 \cdot 10^{-7}$	-
158	$\leq 1.0 \cdot 10^{-6}$	$(6.5 \pm 1.9) \cdot 10^{-7}$	$(9.2 \pm 3.7) \cdot 10^{-9}$
186	$\leq 2.6 \cdot 10^{-6}$	$\leq 2.6 \cdot 10^{-6}$	$\leq 2.6 \cdot 10^{-6}$
215	$\leq 1.4 \cdot 10^{-6}$	$\leq 1.4 \cdot 10^{-6}$	-
259	$\leq 2.6 \cdot 10^{-6}$	$\leq 2.6 \cdot 10^{-6}$	$\leq 1.3 \cdot 10^{-7}$
291	$\leq 2.2 \cdot 10^{-6}$	$\leq 2.2 \cdot 10^{-6}$	$\leq 2.2 \cdot 10^{-6}$
323	$\leq 2.2 \cdot 10^{-6}$	$\leq 2.2 \cdot 10^{-6}$	$\leq 2.2 \cdot 10^{-6}$
334	$\leq 3.0 \cdot 10^{-6}$	$\leq 3.0 \cdot 10^{-6}$	$\leq 3.0 \cdot 10^{-6}$
369	-	-	$\leq 6.0 \cdot 10^{-4}$
394	-	-	$\leq 6.0 \cdot 10^{-4}$

The aim of the test was to study the possible sources of beam induced background, and to have some hints on the sensitivity to the ^{22}Ne+p resonant cross section.
Despite the use of non enriched gas (only 9.25% ^{22}Ne) and a setup which was not optimized for this measurement, gamma-rays from the 186 keV resonance have been observed during the test in a 12 hours run. This resonance was never observed in previous experiments, and the literature upper limit for the resonance strength is $\omega\gamma < 2.6 \cdot 10^{-6}$ eV. Thanks to the LUNA observation it was possible to provide a lower limit $\omega\gamma \geq 0.12 \cdot 10^{-6}$ eV for the 186 keV resonance strength and thus to improve the literature information on this resonance [23].
Following the feasibility test, the characterization of the setup for the first experimental campaign on the ^{22}Ne(p,γ)^{23}Na resonances was started.
Using an extended gas target system requires to accurately know the gas density profile along the beam path. The gas density without beam was deduced from pressure and temperature values measured at different positions inside the target chamber. For this measurement, a dedicated setup was used with a target chamber of the same geometry as the one used for the cross section measurement, but equipped with several flanges along the beam axis allowing to connect pressure or temperature gauges. Then the beam heating effect in natural neon gas has been measured for the first time with the resonance scan technique, using the intense and well known ^{21}Ne(p,γ)^{22}Na resonance at 271.6 keV beam energy [22] and a collimated NaI detector [23].
Once the gas target characterization was completed, the study of the ^{22}Ne(p,γ)^{23}Na low-energy resonances started. For this experimental campaign, a proton beam was delivered to the windowless gas target filled with 99.9% enriched ^{22}Ne.
The gamma-rays emitted in the de-excitation of ^{23}Na were detected by two HPGe detectors collimated at 55° (where the second order legendre polynomimal is zero and possible angular distribution effects are minimal) and 90° with respect to the beam direction (fig. 1). The use of two detectors looking at different angles allows to estimate the effect of the gamma ray angular distribution on the resonance strength. In order to reduce the environmental background at gamma ray energies below 2.6 MeV, the two detectors were surrounded by a copper and lead shielding. This shielding ensured about four orders of magnitude background reduction for γ

ray energies below 3 MeV.

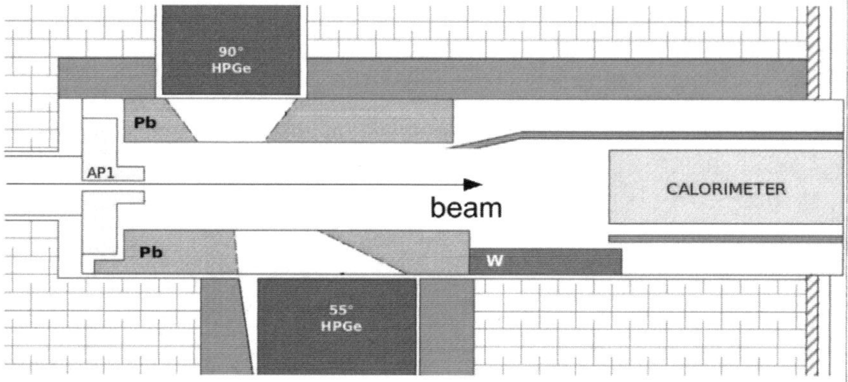

Figure 1. Sketch of the experimental setup used to study the ^{22}Ne(p,γ)^{23}Na low energy resonances at LUNA.

The gamma detection efficiency for the two detectors has been measured at several positions along the target chamber with ^{7}Be, ^{137}Cs, ^{60}Co and ^{88}Y point-like sources. The efficiency curve was then extended up to 6.79 MeV using the intense ^{14}N(p,γ)^{15}O resonance at 278 keV [3]. During about five months of data taking, all the resonances between 70 and 334 keV have been investigated. The resonances at 158, 186 and 259 keV have been observed for the first time in a direct experiment. For these resonances, the complete excitation function was measured and then a long run at the energy of maximum yield was performed. New gamma decay modes have also been observed for the three resonances observed. Moreover, for the 186 and 259 keV resonances the statistics is high enough to allow the study of gamma-gamma coincidences between the two detectors. Fig. 2 shows the long run spectrum over the 259 keV resonance. The observed transitions are also indicated.

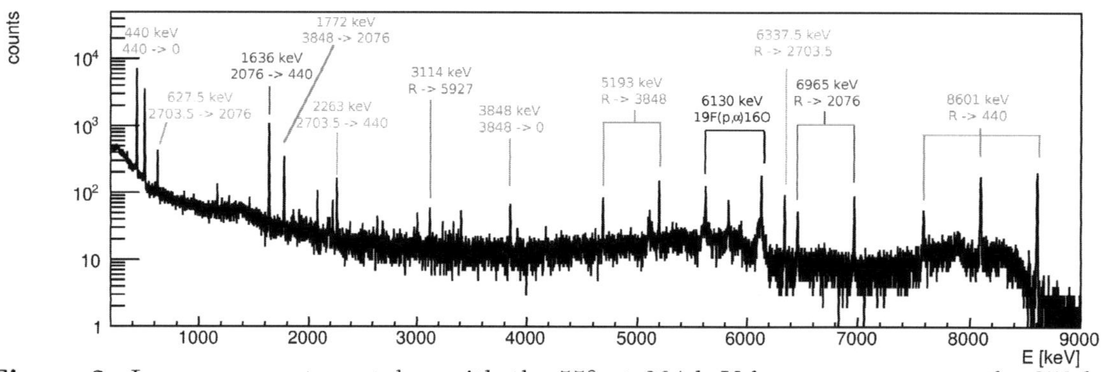

Figure 2. Long run spectrum taken with the 55° at 264 keV beam energy, over the 259 keV resonance.

For the non-detected resonances new upper limits have been measured. The new upper limits are two to three orders of magnitude lower than the previous direct measurement, demonstrating the improvement in sensitivity that can be achieved in underground experiments.

IOP Conf. Series: Journal of Physics: Conf. Series **940** (2017) 012026 doi:10.1088/1742-6596/940/1/012026

3.2. Ongoing measurements

A new experimental campaign on the ^{22}Ne(p,γ)^{23}Na reaction is currently ongoing at LUNA. The aim of this phase is to measure the direct capture contribution to the ^{22}Ne + p cross section and to increase the sensitivity to the tentative resonances at 71, 105 and 215 keV. In this phase, the windowless gas target chamber is surrounded by a high-efficiency BGO detector covering most of the solid angle around the target.
In parallel, the reactions ^{23}Na(p,γ)^{24}Mg of the NeNa cycle and ^{18}O(p,γ)^{19}F of the hot CNO cycle are under investigation on the solid target beam line.

4. Future perspectives

A new scientific program for the LUNA 400 kV accelerator has been approved. The program will last until 2018 and includes six reactions relevant for Big Bang nucleosynthesis and stellar hydrogen and helium burning.
The first of those reactions to be investigated is the ^{2}H(p,γ)^{3}He, which represents the main uncertainty on ^{2}H abundance from standard Big Bang nucleosynthesis calculations.

The future of LUNA also involves the installation of a new higher voltage machine, in the MV range. This machine will be devoted to the study of nuclear reactions that take place at higher temperatures than those occurring during the hydrogen-burning processes studied so far with the 400 KV accelerator.
This project, named "LUNA-MV", involves the realization of a unique facility inside the underground Gran Sasso Laboratory centred on a 3.5 MV single-ended accelerator providing high intensity hydrogen, helium and carbon beams. Two different beam lines are foreseen devoted to solid and gas target experiments, respectively. This will allow us to study the key reactions of helium burning (namely the ^{12}C(α,γ)^{16}O reaction and (α,n) reactions on ^{12}C and ^{22}Ne) and to re-investigate the ^{14}N(p,γ)^{15}O and the ^{3}He(^{4}He,γ)^{7}Be reactions over a wide energy range in order to further diminish its experimental uncertainties.
The LUNA-MV project was approved by the LNGS Scientific Committee in 2010 and it obtained financial support from the Italian Research Ministry in 2012. The tender for the accelerator has been published in 2015.
A key issue for LUNA-MV is the realization of a neutron shielding to the rest of the laboratory and to the internal rock surfaces to preserve the low background characteristic of LNGS. In order to evaluate the neutron flux produced by the reactions above mentioned, a series of GEANT4 simulations of the LUNA-MV building are currently being finalized.
The construction of the building and the infrastructure is planned to start in 2017, while the commissioning is expected to start in 2018.

References

[1] Rolfs C and Rodney W S, 1988 *Cauldrons in the Cosmos* (Chicago and London: University of Chicago Press)
[2] Broggini C et al. 2010 *Annu. Rev. Nucl. Part. Sci.* **60**, 53
[3] Formicola A et al. 2004 *Phys. Lett.* B **591** 61.
[4] Imbriani G et al. 2005 *Eur. Phys. J.* A **25** 455
[5] Bemmerer D et al. 2006 *Phys. Rev. Lett.* **97** 122501
[6] Gyürky G et al. 2007 *Phys. Rev.* C **75** 35805
[7] Confortola F et al. 2007 *Phys. Rev.* C **75** 065803
[8] Strieder F et al. 2012 *Phys. Lett.* B **707** 60
[9] Di Leva A et al. 2014 *Phys. Rev.* C **89** 015803
[10] Anders M et al. 2014 *Phys. Rev. Lett.* **113** 042501
[11] Bruno C G et al. 2015 *Eur. Phys. J.* A **51** 94
[12] Carretta E et al. 2009 *Astron. Astrophys.* **505** 117
[13] Izzard R G et al. 2007 *Astron. Astrophys.* **466** 641
[14] Iliadis C et al. 2002 *Astrophys. J. Suppl. Ser.* **142**, 105

[15] Görres J et al. 1982 *Nucl. Phys.* A **385** 57
[16] Powers J R et al. 1971 *Phys. Rev.* C **4** 6
[17] Hale S E et al. 2001 *Phys. Rev.* C **65** 015801
[18] Jenkins D G et al. 2013 *Phys. Rev.* C **87** 064301
[19] Angulo C et al. 1999 *Nucl. Phys.* A **656** 3
[20] Iliadis C et al. 2010 *Nucl. Phys.* A **841** 251
[21] Anders M et al. 2013 *Eur. Phys. J.* A **49** 28
[22] Becker H W et al. 1992 *Z. Phys.* A **343** 361
[23] Cavanna F et al. 2014 *Eur. Phys. J.* A **50** 179

IOP Conf. Series: Journal of Physics: Conf. Series **940** (2017) 012027 doi:10.1088/1742-6596/940/1/012027

Direct underground measurement of the $^{17}O(p,\alpha)^{14}N$ reaction at LUNA

C G Bruno for the LUNA Collaboration

School of Physics and Astronomy and SUPA, Edinburgh University, UK

E-mail: `carlo.bruno@ed.ac.uk`

Abstract. The $^{17}O(p,\alpha)^{14}N$ reaction plays a role in a number of stellar sites, including giant branch stars and novae. In particular, a narrow and isolated resonance at E_p=70 keV dominates the reaction rate at stellar temperatures around T_9=0.01-0.1. A direct measurement of the $^{17}O(p,\alpha)^{14}N$ reaction has been recently performed at the underground LUNA-400 accelerator in the Gran Sasso Laboratory, Italy. Protons were accelerated on a solid Ta_2O_5 target and the alpha particles emitted were detected with an array of silicon detectors. The background reduction afforded by the underground environment was instrumental in successfully carrying out the measurement. The setup was simulated and commissioned exploiting two well-known resonances at E_p=151 and 193 keV in $^{18}O(p,\alpha)^{15}N$ and $^{17}O(p,\alpha)^{14}N$ respectively. The strengths of the two resonances were found to be respectively $\omega\gamma(151)=164.2 \pm 0.9_{\text{stat}}\,^{+12.1}_{-11.7}{}_{\text{syst}}$ meV and $\omega\gamma(193)=1.68 \pm 0.03_{\text{stat}} \pm 0.12_{\text{syst}}$ meV in good agreement with the literature. Analysis of data acquired on the E_p=70 keV resonance in $^{17}O(p,\alpha)^{14}N$ reaction is currently underway.

1. Introduction

The rare, stable ^{17}O isotope is ideally suited to trace and constrain mixing processes occurring in giant branch stars [1]. Its abundance in stellar models currently resents of the uncertainties in the reaction rate of the $^{17}O(p,\alpha)^{14}N$ reaction (Q_{val}=1.2 MeV). Experimentally, the abundance of ^{17}O on the surface of these types of stars can be estimated precisely from solar grains [2]. Recent studies [3, 4] have found discrepancies between the abundances predicted by models and those experimentally observed for some stellar sites. These inconsistencies could point to additional mixing processes occurring in giant branch stars or issues in the nuclear physics input employed. At temperatures of astrophysical interest (T_9=0.01-0.1), the reaction rate of the $^{17}O(p,\alpha)^{14}N$ nuclear reaction is dominated by two narrow resonances at energies[1] E_p=70 and 193 keV. The latter resonance is quite well-known (e.g. ref. [5, 6]), but direct masurements and the frequent re-analyses of the E_p=70 keV [7, 8, 9, 10] resonance strength paint an incomplete picture in the literature. Because of the weakness ($\omega\gamma \approx$ neV) and importance of this resonance, indirect methods have been recently employed in its study [11, 12] as well. A campaign aimed at precisely measuring the strength of the E_p=70 keV resonance in $^{17}O(p,\alpha)^{14}N$ has been recently completed at the underground LUNA-II accelerator [13].

[1] In this work, E_p indicates the energy of the proton beam in the laboratory frame

IOP Conf. Series: Journal of Physics: Conf. Series **940** (2017) 012027 doi:10.1088/1742-6596/940/1/012027

2. Experimental approach

The aim of the experimental campaign was the precise, direct measurement of the E_p=70 keV resonance strength in $^{17}O(p,\alpha)^{14}N$ under thick-target yield conditions [14]. Protons were accelerated at typical beam currents of around 100μA on a solid Ta_2O_5 target [15] enriched in ^{17}O. The alpha particles produced were detected by an array of eight silicon detectors, placed at backward angles (four at 102.5° and four at 135°) with respect to the beam axis, as shown in Fig.1.

Figure 1. The reaction chamber used during the campaign. The beam enters from the top and is completely by a solid Ta_2O_5 target placed in the centre of the aluminium dome pictured. Eight silicon detectors arranged in two rows at 135 and 102.5 degrees are housed on the aluminium dome. Detectors were protected from elastically scattered protons by thin ($\approx 2.4\mu m$) aluminised Mylar foils mounted on an inner concentric dome (not shown).

In order to shield the detectors from the intense flux of elastically scattered protons, thin ($\approx 2.4\mu m$) aluminised Mylar foils were placed in front of each detector. Great care was taken to optimise the thickness of these foils. They had to be thick enough to stop scattered protons up E_p=200 keV yet thin enough to let the alpha particles through with a detectable energy. For the final thickness chosen, the alpha particles generated by the E_p=70 keV resonance in $^{17}O(p,\alpha)^{14}N$, produced with an initial energy of E_α=1 MeV, reached the detectors with only 200 keV. This low energy, coupled with the low yield expected from this resonance (few counts per Coulomb of collected charge on target) made the measurement especially challenging.

3. Commissioning

An in-depth description of the setup and the commissioning may be found in ref. [16]. We exploited the well-known resonances at E_p=151 and 193 keV in the $^{18}O(p,\alpha)^{15}N$ reaction (Q_{val}=4.0 MeV) and the $^{17}O(p,\alpha)^{14}N$ reaction, respectively. The E_p=151 keV resonance is known with a high precision and accuracy [17] and was used both to validate a GEANT4 simulation [18] of the setup and to periodically check the degradation of the solid targets used during the measurements. We found its resonance strength to be $\omega\gamma(151)=164.2 \pm 0.9_{\rm stat}\,^{+12.1}_{-11.7}{}_{\rm syst}$ meV, in excellent agreement with previous measurements [17].
The E_p=193 keV resonance is also relatively well known [5, 6]. We measured its strength to be $\omega\gamma(193)=1.68 \pm 0.03_{\rm stat} \pm 0.12_{\rm syst}$ meV, once again in excellent agreement with previous values. The E_p=193 keV resonance has a strength at least five orders of magnitude larger than the 70 keV resonance, and we could see a very clear alpha peak in our spectrum (see Fig.2). The position and width of this peak were used to estimate the expected energy of the alpha peak from the 70 keV resonance. Measuring this latter resonance was the final objective of our campaign (for a detailed analysis procedure see refs. [16, 19]). We were able to observe a signal

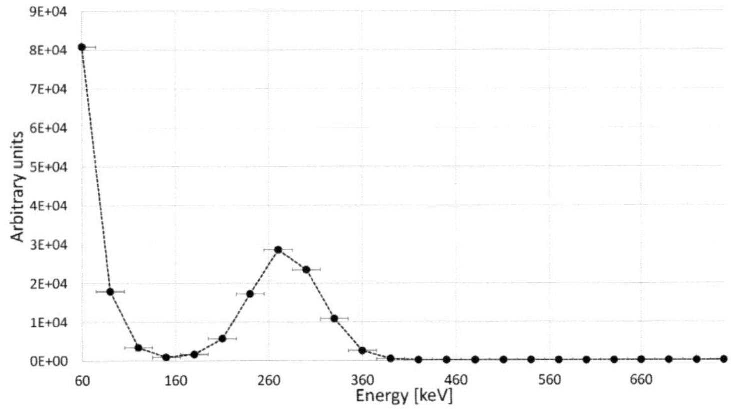

Figure 2. Typical spectrum for the resonance at E_p=193 keV. A clear alpha peak is visibile around 250 keV of detected energy, after passage through protective aluminised Mylar foils. Horizontal error bars indicate bin size, vertical error bars are too small to be seen.

for the 70 keV resonance, thanks to the background reduction afforded by the underground environment. Data analysis is currently in progress.

4. Background suppression

While the reduction in natural background afforded by underground environment is well documented for γ-spectroscopy experiments (e.g. [13, 20]), the benefits of moving underground in α-spectroscopy experiments are less well-established. We carried out a systematic study of the background reduction in the eight silicon detectors we used in the course of the campaign. Background was acquired underground in Gran Sasso Laboratory, Italy, with and without a partial lead shield around the scattering chamber. After the campaign was over, we moved the entire setup to Edinburgh, UK, where background was acquired again with and without a lead shield. A comparison between total spectra in the four experimental conditions in shown in Fig.3.

Three broad energy regions can be identified. Above 4.5 MeV, there is essentially no difference between the background rates in the four conditions. This suggests the origin of the background is intrinsic in the detectors or in the setup. It could be due to, e.g., radioactive elements present in the detectors' bulk material or activity transferred to detectors from radioactive sources (*e.g.* during calibration in previous experiments). In the second energy region, from about 1.5 MeV to 4.5 MeV, the underground environment leads to a significant background reduction, albeit with little difference with or without the shielding. The main source of the background in this region is not obvious. It is reduced by the underground environment, but not by a lead shield. Note the tails from the peaks at 5 MeV play a role in this energy region as well. In the third energy region, below 1.5 MeV, both the lead shielding and the underground environment have a significant effect on the background rate. Compton-generated electrons produced by gamma rays interacting with the silicon detectors probably play a role here. Gamma rays are reduced both by the underground environment and by a lead shielding, leading to a reduction in the Compton-generated electrons.

These results confirm the advantage of moving underground for alpha spectroscopy experiments. In particular, in the region of interest for the 70 keV resonance, the reduction in natural background is as much as a factor 15 between unshielded overground and shielded underground, thus allowing for this challenging experiment.

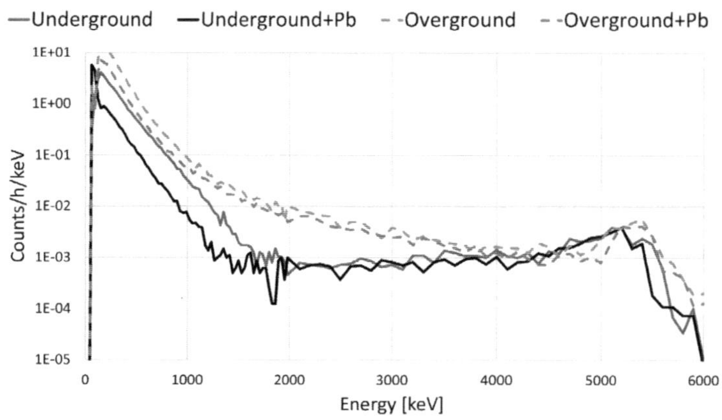

Figure 3. Background spectra acquired underground in Gran Sasso Laboratory, Italy, with (black line) and without (blue) lead, compared with background spectra acquired overground in Edinburgh, UK, with (dashed green) and without (dashed red) lead. See text for details.

5. Conclusions and future prospects

A purpose-built alpha spectroscopy setup was commissioned at the underground LUNA accelerator in the Gran Sasso Laboratory, Italy. The E_p=151 and 193 keV resonances in $^{18}O(p,\alpha)^{15}N$ and $^{17}O(p,\alpha)^{14}N$ respectively were measured. Their strengths were found to be respectively $\omega\gamma(151)=164.2 \pm 0.9_{\rm stat}\,{}^{+12.1}_{-11.7}{}_{\rm syst}$ meV and $\omega\gamma(193)=1.68 \pm 0.03_{\rm stat} \pm 0.12_{\rm syst}$ meV in good agreement with the literature. A region of interest for the alpha peak of the E_p=70 keV resonance in $^{17}O(p,\alpha)^{14}N$ was precisely defined. The significant background reduction afforded by the underground environment will allow us to measure the strength of the E_p=70 keV resonance in $^{17}O(p,\alpha)^{14}N$ with an high precision. Analysis of the data is ongoing and the results will be published in a forthcoming paper [21].

6. Bibliography

[1] K. M. Nollett *et al.*, Astrophys. J. **582** (2003) 1036-1058
[2] S. Palmerini *et al.*, Astrophys. J. **764** (2013) 128
[3] M. Lugaro *et al.*, A&A **461** (2007) 657-664
[4] G. M.Halabi and M. El Eid, MNRAS **451** (2015) 2957-2967
[5] A. Chafa *et al.*, Phys. Rev. C **75** (2007) 035810-1 15
[6] J. Newton *et al.*, Phys. Rev. C **81** (2010) 045801
[7] M. Berheide *et al.*, Z. Phys. A **343**, 483-487 (1992)
[8] S. Niemeyer, Diplomarbeit, Ruhr-Universitaet Bochum, Germany (1996)
[9] J. Blackmon *et al.*, Phys. Rev. Lett. **74** (1995) 2642-2645
[10] M.D. Hannam and W.J.Thompson, Nucl. Inst. Meth. A **431** (1999) 239-251
[11] M.L. Sergi *et al.*, Phys. Rev. C **82** (2010) 032801(R)
[12] M.L. Sergi *et al.*, Phys. Rev. C. **91** (2015) 065803
[13] C. Broggini *et al.* Annu. Rev. Nucl. Part. Sci. **60** (2010) 53-73
[14] C. Iliadis, 2007: *Nuclear Physics of Stars* Wiley-VCH Verlag GmbH
[15] A. Caciolli *et al.*, Eur. Phys. Journ. A **48** (2012) 144
[16] C.G. Bruno *et al.* Eur. Phys. J. A **51** (2015) 94
[17] H.W. Becker *et al.*, Z. Phys. A **351** (1995) 453
[18] GEANT 4.9.4.p02, from URL http://geant4.cern.ch
[19] C.G. Bruno, PhD thesis, The University of Edinburgh, UK, *in preparation*
[20] T. Szücs *et al.*, Eur. Phys. J. A **44** (2010) 513
[21] C.G.Bruno *et al.*, *in preparation*

IOP Conf. Series: Journal of Physics: Conf. Series **940** (2017) 012028 doi:10.1088/1742-6596/940/1/012028

The Feasibility of Studying ^{44}Ti(α, p)^{47}V Reaction at Astrophysical Energies

Tariq Al-Abdullah[1,2], **D. Bemmerer**[1], **Z. Elekes**[3], **D. Schumann**[4]

[1]Institute of Radiation Physics, Helmholtz-Zentrum Dresden-Rossendorf, D-01328 Dresden, Germany
[2]Physics Department, The Hashemite University, P.O. Box 150459, Zarqa 13115, Jordan
[3] Institute for Nuclear Research of the Hungarian Academy of Sciences (MTA ATOMKI), H-4026 Debrecen, Bem tér 18/c, Hungary
[4]Laboratory of Radiochemistry and Environmental Chemistry, Paul Scherrer Institute, CH-5232 Villigen PSI, Switzerland

E-mail: abdullatq@gmail.com

Abstract. The gamma-ray lines from the decay of ^{44}Ti have been observed by space-based gamma-ray telescopes from two supernova remnants. It is believed that the ^{44}Ti(α,p)^{47}V reaction dominates the destruction of ^{44}Ti. This work presents a possible technique to determine its reaction rate in forward kinematics at astrophysically relevant energies. Several online and offline measurements in parallel with Monte Carlo simulations were performed to illustrate the feasibility of performing this reaction. The results will be discussed.

1. Introduction

Core-collapse supernovae are believed to enrich the interstellar medium with heavy elements. Because of its relatively long half life ($T_{1/2} = 59.1 \pm 0.3$ y) [1], the detection of ^{44}Ti isotope may reveal information about the nucleosynthesis processes and address the explosion mechanisms.

The remnant of Cassiopeia A supernova gives clear evidences about its production, where emitted $\gamma-$lines first from ^{44}Ti $\rightarrow$ ^{44}Sc at 67.9 and 78.4 keV, and then to ^{44}Ca at 1157 keV have been detected by INTEGRAL [2] (Also BeppoSAX [3]) and COMPTEL telescopes [4] respectively. These observations are confirmed from the remnant of supernova SN1987A [5]. However, the same γ lines could not be detected by other supernova portions or might be overestimated by theoretical predictions. Including Cas A, the asymmetry between models and observations raised the necessity to explore such stellar explosions.

The ^{44}Ti(α, p)^{47}V reaction is considered the major one that has a direct impact on the synthesis of ^{44}Ti [6]. Its importance raised from its small Q-value ($Q \approx -$ 0.410 MeV). The reaction was studied in inverse kinmatics by Sonzogni *et al* [7]. The measured ^{44}Ti beam energies were higher than the Gamow windows as shown in fig. 1. A similar experiment was conducted at the REX-ISOLDE facility, CERN [8], see fig. 1. The results were able only to establish an upper limit on the cross section of the ^{44}Ti(α, p)^{47}V reaction at $E_{c.m.} = 4.15$ MeV. Trying to improve the reaction rate, we report here a study that investigates the feasibility of performing the ^{44}Ti(α, p)^{47}V reaction in forward kinematics using ^{44}Ti as a target.

IOP Conf. Series: Journal of Physics: Conf. Series **940** (2017) 012028 doi:10.1088/1742-6596/940/1/012028

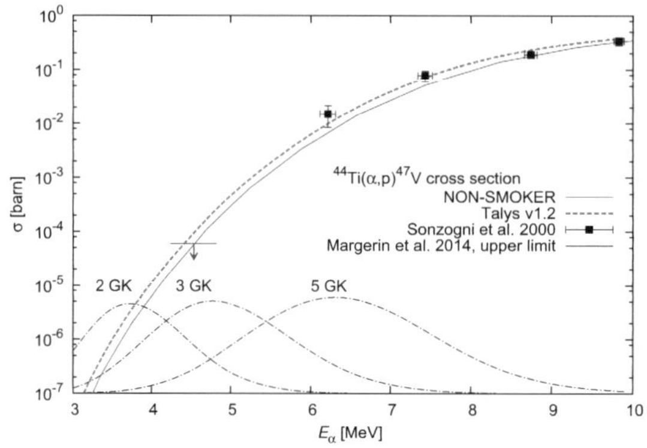

Figure 1. The points represent the measured cross-section values for ^{44}Ti$(\alpha, p)^{47}$V reaction, which are fitted with NON-SMOKER and TALYS codes. The in Ref. [9] suggested Gamow peaks show that data are needed for $E_{c.m.}$ below 6.0 MeV.

2. Production of ^{44}Ti target

^{44}Ti can be produced in great amounts via spallation reactions induced by high-energetic protons in metals with mass numbers slightly hevier than titanium. One possibility to obtain this isotope is to extract it from accelerator waste, as was proposed and carried out by the RadWasteAnalytics group at PSI [10]. Stainless steel samples, irradiated for two years in the high-power target of the spallation neutron source at PSI (SINQ) are best suited for manufacturing of highly active ^{44}Ti samples.

For the target manufacturing, electro-deposition was implemented. Our preparatory studies were aimed to produce a target with around 10 MBq, but for the development of the method we used lower concentrations. An electro-deposition cell with a volume of 27 mL made of Teflon was prepared. As backing material stainless steel disks with a diameter of 27 mm and a thickness of 0.5 mm with a deposition spot of 14 mm were applied. The chemical procedure to produce the target is explained in [11]. Producing a radioactive target in the range of few hundreds of kBq was successful. For the experimental target an aliquot of 10 MBq ^{44}Ti was prepared, but the deposition was thick, porous, loose and tending to disperse. Obviously, besides the stable titanium also parts of the contaminants like Fe, Ca, Mn, Cr and others tend to deposit during the electrolysis, making the target inapplicable for the planned experiment, even assuming that it will be covered by a thin gold layer.

A further ultra-purification of the electrolyte would be mandatory to produce a suitable target. An alternative solution of the problem could be the production of a ^{44}Ti beam and following implantation into a suitable backing. This method would not only have the advantage in getting rid of the contaminating elements, but also obtaining a separation from the stable titanium isotopes.

3. Experimental Setup

The experiment is planned to be performed using the ^{4}He^{++} beam provided by the 3 MV Tandetron accelerator at Helmholtz-Zentrum Dresden-Rossendorf (HZDR). With an intensity of $1 - 2$ μA, the beam energy will be from 0.1 to 3.3 MV. Four totally depleted silicon surface barrier detectors, each with an active area of 300 mm^2 and thickness of 50 μm, are planned to be placed on the same plane around the target. They will be equally spaced and placed at 61°, 93°, 125° and 157° angles with respect to the incident α-beam as shown in Fig. 2. All four detectors will be backed by 300-μm-thick partially Passivated Implanted Planar Silicon

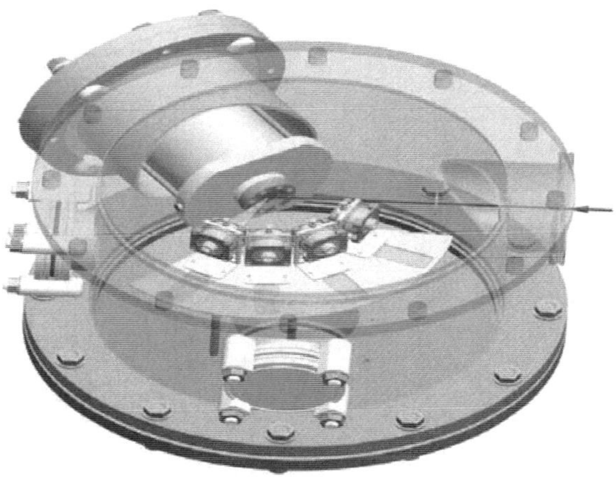

Figure 2. Schematic diagram of the experimental setup shows the path of the α-beam (red line from the right) and the positions of the telescopes (yellow) with respect to the target.

(PIPS) detectors. Together, the 50 μm and 300 μm detectors form a particle telescope providing particle identification information (ΔE,E). Each telescope will sit on a trail track to vary its distance from the target. Aluminum foils of thicknesses range from 15 – 80 μm will be inserted in front of the telescopes to block the scattered alpha particles. The emitted γ rays during the bombardment will be identified using a high-purity germanium (HPGe) detector placed behind the target at 55° with respect to the beam axis. Monitoring γ rays gives additional information about the stability of the target and how the target evolves under the intense beam.

4. Sputtering Radioactive Target

One of the fundamental concerns during the measurements is filling the experimental setup with sputtered ^{44}Ti, especially if the activity of the sputtered ^{44}Ti exceeds the safety regulation limit in the lab. This issue was investigated by online and offline studies using ^{48}Ti as a target to avoid any contamination.

Stable $^{48}Ti^{+}$ ions was implanted at 60 keV energy into a thin Ta disk at room temperature with a fluence of 10^{16} cm^{-2}. The target was then mounted in an apparatus. The surface of the device was tilted to allow the ion beam to impinge on the target at 55°. After preparation, the target was installed in a vacuum chamber pumped down to a pressure of a few times 10^{-7} mbar. A 5.0-MeV α beam with an intensity of 1 μA was then sent on the target for 11 hours. The beam spot was approximately 3.0 mm in diameter and its position was varied to irradiate the whole surface. The ejected atoms were collected by polyester foil which surrounded the target and was thick enough to stop the sputtered atoms. Using the ICP-MS technique, the number of sputtered Ti from the target relative to the beam's current is around $(10 \pm 3) \times 10^{-5}$ atoms/ion.

The above sputtering process was simulated using the SRIM 2013 version of the TRIM [12] and TRIDYN [13] codes. Since the sputtering is negligible deep in the target, the depth of the target is chosen to be 0.1 μm. Sufficient statistical qualities and precision could be achieved with 5.4×10^{18} cm^{-2} total fluence and 4×10^{7} corresponding number of pseudoprojectiles. The SPYL of Ti and Ta atoms were determined to be 6.35×10^{-5} and 1.79×10^{-3} atoms/ion, respectively. Neither varying the width of the slabs nor oxidizing the whole target changed the above results by more than 20%. Both codes agree within uncertainty of 20%. The results are shown in Fig. 3.

The simulated calculation and ICP-MS results coincide very well. This infers that TRIDYN and TRIM codes are reliable to predict the SPYL when the beam irradiates the ^{44}Ti target. If 1.0 μA of α particles bombarding the radioactive target for 10 days, the yield will be 5×10^{-5}

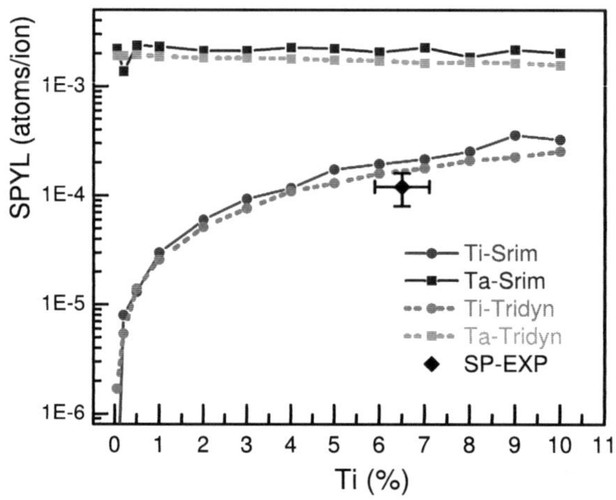

Figure 3. A comparison between the sputtered yield of Ti and Ta atoms obtained by the TRIM and TRIDYN codes, as a function of initial maximum Ti concentration in the bulk, at $E_\alpha = 5$ MeV. The single point (black) represents the sputtering experiment.

atoms/ion. That means there will be less than 100 kBq ejected activity in the scattering chamber from a 1.0 MBq target. Since higher α-beam energies are related to lower SPYL, the sputtered activities are expected to be lower than the above value. According to the legal regulations in Germany, 100 kBq ^{44}Ti/^{44}Sc in secular equilibrium is the activity limit below which a source is not regulated when outside a radiation controlled area.

5. Offline Measurements and Simulations

One of the concerns regarding the reaction products of α + ^{44}Ti is to estimate the continuum background from β^+ decay of the ^{44}Ti targets daughter, and their consequences on the detection system. The experiment will be performed with incident α-beam at $3-8$ MeV. The kinematics give the ejected protons at 125° energies from $2.16-6.42$ MeV before the telescope. The possible interference between the proton peaks and the β^+ background produced is investigated. For this study, a mixed-radioactive source containing ^{239}Pu, ^{241}Am and ^{244}Cm with main energies of $\alpha-$particles of 5.157, 5.485, and 5.805MeV, respectively, was first inserted below a detector telescope. The telescope consisted of two PIPS detectors; 25 m and 300 m thick Ortec ED-100-450-25 and CU-023-600-300 used as ΔE and E detectors, respectively. All the tests were performed under 10^{-6} mbar vacuum. A ^{44}Ti/^{44}Sc sample was placed below the telescope and near the mixed-source. The activity of the sample is 83.1 $\pm$ 1.2 kBq. Several test runs were performed using these α and β sources. Data were collected when both were seen by E-detector, when only one of them was placed in the chamber and when none of them were close to the detector. The energy calibration was achieved by identifying the α peaks in the spectrums of ΔE versus E + ΔE. The data for the three peaks is shown in Fig. 4.

To count for the 1 MBq activity, the spectrum from ^{44}Ti will be increased by almost one order of magnitude. Using of the telescope as estimated in Fig. 4 shows a very good separation between proton spots and continuum background radiations due to β^+ emissions for all incident E_α. Hence, those protons are possible to be observed with the suggested experimental setup.

On the other hand, naturally occurring Ti, which is already present in accelerator waste, will act as a carrier during the chemical separations. Five Ti isotopes (abundance), ^{46}Ti(8.0%), ^{47}Ti(7.3%), ^{48}Ti(73.8%), ^{49}Ti(5.5%), and ^{50}Ti(5.4%), are considered as impurities in the ^{44}Ti target. Because of the small variations in the $Q-$values when $^{44-49}$Ti are involved, and very

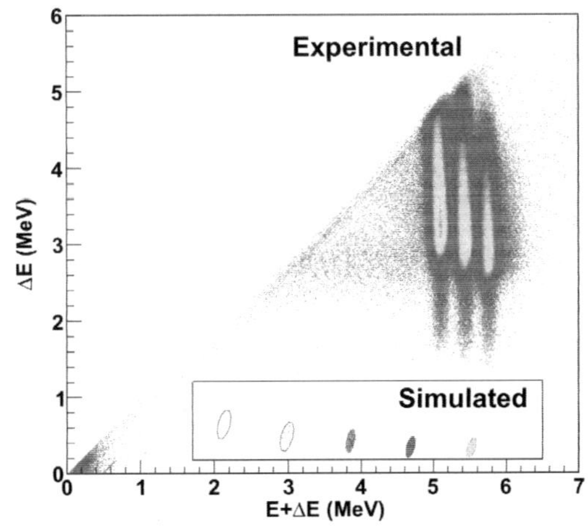

Figure 4. ΔE vs. $E + \Delta E$ plot of the experimental results when both sources are below the telescope. The data inside the box are the result of simulations.

comparable cross-sections, on average 0.57 mb, there will be a possible interference in the reaction products.

A Monte Carlo simulation using the GEANT4 toolkit was carried out to model the experimental setup. The code included those parts of the system which might effect the trajectory, the angular and energy straggling of the ejected particles, namely the target, the Si detectors with their housings and the Al degrader foils in front of the telescopes. The protons were ejected according to two-body kinematics using a monochromatic $\alpha-$beam. Assuming the worst-case scenario the particles were started from a depth of 600A while the lateral distribution was chosen to be 3 mm which resembled the expected beam spot size. No excitation of the heavy ions leaving the reaction was considered. For the different isotopes, separate runs were carried out presuming 100% purity. For each case the same number of initial particles were initiated. As a result, a clear separation between the protons originating from (α, p) reactions on different isotopes in the target can be seen. The above simulation assumed similar concentrations to the radioactive and stable Ti in the target. To observe the $^{44}\mathrm{Ti}(\alpha, p)^{47}\mathrm{V}$ reaction, the amount of impurities in the target must be low.

6. Conclusion

We have discussed the possibility of studying $^{44}\mathrm{Ti}(\alpha, p)^{47}\mathrm{V}$ reaction directly in forward kinematics. The proposed experimental setup consists of a 1 MBq ^{44}Ti target and four detector telescopes at specific backward angles. The biggest challenge in the study is to prepare a radioactive target with minimum amount of naturally occurring Ti and high ^{44}Ti activity. Estimating the continuum of the background radiation, mainly due to the β^+ emission, by offline measurements using E and ΔE PIPS-detectors and radioactive sources show that events from $^{44}\mathrm{Ti}(\alpha, p)^{47}\mathrm{V}$ are well separated, and thus can be observed. Summarizing, the experiment is found to be safe and feasible.

Acknowledgment

Support from DFG (BE 4100/2-1, BE 4100/3-1) and from the Helmholtz Association Nuclear Astrophysics Virtual Institute (NAVI, VH-V1-417) and by Euro GENESIS is gratefully acknowledged.

Referrences

[1] I. Ahmad, J. P. Greene, E. F. Moore, S. Ghelberg, A. Ofan, M. Paul, and W. Kutschera 2006, Phys. Rev. C **74**, 065803
[2] M. Renaud *et al.* 2006, Astrophys. J **647**, L41
[3] J. Vink *et al.* 2001, Astrophys. J. **560**, L79
[4] A. F. Iyudin *et al.* 1994, Astron. Astrophys. **284**, L1
[5] A. Jerkstrand, C. Fransson, and C. Kozma 2011, Astron. Astrophys. **530**, A45
[6] G. Magkotsios, F. X. Timmes, A. L. Hungerford, C. L. Fryer, P. A. Young, and M. Wiescher 2010, Astrophys. J. Suppl. Ser. **191**, 66
[7] A. A. Sonzogni *et al.* 2001, Phys. Rev. Lett. **84**, 1651
[8] V. Margerin *et al.* 2014, Phys. Lett. B **731**, 358
[9] T. Rauscher 2010, Phys. Rev. C **81**, 045807
[10] D. Schumann and J. Neuhausen 2008, J Phys. G: Nucl. Part. Phys. **35**, 014046
[11] T. Al-Abdullah *et al.* 2014, Eur. Phys. J. A **50**, 140
[12] J. F. Ziegler, M. D. Ziegler, and J. P. Biersack 2010, Nucl. Instrum. Methods Phys. Res. B **268**, 1818
[13] W. Möller and W. Eckstein 1984, Nucl. Instrum. Methods Phys. Res. B **2**, 814.

IOP Conf. Series: Journal of Physics: Conf. Series **940** (2017) 012029 doi:10.1088/1742-6596/940/1/012029

Investigation of interaction processes of light nuclei with ^{14}N

D.Alimov[1], N.Burtebayev[2], I.Boztosun[3], Z.Kerimkulov[2], J.Burtebayeva[2], M.Karakoc[3], Y.Mukhamejanov[1], D.Janseitov[4], A.Bahtibayev[5], A.Pattayev[5] and Sh.Hamada[6]

[1] Al-Faraby Kazakh National University, 050032, Almaty, Kazakhstan,
[2] Institute of Nuclear Physics, Almaty, Kazakhstan,
[3] Akdeniz University, Faculty of Science, Department of Physics, 07058, Antalya, Turkey,
[4] Eurasian National University, Astana, Kazakhstan,
[5] A. Yesevi IKTU, Turkestan, Kazakhstan,
[6] Tanta University, Faculty of Science, Tanta, Egypt

E-mail: Diliyo@mail.ru

Abstract. Elastic scattering of ^{3}He on ^{14}N as an example of 1p-shell nuclei at 50 and 60 MeV has been investigated within the framework of Optical Model. In the Optical Model analysis, both microscopic double-folding and phenomenological potentials for the real part of the complex nuclear potential have been used. For both microscopic double-folding and phenomenological analyses, the imaginary potential has taken to be WS volume shaped. It is noticed that while a normalization of the strength of the double-folding real potential is needed to explain the structure observed in the experimental data, a good agreement between experimental data is obtained for the phenomenological potential case.

1. Introduction

The study of the interaction of ^{3}He with ^{14}N nuclei at low energies is of interest both from the point of view of establishing reliable values of the parameters of the internuclear potential interaction of ^{3}He and to study the mechanisms of cluster effect in scattering processes [1].

The elastic and inelastic experimental data of ^{3}He+^{14}N system has been measured in INP Almaty as a part of larger experimental research program of the scattering of ^{3}He on light-heavy nuclei with a mass range between 4 and 28. Experiment on the elastic and inelastic scattering of ^{3}He ions with energies 50 and 60 MeV on ^{14}N were carried out on the extracted beam of the isochronous cyclotron U-150M in INP Almaty, Kazakhstan. The energy of the accelerated ions was determined by comparing the energy spectrum of the particles scattered from a thin Au target at a small angle, the spectra of the α - particle radioactive source. The spread of the beam energy was less than 1%. The targets were thin films and foils (1 mg/cm^2) of enriched isotope of corresponding elements. The thickness of the gas target was determined by weighing or by energy loss of α - particles from a radioactive source with exactly $6-9\%$. Nuclear reaction products were detected by using telescope detectors consisting of surface-barrier silicon counters with a thickness - 10, 18, 33, 50, 100 micrometers or ionization chamber, the effective thickness may be smoothly changed in a broad range. In both variants, telescopes as E-counter used diffusion-drift silicon detectors with a sensitive layer thickness of 2 mm. The angular distributions of scattered

IOP Conf. Series: Journal of Physics: Conf. Series **940** (2017) 012029 doi:10.1088/1742-6596/940/1/012029

particles in the investigated nuclei were measured in the angular range of 10°-170° in laboratory coordinate system in steps of 3°. Spectrum analysis has been performed by using the program MAESTRO [2].

Experimental data of $^{3}He+^{14}N$ system measured at INP Almaty has been analysed within the standard optical model and microscopic double-folding potentials using the computer code FRESCO. In the next section, we present our optical model potentials and then show our results in section 3. We also present our summary and conclusion in this section.

2. The Model

We have used both phenomenological deep and double-folding real potentials with a volume type imaginary potential [3, 4, 5, 6, 7, 8, 9, 10, 14]. Our total real potential for these cases consists of the nuclear, $V_{Nuclear}$, Coulomb and centrifugal potentials, V_{Coul}, V_{Cent} respectively.

$$V_{total}(r) = V_{Nuclear}(r) + V_{Coulomb}(r) + V_{Centrifugal}(r) \tag{1}$$

The phenomenological nuclear potential is assumed to have the square of a Woods-Saxon shape and Coulomb potential [10, 11] distributed uniformly over a sphere of radius R_C, is also added.

Table 1. Phenomenological and Double Folding potential parameters with N_r.

E_b(MeV)	V_0(MeV)	r_0(fm)	a_0(fm)	W_V(MeV)	r_V(fm)	a_V(fm)	N_r	R_C(fm)
50	60	1.2	0.823	8.8	1.412	0.747		4.62
DF				8.8	1.412	0.747	0.9	
60	55	1.2	0.823	9.8	1.318	0.747		4.62
DF				9.8	1.318	0.747	0.9	

Double folding potential [12, 13] is calculated by using the nuclear matter distributions of both projectile and target nuclei together with an effective nucleon-nucleon interaction potential (ν_{NN}) are used. Thus, the double folding potential is

$$V_{DF}(\mathbf{r}) = \int d\mathbf{r}_1 \int d\mathbf{r}_2 \rho_p(\mathbf{r}_1)\rho_t(\mathbf{r}_2)\nu_{NN}(\mathbf{r}_{12}) \tag{2}$$

where $\rho_p(\mathbf{r}_1)$ and $\rho_t(\mathbf{r}_2)$ are the nuclear matter density of projectile and target nuclei, respectively. Gaussian density distributions defined as following have been used for both nuclei

$$\rho(r) = \rho_0 exp(-\beta r^2) \tag{3}$$

where β is adjusted to reproduce the experimental value for the rms radius of the ^{14}N=2.58 fm and ^{3}He=1.877 fm [14]. ρ_0 values can be obtained from the normalization condition

$$\int \rho(r) r^2 dr = \frac{A}{4\pi} \tag{4}$$

where A is the mass number. For the nucleon-nucleon interaction potential (ν_{NN}), we have used the following M3Y nucleon-nucleon realistic interaction with the relevant exchange correction term due to the Pauli principle [12, 13], given by

$$\nu_{NN}(r) = 7999\frac{exp(-4r)}{4r} - 2134\frac{exp(-2.5r)}{2.5r} + J_{00}(E)\delta(r)MeV, \tag{5}$$

where

$$J_{00}(E) = 276[1 - 0.005E_{Lab}/A_p]MeV fm^3 \tag{6}$$

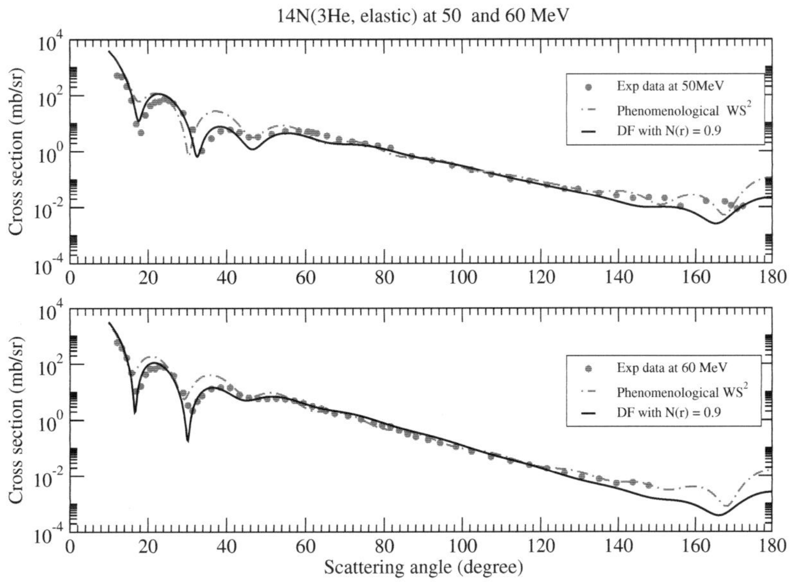

Figure 1. The results of DF and phenomenological WS^2 potentials in comparison with experimental data for $^3He+^{14}N$ system at 50 and 60 MeV.

3. The Results and Conclusion

The results obtained by using the microscopic double-folding and phenomenological nuclear potentials with the above-described models are shown in Fig.1 in comparison with the experimental data. As it can be seen from this figure that the real part of the double folding potential requires a normalization in order to obtain a reasonable result. Without this normalization, $N_r = 0.9$, we could not get an agreement with the experimental data. The phenomenological Woods-Saxon typed potential has provided a better agreement with the experimental data as seen in the same figure with dashed line.

Double-Folding analysis could effectively fit the experimental data at forward (angles lower than 70°), while phenomenological analysis could fit the experimental data at backward angles. The resulting potentials can be used for model calculations of yields of nuclear reactions necessary for astrophysical applications.

[1] Burtebayev N, Hamada Sh, Kerimkulov Zh, Alimov D.K and Yushkov A.V, 2014 *WASET*, **86** 9997511.
[2] MAESTRO-32 MCA Emulator for Microsoft Windows 98, 2000, NT, and XP A65-B32 Software.
[3] Burtebayev N, Alimov D, Boztosun I et.al 2015 *J. Phys.: Conf. Series* **590** 012055.
[4] Kucuk Y and Boztosun I, *Nucl. Phys.* A 764 (2006) 160.
[5] Freer M et al. 2007 *Phys. Rev.* C **76** 034320.
[6] Boztosun I and Rae W.D.M. 2001 *Phys. Rev.* C **63** 054607.
[7] Boztosun I and Rae W.D.M. 2001 *Phys. Rev.* C **64** 054607.
[8] Boztosun I and Rae W.D.M. 2002 *Phys. Rev.* C **65** 024603.
[9] Kucuk Y, Boztosun I and Topel T 2009 *Phys. Rev.* C **80** 054602.
[10] Boztosun I, 2002 *Phys. Rev.* C **66** 024610.
[11] Boztosun I and Rae W.D.M, 2001 *Phys. Lett.* B **518** 229.
[12] Karakoc M and Boztosun I, 2006 *Phys. Rev.* C **73** 047601.
[13] Karakoc M and Boztosun I, 2006 *J. of Mod. Phys. E-Nucl. Phys.* **15** 1317.
[14] Vries H.De, Jager C.W.De, and Vries C.De, 1987 *At. Data and Nucl. Data Tables* **36** 495.

IOP Conf. Series: Journal of Physics: Conf. Series **940** (2017) 012030 doi:10.1088/1742-6596/940/1/012030

Nucleosynthesis of Mo and Ru isotopes in neutrino-driven winds

Julia Bliss[1,*], **Almudena Arcones**[1,2]
[1]Institut für Kernphysik, Technische Universität Darmstadt
[2]GSI Helmholtzzentrum für Schwerionenforschung GmbH
E-mail: jbliss@theorie.ikp.physik.tu-darmstadt.de

Abstract. The solar system origin of the p-isotopes ^{92,94}Mo and ^{96,98}Ru is a long-lasting mystery. Several astrophysical scenarios failed to explain their formation. Moreover, SiC X grains show a different abundance ratio of ^{95,97}Mo than in the solar system. We have investigated if neutrino-driven winds can offer a solution to those problems.

1. Introduction

Neutrino-driven winds from nascent neutron stars are an exciting astrophysical site for the synthesis of elements between Sr-Ag. Observations of UMP stars (see e.g., [1]) have shown that those elements cannot be only produced by the r-process but rather at least one additional contribution is required [2, 3]. This astrophysical site may be the neutrino-driven wind where elements from Sr up to Ag are formed (see e.g., [4]) either by the weak r- [5] or νp-process (see e.g., [6]). It is not clear if the conditions in the wind are neutron- or proton-rich. Therefore, we investigate the nucleosynthesis for both conditions and systematically explore the impact of the wind parameters (electron fraction Y_e, entropy, expansion time scale) on the abundances (see [7]). Molybdenum and ruthenium have raised interest since several astrophysical scenarios failed to reproduce the solar system (SoS) abundance ratio of ^{92,94}Mo and ^{96,98}Ru. Besides, available data of silicon carbide grains of type X (SiC X) exhibit different isotopic ratios of ^{95,97}Mo than in the solar system. We have investigated whether the neutrino-driven wind can explain the SoS $Y(^{92}\text{Mo})/Y(^{94}\text{Mo})$ and the origin of $Y(^{95}\text{Mo})/Y(^{97}\text{Mo})$ found in SiC X.

2. Solar system abundances of ^{92,94}Mo and ^{96,98}Ru

We have performed a systematic nucleosynthesis study to identify the necessary conditions for the formation of the different Mo and Ru isotopes in neutrino-driven winds. Fig. 1 (left panel) shows the abundances of ^{92}Mo for different Y_e – entropy combinations. The color contours correspond the abundances in log scale. ^{92}Mo can be produced in neutron- ($Y_e < 0.5$) and proton-rich ($Y_e > 0.5$) conditions for a significant range of different entropies. The abundance pattern of ^{94}Mo is similar but the abundances are silghtly smaller for the same conditions. In neutron-rich winds, larger abundances of ^{92,94}Mo are obtained for smaller entropies, i.e., lower neutron-to-seed ratio. To reach ^{92,94}Mo in neutron-rich conditions the nucleosynthesis path has to move along the valley of stability. If the neutron abundance is too large the path moves

*Supported by the WE-Heraeus Foundation and the HGS-HIRe Graduate School.

farther away from stability and then ^{92,94}Mo are blocked due to stable ^{92,94}Zr, respectively. Therefore, it is a charged-particle process instead of a neutron capture process that synthesizes the Mo p-isotopes. In proton-rich winds, the abundances grow with increasing entropy and Y_e [8]. The νp-process forms elements via (n,p) and (p,γ) reactions which both depend on the proton abundance. Thus, with increasing proton-to-seed ratio, i.e., larger Y_e and entropy, the νp-process yields heavier elements.

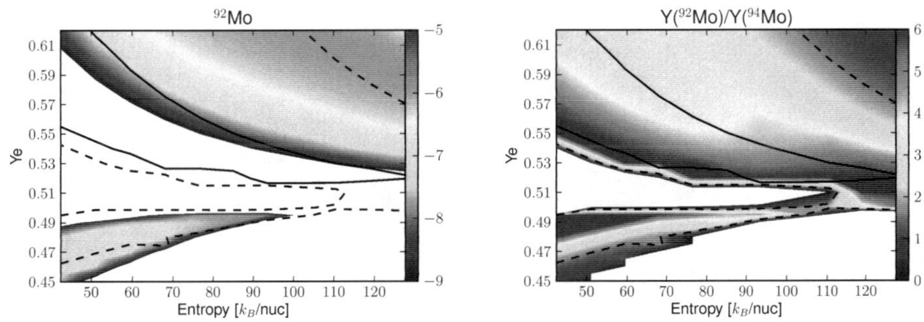

Figure 1. Left panel: Color contours show the abundances of ^{92}Mo in log scale for different Y_e and entropy. Right panel: Abundance ratio of ^{92,94}Mo for the same Y_e-entropy space. In both panels the regions where the abundance ratio of the Mo and Ru p-isotopes in the wind reaches the SoS values are indicated by dashed and solid lines, respectively.

Unlike the Mo p-isotopes, ^{96,98}Ru can only be formed in proton-rich conditions. The reason for this is that the nucleosynthesis path has to pass ^{92,94}Mo to reach the Ru p-isotopes in neutron-rich conditions, but $Y(^{92,94}\mathrm{Mo})$ are not large enough. In proton-rich conditions, the synthesis is similar to the one of ^{92,94}Mo.

To investigate if neutrino-driven winds can contribute to the SoS abundances of the Mo and Ru p-isotopes we use their SoS abundance ratios. Fig. 1 (right panel) shows the abundance ratio of ^{92,94}Mo. The dashed lines illustrate the region where the Mo p-isotope ratio in the wind is the same as in the SoS ($Y_\odot(^{92}\mathrm{Mo})/Y_\odot(^{94}\mathrm{Mo}) = 1.60$ [9]) and the solid lines mark the regions where the Ru p-isotope ratio in the wind equals the one in the SoS ($Y_\odot(^{96}\mathrm{Ru})/Y_\odot(^{98}\mathrm{Ru}) = 2.97$ [9]). The fact that solid and dashed lines do not cross implies that there is no set of wind parameters that simultaneously reproduce the solar Mo and Ru p-isotope ratios. This means that other astrophysical sites, e.g., type Ia supernovae [10], could contribute. Only with a combination of wind parameters it is possible to explain the ratios only due to the wind. A possiblity to investigate this is the production factor (see e.g., [8]). In order not to overproduce elements associated with core-collapse supernovae the production factor should be ~ 10 (see e.g., [11]). The wind fulfills the Mo p-isotope ratio in slightly neutron-rich conditions [12, 13] and $Y(^{92,94}\mathrm{Mo})$ within these conditions (Fig. 1 left panel) are sufficient. Moreover, there is an overproduction of Sr, Y, Zr due to the shell closer at $N = 50$ [14, 13]. In proton-rich winds there are two possible sites to obtain the SoS Mo p-isotope ratio. For slightly proton-rich winds [15] $Y(^{92,94}\mathrm{Mo})$ are too low. ^{92,94}Mo are produced in enough amounts for high Y_e. Here the problem is due to the Ru p-isotopes since their production factors are larger and their abundance ratio is not the solar one. Therefore, even with a combination of wind trajectories it is very challenging to explain the Mo and Ru p-isotopes in the SoS based only on neutrino-driven winds.

3. Solar system and SiC X abundances of ^{95,97}Mo

The s-, r-nuclides ^{95,97}Mo can be formed within slightly neutron-rich conditions and their abundances grow with increasing neutron-to-seed ratio, i.e., lower Y_e and higher entropy (see left panel in Fig. 2). ^{95,97}Mo are overabundant in SiC X grains in comparison to the SoS [16]. The grains indicate a mixture of SoS ^{95,97}Mo and an unknown component. Besides, SiC X grains do not exhibit an excess of ^{96,98,100}Mo that differs from a pure s- or r-process enrichment. We have investigated if neutrino-driven winds can explain the SoS abundance ratio without s-process contribution ($Y_{\odot-s}(^{95}\mathrm{Mo})/Y_{\odot-s}(^{97}\mathrm{Mo}) = 1.88$ [9]) that is similar to the abundance ratio found in SiC X grains ($Y_{\mathrm{SiC\ X}}(^{95}\mathrm{Mo})/Y_{\mathrm{SiC\ X}}(^{97}\mathrm{Mo}) = 1.83$ [16]). Fig. 2 (right panel) shows the abundance ratio of ^{95,97}Mo. The solid lines mark the regions for which $Y(^{95}\mathrm{Mo})/Y(^{97}\mathrm{Mo})$ in the wind is the same like in SiC X. The SoS and SiC X ratio can be obtained in the wind for a wide range of entropies and Y_e. The solid lines in Fig. 2 (left panel) show that the abundances of ^{95,97}Mo within these conditions are sufficient. Since the production factor becomes larger than 10 for those wind parameters, not all winds can have such conditions, otherwise there would be overproduction of ^{96}Zr in the SoS.

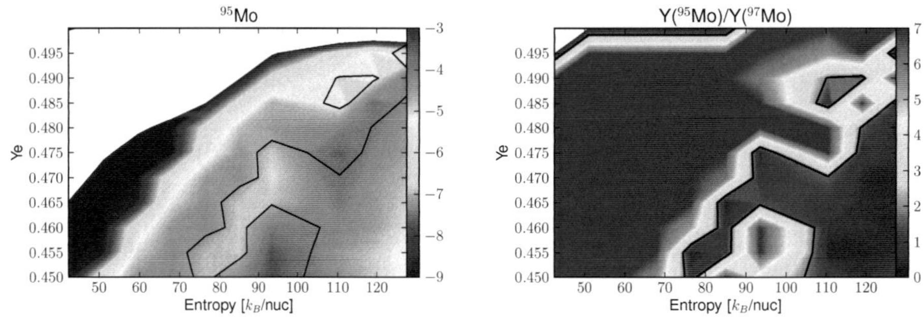

Figure 2. Left panel: Abundances of ^{95}Mo in log scale. Right panel: Abundance ratio of ^{95,97}Mo. In both panels the solid lines show the regions where $Y(^{95}\mathrm{Mo})/Y(^{97}\mathrm{Mo})$ in the wind is the same as in SiC X grains.

4. Summary

Nucleosynthesis studies like the one here combined with observations of UMP stars can help us to understand better the synthesizes of elements between Sr and Ag, and to put constraints on the wind parameters. We have shown that neutrino-driven winds can produce the SoS $Y(^{92}\mathrm{Mo})/Y(^{94}\mathrm{Mo})$ in neutron- and proton rich conditions. The SoS $Y(^{96}\mathrm{Ru})/Y(^{98}\mathrm{Ru})$ can be formed in proton-rich winds. Therefore, neutrino-driven winds may be important for the origin of SoS ^{92,94}Mo and ^{96,98}Ru but other sites e.g., type Ia supernovae [10] cannot be neglected. Besides, the wind can explain the SoS and SiC X abundances of ^{95,97}Mo.

References

[1] Cowan J J and Sneden C 2006 *Nature* **440** 1151–1156
[2] Qian Y Z and Wasserburg G J 2007 *Physics Reports* **442** 237–268
[3] Travaglio C, Gallino R, Arnone E, Cowan J, Jordan F and Sneden C 2004 *ApJ* **601** 864–884
[4] Arcones A and Montes F 2011 *ApJ* **731** 5
[5] Truran J W and Cowan J J 2000 *Nuclear Astrophysics, 2000* ed Hillebrandt W and Müller E p 64
[6] Fröhlich C, Martínez-Pinedo G, Liebendörfer M, Thielemann F K, Bravo E, Hix W R, Langanke K and Zinner N T 2006 *Physical Review Letters* **96**
[7] Arcones A and Bliss J 2014 *J. Phys. G: Nucl. Phys.* **41** 044005
[8] Pruet J, Hoffman R D, Woosley S E, Janka H T and Buras R 2006 *ApJ* **644** 1028–1039
[9] Lodders K 2003 *ApJ* **591** 1220–1247
[10] Travaglio C, Gallino R, Rauscher T, Dauphas N, Röpke F K and Hillebrandt W 2014 *ApJ* **795** 141
[11] Timmes F X, Woosley S E and Weaver T A 1995 *ApJS* **98** 617–658
[12] Farouqi K, Hallmann O, Kratz K L and Ott U 2010 *Nuclei in the Cosmos* p 81
[13] Hoffman R D, Woosley S E, Fuller G M and Meyer B S 1996 *ApJ* **460** 478–488
[14] Hoffman R D, Woosley S E and Qian Y Z 1997 *ApJ* **482** 951–962
[15] Fisker J L, Hoffman R D and Pruet J 2009 *ApJL* **690** L135–L139
[16] Pellin M J, Davis A M, Lewis R S, Amari S and Clayton R N 1999 *Lunar and Planetary Inst. Technical Report* vol 30 p 1969

IOP Conf. Series: Journal of Physics: Conf. Series **940** (2017) 012031 doi:10.1088/1742-6596/940/1/012031

Backgrounds in a BGO detector underground

A Boeltzig[1], A Best[2], A Di Leva[3,4], G Imbriani[3,4] and M Junker[1,2] for the LUNA collaboration

[1] Gran Sasso Science Institute, L'Aquila, Italy
[2] Laboratori Nazionali del Gran Sasso, Assergi, Italy
[3] Università degli Studi di Napoli Federico II, Napoli, Italy
[4] INFN Sezione di Napoli, Napoli, Italy

E-mail: axel.boeltzig@gssi.infn.it

Abstract. For the measurement of reactions with small cross sections as aimed for in the Laboratory for Underground Nuclear Astrophysics (LUNA) at the Gran Sasso National Laboratory (LNGS), low detector background levels are compulsory. Background measurements with a BGO detector used at LUNA have been evaluated to develop a model of the contributions in the region of interest for ongoing and planned measurements. Conclusions from this model and implications for further background reduction for this detector setup are discussed. A customized lead shielding against environmental backgrounds is described.

1. Challenges in direct measurements of small cross sections

The relevant energy region for nuclear reactions in stellar scenarios is given by the Gamow window. Direct measurements of nuclear reactions in or close to this energy region are often very challenging as the cross sections for these processes fall rapidly with decreasing energy. The Laboratory for Underground Nuclear Astrophysics (LUNA), located at the Gran Sasso National Laboratory (LNGS) is dedicated to measurements of such reactions. Two factors are crucial to achieve this: a large detection efficiency for the signal of the reaction and a small background in the signal region of interest.

2. The Bismuth Germanium Oxide (BGO) detector at LUNA

An important tool for the detection of (p, γ) reactions at LUNA is a BGO detector [1]. BGO efficiently absorbs gamma rays thanks to its density and the large atomic number of bismuth. Its scintillation properties allow the subsequent detection of the absorbed radiation. A disadvantage to other gamma ray detection techniques is the moderate energy resolution. The detector at LUNA has six crystal segments that are aligned axially around the borehole for the beamline. The setup allows to place the target in the center of the detector and achieve a close to 4π solid angle coverage. The crystals are optically isolated and coupled to PMTs that are read out individually to obtain six single crystal spectra. When coincident events in the individual crystals are added, a total detector spectrum can be obtained. This spectrum imitates one large crystal and maximizes the efficiency for detection of the complete gamma ray energies. Hence, the region of interest in the total detector spectrum for a (p, γ) reaction is usually given by the Q value of the reaction under study. The single crystal spectra can yield additional information on the individual gamma rays emitted in the reaction.

IOP Conf. Series: Journal of Physics: Conf. Series **940** (2017) 012031 doi:10.1088/1742-6596/940/1/012031

3. Background contributions

Without dedicated efforts, the background signal in the detector can easily be the limiting factor for the sensitivity of a measurement. An understanding of the different background contributions is necessary to choose an appropriate background reduction strategy. Two primary sources contribute to the background in a radiation detector: radioactive decays and cosmic rays.

The radioactive component stems primarily from natural radioactivity in the form of primordial decay chains and ^{40}K. It contributes to the detected environmental gamma ray background, but also to the ambient neutron flux via (α, n) reactions. Particle (α and β) radiation will typically not be detected from sources outside the detector, due to its limited range. It can be relevant, however, for intrinsic radioactivity inside or on the scintillator.

Cosmic rays, in particular penetrating muons, contribute by direct detection or via secondary processes that produce gamma rays or neutrons. At LNGS, the cosmic component is dominated by muons, whose flux is about a factor 10^6 lower than on surface. This reduction is also directly reflected in the muon-induced backgrounds, e. g. for high-energy gamma rays (see [2] for a comparison with this detector on surface and underground).

Thermal neutrons constitute a large fraction of the environmental neutron flux at LNGS (cf. e. g. [3]). For the studied detector these neutrons can contribute significantly to the background via (n, γ) reactions, typically with large Q values, in or close to the active detector volume.

4. Modeling of a background spectrum acquired underground

Background spectra were acquired during multiple runs in the course of the ^{25}Mg(p, γ) campaign at LUNA (cf. [4]) with an accumulated real time of approximately one month. These data, acquired without additional shielding around the detector, were used as the basis for the development of the background model. Focus of the model was the energy region above 6 MeV given by the Q values of the reactions to be studied.

Simulations of the detector setup are required to determine the energy deposition in the detector caused by different background sources. They have been performed with a simulation based on Geant4 [5]. The detector geometry had been implemented and tested previously (cf. e. g. [4]). The radioactive decay process provided by Geant4 was used to generate the primary particles from the decays of the different radioactive nuclides. A custom primary particle generator was used for the radiative capture of thermal neutrons, creating the deexcitation gamma cascades according to the level schemes of the daughter nuclei as far as the branchings were known. Contributions from the radioactive nuclides were simulated homogeneously distributed throughout the surrounding rock, the steel parts of the detector and the BGO crystals. The radiative neutron captures were simulated homogeneously throughout the volumes, neglecting the self-shielding of the detector. For each simulated contribution, the energy deposition in the BGO crystals was saved, separately for electrons and for α particles.

The detector response to the energy deposition has to be modeled to derive a simulated signal that can be compared to the measured spectrum. The quenching effect, leading to a lower scintillation light yield of α particles compared to electrons of the same energy, was modeled using Birks' law as described in [6]. To account for the detector resolution, the visible energy deposition after quenching was smeared with a Gaussian response with an energy-dependent resolution σ. A quadratic relation between the visible energy deposition and the detector signal (ADC channel) was assumed.

Several assumptions had to be made to limit the number of free parameters when combining the large number of individual contributions to obtain the total spectrum. The decay chains were assumed to be in equilibrium. Intrinsic radioactivity was treated only as a function of material, possible differences in radiopurity between separate parts have been neglected. Pile-up, i. e. the random summing of energy depositions from independent decays or reactions, has been considered for each combination of two or three physical events occurring within a certain time window for a single detected event, based on the rates of the contributions.

As a result, neutrons from (n, γ) reactions in BGO and the steel parts of the setup have been identified as the main contribution to the backgrounds for energies of 6-12 MeV. In the range of circa 4-6 MeV, pile-up events were the largest contribution.

5. Customized lead shielding

A lead shielding around the detector can suppress the detection of environmental gamma rays. Even though many of the reactions studied with the BGO detector at LUNA have Q values above the direct contribution of environmental gamma rays, the region of interest for these reactions can be affected by pile-up from random coincidences of environmental background events.

The use of a stationary shielding using lead bricks is not practical for this setup, as the detector has to be moved away regularly to exchange the irradiated target. A customized shielding has been built, consisting of few massive pieces that house the detector and are mounted on rails, such that the whole detector setup can be moved along the beamline to access the target. The setup provides 10 cm of lead around the detector (except for openings for the beamline and for the cables). It is also manufactured to house a HPGe detector (in two different positions) with 15 cm of lead around.

First measurements with the shielding in place showed a reduction of approximately 1 order of magnitude in the count rate around 1 MeV. With the shielding, in combination with a new DAQ that allowed shorter coincidence intervals for the total detector spectrum, the pile-up contribution could be reduced significantly, and the region where neutron-capture reactions are predominant extended to lower energies. A reduction in the number of detected neutron-induced events has been observed as well, presumably mostly caused by the absorption of thermal neutrons on ^{206}Pb in the shielding.

6. Conclusions and outlook

A model for the background contributions was developed for the LUNA BGO setup, focusing on the high-energy part of the spectrum. A reduction of pile-up events and neutron induced events has been achieved using a customized lead shielding. The identification of radiative neutron capture as the main background contribution at large energies motivates the consideration of an additional neutron shielding.

Updates of the model are ongoing to improve the description over a larger energy range and incorporate the information from shielded background measurements, aiming at a better distinction between external and intrinsic backgrounds. Ultimately the final model will be used to assess the feasibility of a further background reduction to enhance the sensitivity of the presented setup, and therefore allow the measurement of reactions at lower beam energies or reactions with lower Q values.

Acknowledgments

The customized lead shielding was designed and installed by L. Roscilli (INFN Section of Naples) and the technical division of the University of Naples, with support from the mechanical workshop at LNGS.

References

[1] Casella C *et al.* 2002 *Nucl. Instr. Meth. Phys. Res.* A **489** 160–169
[2] Bemmerer D *et al.* 2005 *Eur. Phys. J.* A **24** 313–319
[3] Wulandari H, Jochum J, Rau W and von Feilitzsch F 2004 *Astropart. Phys.* **22** 313–322
[4] Strieder F *et al.* 2012 *Phys. Lett.* B **707** 60–65
[5] Agostinelli S *et al.* 2003 *Nucl. Instr. Meth. Phys. Res.* A **506** 250–303
[6] Glasow R, Kampert K H, Löhner H, Beckmann P and Santo R 1985 *Nucl. Instr. Meth. Phys. Res.* A **228** 354–358

IOP Conf. Series: Journal of Physics: Conf. Series **940** (2017) 012032 doi:10.1088/1742-6596/940/1/012032

Anharmonic oscillator potentials in the prolate γ-rigid regime of the collective geometrical model

R Budaca

Horia Hulubei National Institute of Physics and Nuclear Engineering, Bucharest-Măgurele RO-077125, Romania

E-mail: rbudaca@theory.nipne.ro

Abstract. Based on higher order JWKB approximation, an analytical formula is derived for the energy spectrum of the prolate γ-rigid Bohr-Mottelson Hamiltonian with an oscillator potential and a quartic or sextic anharmonicity in β shape variable. The model is found to provide good description for near vibrational nuclei situated in vicinity of a closed neutron shell.

1. Introduction

An analytical expression for the energy spectrum of the ground and β bands was obtained in the axially symmetric γ-rigid regime of the Bohr Hamiltonian with a harmonic oscillator potential in the β shape variable amended with a higher order anharmonicity term. The anharmonic terms are considered of quartic [1] and sextic [2] types. The particular structure of the model space facilitates the exact separation of angular variables from the β variable, leading to a differential Schrödinger equation with an anharmonic potential and a centrifugal-like barrier. The Schrödinger equation for such potentials is not exactly solvable. Thus, the corresponding eigenvalues are approximated by an analytical formula derived on the basis of the JWKB approximation [3, 4], which depends on a single parameter except the scale. The experimental realization of the model is found in a variety of vibrational-like nuclei exhibiting some regularity regarding the order of the considered anharmonicity.

2. Prolate γ-rigid collective Hamiltonian

The Hamiltonian associated to a prolate γ-rigid nucleus is [1, 2, 5, 6, 7]:

$$H = -\frac{\hbar^2}{2B}\left[\frac{1}{\beta^2}\frac{\partial}{\partial\beta}\beta^2\frac{\partial}{\partial\beta} - \frac{\hat{\mathbf{I}}^2}{3\hbar^2\beta^2}\right] + U(\beta), \tag{1}$$

where $\hat{\mathbf{I}}$ is the angular momentum operator from the intrinsic frame of reference, while B is the mass parameter. The Schrödinger equation associated to such a Hamiltonian is solved by separating the β variable from the angular ones which is achieved through the factorization $\Psi(\beta,\theta_1,\theta_2) = F(\beta)Y_{IM}(\theta_1,\theta_2)$, where the angular factor state is a spherical harmonic function. With this, the Schrödinger equation is reduced to a second order differential equation in variable

β, which is brought to a canonical form through the change of function $F(\beta) = \frac{f(\beta)}{\beta}$:

$$\left[\frac{d^2}{d\beta^2} - \frac{I(I+1)}{3\beta^2} + 2(\varepsilon - u(\beta))\right] f(\beta) = 0, \tag{2}$$

where $\varepsilon = BE/\hbar^2$, $u(\beta) = BU(\beta)/\hbar^2$ denote the reduced energy and potential.

3. Quartic and sextic anharmonicities

The eigenvalue problem for potentials:

$$u(\beta) = \frac{1}{2}\alpha_1\beta^2 + \alpha_2\beta^{2m}, \quad \alpha_1 \geqslant 0, \alpha_2 > 0 \text{ and } m = 2, 3, \tag{3}$$

has the following scaling property:

$$\varepsilon(\alpha_1, \alpha_2) = \alpha_2^{\frac{1}{m+1}}\varepsilon(\lambda_m, 1) = \alpha_2^{\frac{1}{m+1}} W^m(\lambda_m), \tag{4}$$

where $\lambda_m = \alpha_1\alpha_2^{\frac{m}{6}-1} \geqslant 0$. This follows from the change of variable $\beta' = \alpha_2^{\frac{1}{2(m+1)}}\beta$ in the differential equation (2). Similar scaling was used in Refs.[7, 8] in the case of a quasi exactly solvable sextic potential. The problem is then reduced to a radial Schrödinger equation with a modified centrifugal term for the scaled potential $\tilde{u}(\beta') = \frac{1}{2}\lambda_n\beta'^2 + \beta'^{2m}$. Adapting the procedure of Refs.[3, 4] one can write the eigenvalue W^m as function of λ_m, n - the β vibration quantum number and the intrinsic angular momentum I, as follows:

$$W^2_{nI}(\lambda_2) = (C_2N_{nI})^{\frac{4}{3}}\sum_{k=0}^{8} G^2_k(\lambda_2, I)\,(C_2N_{nI})^{-\frac{2k}{3}}, \tag{5}$$

$$W^3_{nI}(\lambda_3) = (C_3N_{nI})^{\frac{3}{2}}\sum_{k=0}^{5} G^3_k(\lambda_3, I)\,(C_3N_{nI})^{-k}, \tag{6}$$

where $N_{nI} = 2n+1+(\sqrt{1+4I(I+1)/3})/2$. The functions $G^m_k(\lambda_m, I)$ represent polynomials in λ_m of order k with angular momentum dependent coefficients and are given explicitly in [1, 2] together with the constants C_m.

The energy spectrum of a nucleus described by the Hamiltonian (1) with an oscillator potential with quartic or sextic anharmonicities is then given by:

$$E^m_{nI} = \frac{\hbar^2}{B}\alpha_2^{\frac{1}{m+1}}\left[W^m_{nI}(\lambda_m) - W^m_{00}(\lambda_m)\right]. \tag{7}$$

4. Numerical results

The energy formula depends up to an overall factor on a single free parameter which is bounded by the convergence radius λ^c_m of the considered approximation. The acting space of the model is then restricted to the interval $[0, \lambda^c_m]$ which put in terms of the ratios $R_{4/2}$ and $R_{0/2}$ lies between $[2.29, 2.00]$ and $[2.37, 1.81]$ for quartic and respectively $[2.34, 2.15]$ and $[2.56, 2.11]$ for sextic anharmonicities (See Figure 1). The upper limit corresponds to $\lambda_m = 0$ which is associated to the free parameter models $X(3)$-β^{2m}, while the lower limit is close to the $X(3)$-β^2 predictions [1].

As can be seen from Figure 2, the quartic model is experimentally realized for ^{100}Pd, ^{148}Sm, and ^{222}Th nuclei, while the sextic one describes their next heavier neighbours. The three isotopic pairs also have the same number of neutrons above the corresponding shell closure at magic numbers $N = 50, 82$ and 126.

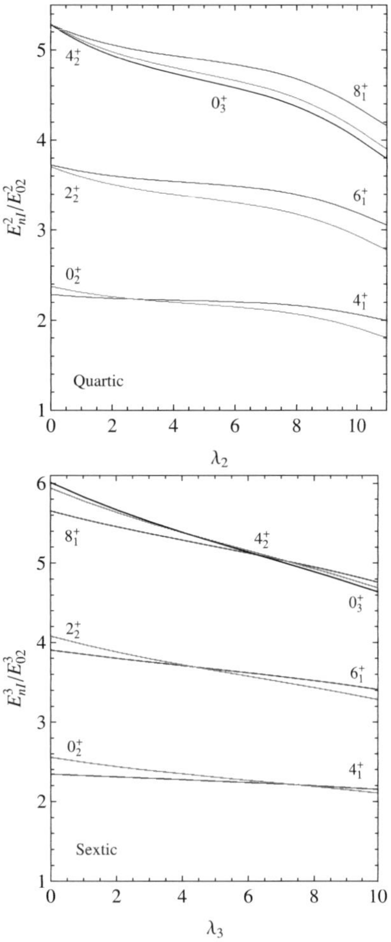

Figure 1. Theoretical low lying energy spectrum for ground and two β excited bands.

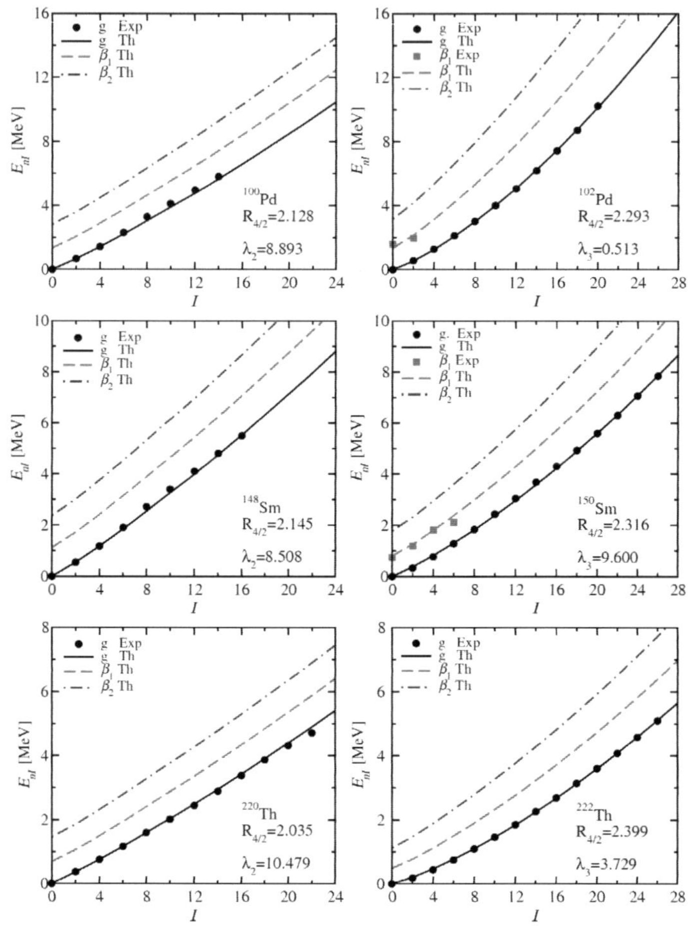

Figure 2. Quartic model description of the lighter isotopes of Pd, Gd and Th [9] in comparison to the heavier ones treated by means of a sextic term.

5. Outlook

The present approximate formulas can be easily translated to the case of four and five dimensional Bohr-Mottelson Hamiltonians with a similar separated potential for the β shape variable.

References

[1] Budaca R 2014 *Eur. Phys. J.* A **50** 87
[2] Budaca R 2014 *Phys. Lett.* B **739** 56
[3] Seetharaman M and Vasan S S 1986 *J. Mat. Phys.* **27** 1031
[4] Vasan S S, Seetharaman M and Sushama L 1993 *Pramana J. Phys.* **40** 177
[5] Bonatsos D, Lenis D, Petrellis D, Terziev P A and Yigitoglu I 2006 *Phys. Lett.* B **632** 238
[6] Budaca R and Budaca A I 2015 *J. Phys. G: Nucl. Part. Phys.* **42** 085103
[7] Buganu P and Budaca R 2015 *J. Phys. G: Nucl. Part. Phys.* **42** 105106
[8] Buganu P and Budaca R 2015 *Phys. Rev.* C **91** 014306
[9] http://www.nndc.bnl.gov/ensdf/

IOP Conf. Series: Journal of Physics: Conf. Series **940** (2017) 012033 doi:10.1088/1742-6596/940/1/012033

Investigation of α-particle scattering from ^{13}C at 29 MeV

Nassrulla Burtebayev[1], Daniyar Janseitov[1,3], Zhambul Kerimkulov[1], Alla Demyanova[2], Andrei Danilov[2], Dilshod Alimov[4] and Yerzhan Mukhamejanov[4]

[1]Institute of Nuclear Physics, 050032 Almaty, Kazakhstan
[2]Kurchatov Institute, 123182 Moscow, Russia
[3]Joint Institute for Nuclear Research, 141980 Dubna, Russia
[4]Kazakh National University, 050040 Almaty, Kazakhstan
E-mail: janseit.daniar@gmail.com

Abstract: We have measured the angular distributions for the elastic and inelastic scattering of α-particles from ^{13}C in the isochronous cyclotron U-150M INP (Almaty, Kazakhstan). The experimental results were analyzed within the framework of both the optical model using different complex potentials and the double folding potential obtained with different density-dependent nucleon-nucleon interactions.

1. Introduction

Neutron-halo nuclei discovered in 1985 by Tanihata et al. [1,2] are exotic systems consisting of a tightly bound core surrounded by a diffuse nuclear cloud. These systems were found in the ground states (g.s.) of some light, neutron-rich radioactive nuclei located close to the neutron dripline. Until now neutron halo was observed almost exclusively in the ground states of some radioactive nuclei. In the framework of the hypothesis of the α-partial condensation, some excited states of ^{13}C may have larger values of the radius.

^{13}C is a good example of a "normal" nucleus well described by the shell model. Its level scheme is reliably determined up to the excitation energies ~ 10 MeV [3]. Analysis [4] of the ^{13}C + α scattering data measured at E (α) = 388 MeV [5] really demonstrated a considerable enhancement of the radius at this particular energy. Consequently, new measurements especially at lower energies are highly desirable.

In the current work we continue the investigation of the nature of ^{13}C excited states at low energies.

2. Experimental Details

The experiment was carried out at the isochronous cyclotron U-150M in Institute of Nuclear Physics (Almaty, Kazakhstan) at energy E(α)=29 MeV. Sets of E-ΔE telescopes for detection of the alpha-particles were used in the experiments. A self-supporting ^{13}C target (0.4 mg/cm^2) with 90% enrichment was used. It contained some impurities of ^{12}C and ^{16}O.

The (E-ΔE) telescopes of silicon semiconductor detectors, which cover the scattering angles of 10–80°, were used in data acquisition. For optimal focusing of the accelerated ions beam on the target, two collimators of diameter 2 and 3 mm were used. The (E-ΔE) method was used in the recording and identification of the reaction products.

3. Results and discussion:

The theoretical calculations for the elastic scattering of α-particles from ^{13}C target at energies 29, 48.7, 54.1 and 65 MeV were performed within the framework of the code Fresco [6] using two different potentials: phenomenological optical potential and semi-microscopic double folding potential. The comparison between the experimental data and the theoretical predictions for $^{13}C(\alpha,\alpha)^{13}C$ at the aforementioned energies is shown in figure 1 using the potential parameters listed in table 1. The table 1 shows: depth (V_0, W), radii (r_v, r_w) and diffuseness (a_v, a_w) of the real and imaginary part of the potential, respectively and the normalization coefficient (N_r), that was used for the calculations in the folding model. Coulomb radius is r_c=1.28 fm.

In the Optical Model (OM) calculations at energies 48.7, 54.1 [7] and 65 MeV [8] we used two different potential sets (A and B), which are a combination of real volume part plus imaginary volume part. The potential set A was used to reproduce the experimental data in the forward hemisphere with the fixed values of radii r_V=1.112 fm and r_W=1.6 fm of real and imaginary part, respectively; the potential set B was used to reproduce the experimental data in the full angular range using the same fixed values for the parameters r_V and r_W as in set A.

Table 1: The optimal potential parameters obtained for elastic scattering of α-particles from ^{13}C at different energies.

E MeV	Potential set	V_0 MeV	r_v fm	a_v, fm	N_r	W MeV	r_w fm	a_w fm	r_c fm
	OM A	147.22	1.112	0.736		12.844	1.6	0.267	1.28
29	OM B	142.23	1.245	0.762		16.75	1.6	0.731	1.28
	DF A				0.97	12.844	1.6	0.267	1.28
	OM A	134.49	1.112	0.79		14.79	1.6	0.639	1.28
48.7	OM B	120.38	1.245	0.73		15.81	1.6	0.639	1.28
	DF A				0.97	14.79	1.6	0.639	1.28
	OM A	129.48	1,112	0.795		14.21	1.6	0.8	1.28
54.1	OM B	117.57	1.245	0.753		15.91	1.6	0.76	1.28
	DF A				0.96	14.21	1.6	0.8	1.28
	OM A	123.07	1.112	0.8		14.97	1.6	0.76	1.28
65	OM B	119.49	1.245	0.784		20.02	1.6	0.728	1.28
	DF A				0.98	14.97	1.6	0.76	1.28

In the double folding (DF) calculations, the real part of the potential was derived by folding the density distributions of the target nucleus and the projectile with the nucleon-nucleon interaction potential. In potential DF at energies 29 and 65 MeV, the normalization coefficient for the real part of the potential equals 0.97 and 0.98 respectively and the phenomenological imaginary volume parameters were fixed to the same values used in potential set A for OM calculations. At energies 48.7 and 54.1 MeV the normalization coefficient for the real part of the potential equals 0.97 and 0.96 respectively.

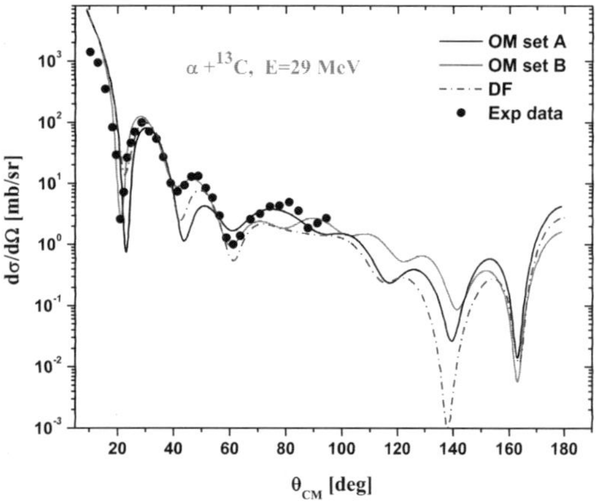

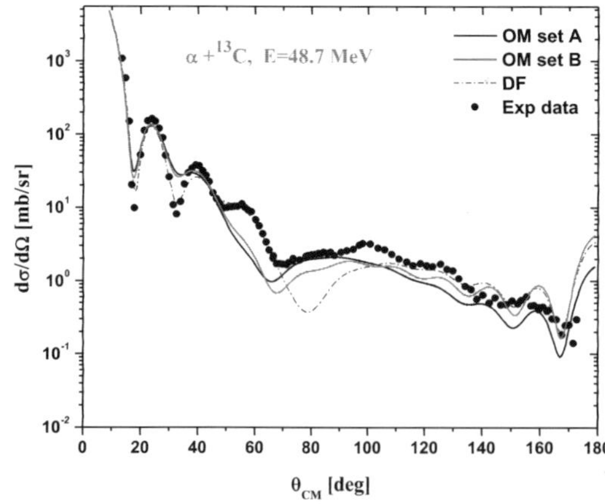

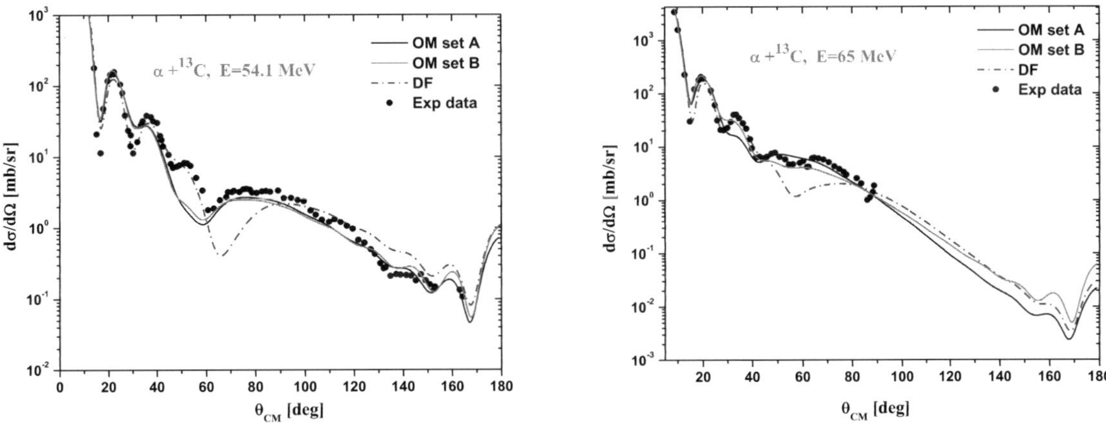

Figure 1. Comparison between the experimental data and the calculated differential cross section for elastic scattering of α-particles from ^{13}C at energies 29, 48.7, 54.1 and 65 MeV using both OM (set A and set B) and DF potentials.

4. Summary

Analysis of data on elastic scattering at energies 29 MeV, together with published data at energies of 65 MeV, 54.1 and 48.7 MeV was carried out within the framework of optical model with the phenomenological optical potentials parameters and semi-microscopic potentials calculated within the framework of double-folding model. Physically reasonable parameters of interaction potentials were obtained in fitting experimental data in a wide energy range of incident particles. The energy dependence of the depth of the real part of the potential was determined.

Analysis of data on inelastic scattering is planned to done with the use of the obtained parameters of optical potentials in the following papers.

References:

[1] I Tanihata *et al.*, *Phys. Lett. B* **160**, 380 (1985).
[2] I Tanihata *et al.*, *Phys. Rev. Lett.* **55**, 2676 (1985).
[3] W von Oertzen, M. Freer, Y. Kanada-En'yo, *Phys. Rep.* **432**, 43 (2006)
[4] A S Demyanova et al., Int. J. Modern. Phys. E **20**, No 4, 915 (2011)
[5] T Kawabata et al., *Journal of Physics: Conference Series,* **111**, 012013 (2008); *Journal of Modern Physics* E **17**, 2071 (2008)
[6] I J Thompson Coupled reaction channels calculations in nuclear physics Comput. *Phys. Rep.* **7** (1988)
[7] H Abele, H J Hauser, A Korber et al. *Zeitschrift fur Physik, Atomic nuclei,* **V.326,** p. 373-381, 1987
[8] A S Demyanova et.al., *EPJ Web of Conferences* **66** 02027 (2014) 201

IOP Conf. Series: Journal of Physics: Conf. Series **940** (2017) 012034 doi:10.1088/1742-6596/940/1/012034

Investigation of the elastic and inelastic scattering of ^{4}He from ^{11}B in the energy range 29-50.5 MeV

N.Burtebayev[1], Zh.Kerimkulov[1], M.Bakhtybayev[1], Sh.Hamada[2], Y.Mukhamejanov[3], M.Nassurlla[3], D.Alimov[3], A.Morzabayev[4], D.Janseitov[4], W.Trzaska[5]

[1] Institute of Nuclear Physics, Almaty, Kazakhstan
[2]Faculty of Science, Tanta University, Tanta, Egypt
[3]Kazakh National University, Almaty, Kazakhstan
[4]Eurasian National University, Astana, Kazakhstan
[5]University of Jyväskylä, Finland
E-mail: y.mukhamejanov@gmail.com

Abstract: We have measured the angular distributions for the elastic and inelastic scattering of ^{4}He from ^{11}B in the isochronous cyclotron U-150 M INP RK. The extracted α-particles beam has been accelerated to energy 29 MeV and then directed to ^{11}B target. The experimental results were analyzed within the framework of both the optical model using different complex potential and the double folding potential obtained with different density-dependent NN interactions which give the corresponding values of the nuclear incompressibility K in the Hartree-Fock calculation of nuclear matter.

1. Introduction

Study of nuclear reaction is of a special interest as it could provide us with useful information about the nuclear structure, potential parameters, deformation, and transition probabilities. The α-nucleus interaction is an essential tool for the understanding of nuclear structure and nuclear reactions. The concept of the α-particle mean field has been widely used to unify the bound and scattering α-particle states in a similar way to use of the nuclear mean field to calculate the properties of bound single particle states and also the scattering of unbound nucleons by nuclei.

2. Experimental Details

The experimental angular distribution measurements for ^{4}He elastically and inelastically scattered from ^{11}B at energy 29 MeV were performed in the isochronous cyclotron U-150M INP located in Almaty, Kazakhstan, which allows accelerating protons up to energy 30 MeV, deuterons up to energy 25 MeV, ^{3}He up to energy 60 MeV and α-particles up to energy 50 MeV. Three (E-ΔE) telescopes of silicon semiconductor detectors, which cover the scattering angles of 10–70° were used in data acquisition. For optimal focusing of the accelerated ions beam on the target, two collimators of diameter 2 mm were used. The (E-ΔE) method was used in the registration and identification of reaction products. In the telescope detectors "E-ΔE", ΔE - detector is a surface-barrier silicon detector-company ORTEC- thick active layer of 30 to 200 μm with thin inlet (~ 40μg/cm^2 Au) and outlet

(~40μg/cm^2 Al) windows. The complete absorption *E* detector is used as a stop detector company ORTEC high-purity silicon; thickness of 2 mm.

3. Results and discussion:

The theoretical calculations for the elastic scattering of α-particles from ^{11}B target at energies 29, 40 and 50.5 MeV were performed within the framework of code Fresco [1] using two approaches (a) phenomenological optical potential and (b) semi-microscopic double folding potential.

In the **Optical Model** calculations at energies 40 and 50.5 MeV [2] we used two different potential sets (A and B), which is a combination of real volume part plus imaginary volume part. The potential set A was used to reproduce the experimental data in the forward hemi sphere with fixed values for the parameters r_V and r_W, the potential set B was used to reproduce the experimental data in the full angular range also with the same fixed values for the parameters r_V and r_W as in set A. While, at energy 29 MeV a potential combination of real volume part plus imaginary surface part (set A) was found to be more effective to reproduce the experimental data in the full angular range at this energy.

Table 1: The optimal potential and deformation parameters obtained for elastic and inelastic 5/2$^-$ (4.45 MeV) scattering of ^{4}He from ^{11}B at energies 29, 40 and 50.5 MeV

E MeV	Potential set	V_o MeV	r_v fm	a_v fm	Nr	W_o MeV	r_w fm	a_w fm	W_d MeV	r_d fm	a_d fm	β 5/2$^-$
29	OM A full	79.55	1.2	0.743					23.65	1.76	0.438	0.46
	DF A full				1.234				37.13	1.8	0.365	0.46
	OM A short	77.85	1.245	0.856		19.01	1.57	0.7				0.42
40	OM B full	130.97	1.245	0.76		18.38	1.57	0.64				0.42
	DF A full				1.2	18.38	1.57	0.64				0.4
	DF B full				1.2	30.0	1.262	0.7				
	OM A short	77.22	1.245	0.856		16.85	1.57	0.833				0.35
50	OM B full	120.0	1.245	0.752		18.14	1.57	0.731				0.35
	DF A full				1.2	18.14	1.57	0.731				
	DF B full				1.2	30.0	1.33	0.746				0.35

In the **double folding** calculations, the real part of the potential was derived by folding the density distributions of the target nucleus and the projectile with the NN interaction potential. We used M3Y interactions which were designed to reproduce the *G*-matrix elements of the Reid [3] and Paris [4] free NN potentials in an oscillator basis. At energies 40 and 50 MeV two sets of potential (A and B) were used in the calculations. In potential set A, the normalization coefficient for the real part of the potential equals 1.2 and the phenomenological imaginary volume parameters were fixed to the same values used in potential set B for optical model calculations. While, at energy 29 MeV, the obtained normalization coefficient for the real part of the potential equals 1.234 and the phenomenological imaginary surface parameters were taken as free parameters.

The comparison between the experimental data and theoretical predictions for the state 5/2$^-$ at energies E^4_{He}= 29, 40 and 50.5 MeV are shown in figures 1. The theoretical calculations of the angular distributions for the different excited states were performed using the coupled channel method implemented in code Fresco using different potential sets.

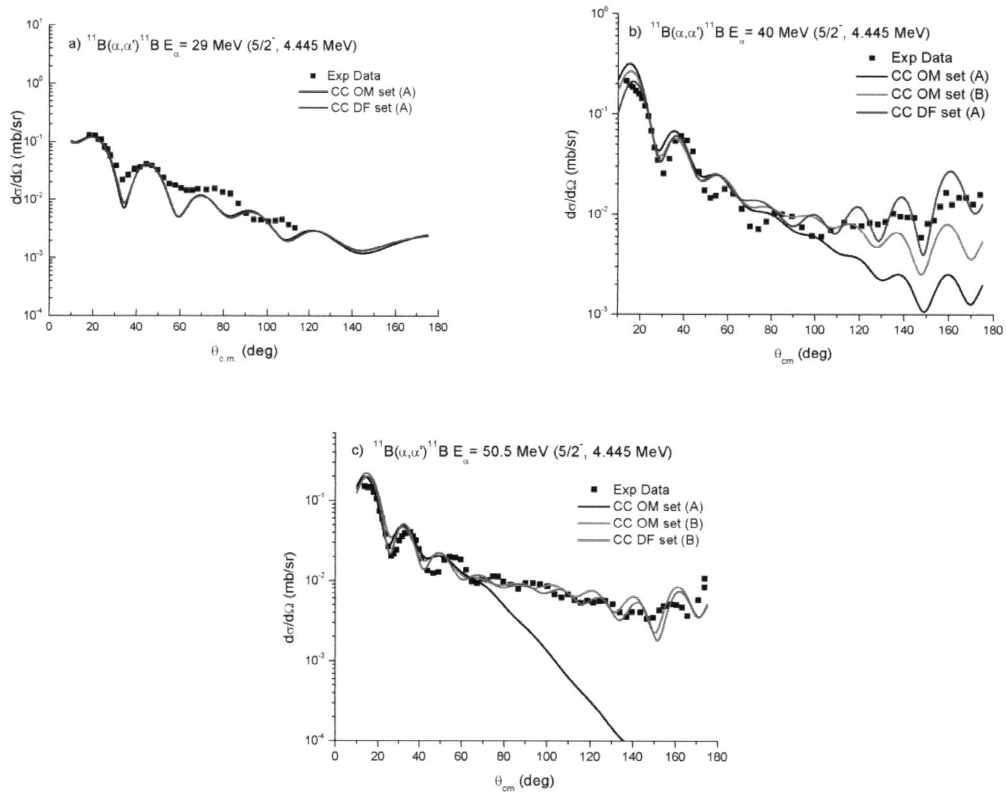

Figure 1. Comparison between the experimental data and the calculated differential cross section for inelastic scattering of ^{4}He from ^{11}B ($5/2^{-}$ state, 4.445 MeV) at energies 29, 40 and 50.5 MeV coupled channel method using different potential sets.

4. Summary

We have measured the angular distributions for the elastic and inelastic $5/2^{-}$ (4.45 MeV) scattering of ^{4}He from ^{11}B target at energy 29, 40 and 50.5 MeV. In addition to our experimental data at energy 29 MeV the elastic scattering of $^{4}He+^{11}B$ were studied at different energies from literature in order to clarify the position of Airy minimum and evolution of rainbow. The data on elastic scattering were analyzed using two approaches optical model (Phenomenological) and double folding (semi-microscopic). The data for the excited states were analyzed within the framework of coupled channel method, the extracted deformation parameters for $5/2^{-}$ state is in a good agreement with the experimental values.

References:

[1] Thompson I J Coupled reaction channels calculations in nuclear physics Comput. *Phys. Rep.* **7** (1988).
[2] Burtebaev N, Baktybaev M K, Duisebaev B A, Peterson R J, Sakuta S B *Physics of Atomic Nuclei* **68** (2005) 8.
[3] Bertsch G, Borysowicz J, McManus H and Love W G *Nucl. Phys.* A **284** 399(1977)
[4] Anantaraman N, Toki H. and Bertsch G *Nucl. Phys.* A **398** 269 (1983).

IOP Conf. Series: Journal of Physics: Conf. Series **940** (2017) 012035 doi:10.1088/1742-6596/940/1/012035

Nuclear Astrophysics at LNL: the ^{25}Mg(α,n)^{28}Si and ^{10}B(p,α)^{7}Be cases

Antonio Caciolli

University and INFN of Padua, Padua, Italy

E-mail: caciolli@pd.infn.it

Abstract. The National Laboratory of Legnaro (LNL) has a long experience in Nuclear Physics measurements. Recently a new effort to perform Nuclear Astrophysics studies has been initiated. This effort started with the collaboration of LNL with the LUNA (Laboratory for Underground Nuclear Astrophysics) collaboration for the study of targets. After that in 2012, thank to a fruitful collaboration between nuclear astrophysicist and nuclear physics groups involved in neutron detection, the study of the ^{25}Mg(α,n)^{28}Si reaction was developed in order to help solving the ^{26}Al puzzle. For the first time the angular distributions of neutrons emitted by this reaction were studied deeply founding discrepancies between the previous studies in literature. In 2014 the study of ^{10}B(p,α)^{7}Be was performed in order to give a precise normalization to the indirect measurements. This study was done by measuring the activated samples and it is still under analysis. A report of the status of the two experiments will be given in this contribution.

1. Introduction

The goal of nuclear astrophysics is to measure the parameters of nuclear reactions involved in astrophysical scenarios. Those parameters, and in particular the cross section, are important for stellar models in order to have precise prediction of stellar nucleosynthesis [1]. Nuclear reaction cross section involving light charged particles are also important in many other fields of research and there is an high effort to study these reactions for applied physics in recent years [2, 3, 4].

The LNL, thanks to their long tradition of nuclear physics studies with different accelerators, offers a perfect environment to perform cross section measurements for nuclear astrophysics and applied physics. At the LNL there are also several facilities for target production and characterizations. Our studies in these laboratories has started exactly with target production and target characterization [5, 6, 7, 8] with ion beam analysis experiments installed at the AN2000 and CN accelerators.

In the present contribution the two recent results achieved at the AN2000 and CN accelerators for the study of two important reactions of nuclear astrophysics are summarized.

2. Study of the ^{25}Mg(α,n)^{28}Si reaction

The observation of ^{26}Al gives us the proof of active nucleosynthesis in the Milky Way. However the identification of the main producers of ^{26}Al is still a matter of debate. Many sites have been proposed, but our poor knowledge of the nuclear processes involved introduces high uncertainties. In particular, the limited accuracy on the ^{25}Mg(α,n)^{28}Si reaction cross section has been identified as the main source of nuclear uncertainty in the production of ^{26}Al in C/Ne explosive burning in massive stars, which has been suggested to be the main source of ^{26}Al in the Galaxy. We studied this reaction through neutron spectroscopy at the CN Van de Graaff accelerator of the Legnaro National Laboratories. The ^{26}Al is produced in the Mg-Al cycle by the well known ^{25}Mg(p,γ)^{26}Al reaction [9, 10]. This reaction has been precisely studied by the LUNA experiment [11] thanks to its underground location [12, 13, 14]. While ^{25}Mg(p,γ)^{26}Al produces ^{26}Al, there are other reactions involved in the destruction of this isotope and its seed, ^{25}Mg: ^{26}Al(n,α)^{23}Na, ^{26}Al(n,p)^{26}Mg, and ^{25}Mg(α,n)^{28}Si. Their impact on the ^{26}Al production during C/Ne explosive burning in massive stars has been recently studied and detailed comparisons between data and models have been performed finding strong discrepancies [15]. In particular, Iliadis and coworkers claimed that the nuclear contribution to the ^{26}Al uncertainty is dominated by the uncertainty on the ^{25}Mg(α,n)^{28}Si reaction rate and they underlined the need of new experimental efforts to reduce the errors on the determination of the cross section value [15]. This reaction has been measured by several authors [17, 16, 18, 19] before 2014, but still there are high discrepancies between the data in literature. A detailed comparison of the existing data can be found in [20], where also the details of the experimental apparatus used at LNL is clearly described. Angular distributions of several branchings of emitted neutrons have been measured at ten different angles by using neutron spectroscopy with the time of flight technique. These results are also reported in [20] where all values are tabulated.

3. Study of the ^{10}B(p,α)^{7}Be reaction

Among the light elements, boron is largely used in many industrial applications and technological solutions, such as semiconductor doping or neutron absorption. Natural boron is composed by two stable isotopes: ^{10}B (19.9%) and ^{11}B (80.1%). Thus it is important to have precise knowledge of nuclear mechanisms involved with both isotopes. As for example, in new concept fusion reactors [21, 22] the (p,α) reaction on ^{10}B produces ^{7}Be. The production of radioactive ^{7}Be could pose serious radiation-safety problems because of its relatively long half-life, $T_{1/2}$=53.22±0.06 d [23]. This calls for precise cross section measurements of the ^{10}B(p,α)^{7}Be reaction in order to understand its impact in future fusion reactor projects.

Boron is also used as a probe of stellar structure for both pre-main sequence [25] and main-sequence stars [26]. Therefore boron plays an important role in astrophysics. These elements are destroyed at different depths in stellar interiors and residual atmospheric abundances can be used to constrain mixing phenomena occurring in such stars [26]. Boron burning is triggered at temperatures $T < 5$ MK via the (p,α) process, with a corresponding Gamow energy centred at about 10 keV. At such energies, the cross section, or equivalently its S-factor, is dominated by the s-wave resonance due to the population of the 8.699 MeV ^{11}C resonant level.

The ^{10}B(p,α)^{7}Be reaction has been measured by several works. Many of them are reported in the NACRE compilation [27]. A recent work has investigated the S-factor at very low energies by means of the Trojan Horse Method (THM) [28]. A discrepancy of about a factor of 2 is evident between the data in [29] and [30]. In addition the data in [31] are characterized by high uncertainties. This would allow first to have new precise data in a region affected by the previously discussed discrepancy and second to provide to THM measurements a more extended energy region for normalisation purposes.

A new measurements of the ^{10}B(p,α)^{7}Be reaction has been done at LNL during December 2014. The cross section has been measured by using the activation technique [1]. Boron samples enriched in ^{10}B up to 93% were irradiated for several hours with a 200-300 nA proton beam. Then the irradiated samples were inserted in the low level counting facility of the LNL [32]. The facility is made of two fully shielded germanium detectors fully shielded with lead and copper. The ^{7}Be decays were observed by detecting the 478 keV gamma line that is emitted with a branching of 10%. The details of the experimental setup and the analysis are described in [33].

IOP Conf. Series: Journal of Physics: Conf. Series **940** (2017) 012035 doi:10.1088/1742-6596/940/1/012035

The S-factor of the new data is a factor of 2 higher than [30] and appears in agreement with the trend of data reported by [31]. In particular the recent data show low uncertainties, lower than 6%, offering a good normalisation for indirect methods.

References

[1] Iliadis C 2007 *Nuclear Physics of Stars* (New York: Wiley-VCH)
[2] Caciolli A, *et al.* 2006 *Nucl. Inst. and Meth. B* **249** 98 – 100
[3] Caciolli A, *et al.* 2006 *Nucl. Inst. and Meth. B* **249** 95 – 97
[4] Caciolli A, *et al.* 2008 *Nucl. Inst. and Meth. B* **266** 1392 – 1396
[5] Depalo R *et al* 2015 *arXiv:1507.03893 [nucl-ex]*
[6] Marta M, *et al.* 2011 *Phys. Rev. C* **83**(4) 045804
[7] Caciolli A, *et al.* 2011 *A& A* **533** A66
[8] Caciolli A, *et al.*, 2012 *Eur. Phys. J. A* **48**, 1
[9] Limata B, *et al.* 2010 *Phys. Rev. C* **82**, 015801
[10] Strieder F, *et al.* 2012 *Physics Letters B* **707**, 60
[11] Broggini C, *et al.* 2010 *Annual Review of Nuclear and Particle Science* **60**, 53
[12] Bemmerer D, *et al.* 2005 *Eur. Phys. J. A* **24**, 313
[13] Caciolli A, *et al.* 2009 *Eur. Phys. J. A* **39**, 179
[14] Szücs T, *et al.* 2010 *Eur. Phys. J. A* **44**, 513
[15] Iliadis C *et al.*, 2011 *The Astrophysical Journal Supplement Series* **193**, 16
[16] K.W.G. L. Van der Zwan, 1981 *Nucl. Sci. Eng.* **79**(2), 197
[17] Anderson M, *et al.* 1983 *Nuclear Physics A* **405**(1), 170
[18] Wieland O, *Diploma Thesis* (IFS-University of Stuttgart, 1995)
[19] Falahat S, *Ph.D. Thesis* (University of Mainz, 2010)
[20] Caciolli A, *et al.* 2014 *Eur. Phys. J. A* **147** 50
[21] Stave S, *et al.* 2011 *Phys. Lett. B* **696** 2629
[22] Spraker M C *et al.* 2012 *J. Fusion Energy* **31** 357367
[23] Tilley D *et al.* 2002 *Nuclear Physics A* **708** 3
[24] Rauscher T, *et al.* 1996 *Phys. Rev. C* **53** 2496
[25] Tognelli E *et al.* 2011 *Astronomy and Astrophysics* **533** A109
[26] Boesgaard A M *et al.* 2005 *Astr. Phys. Journ.* **621** 991
[27] Angulo C *et al.* 1999 *Nucl. Phys. A* **656** 3
[28] Spitaleri C *et al.* 2014 *Phys. Rev. C* **90** 035801
[29] Angulo C *et al.* 1993 *Z. Phys. A* **187** 345
[30] Youn M *et al.* 1991 *Nucl. Phys. A* **533** 321
[31] Roughton N A *et al.* 1979 *At. Data Nucl. Tables* **23** 177
[32] Xhixha G *et al.* 2013 *Journ. Rad. and Nucl. Chem.* **295** 445
[33] Caciolli A *et al.* 2015 *Submitted to PLB*

IOP Conf. Series: Journal of Physics: Conf. Series **940** (2017) 012036 doi:10.1088/1742-6596/940/1/012036

Detection of solar neutrinos with a torsion balance with sapphire crystal

M Cruceru and G Nicolescu

Horia Hulubei National Institute for R&D in Physics and Nuclear Engineering, Magurele, Jud.Ilfov, RO-077125, ROMANIA

E-mail: madalina.cruceru@nipne.ro

Abstract: The solar neutrinos (antineutrinos) are detected with a dedicated torsion balance in the case when they interact coherently on stiff crystals (sapphire with high Debye temperature ~1000K and lead with ~100K Debye temperature). The balance consists in two equal masses of lead and sapphire, of 25g. An autocollimator coupled to this balance measures small rotation angles of the balance. The force with which neutrino flux interacts with these crystals is between 10^{-5}dyn and 10^{-8} dyn, comparable with that reported in Weber's experiments [1]. A diurnal effect is observed for solar neutrinos due to the rotation of the Earth around its own axes. The solar neutrino flux obtained at the site of our experiment is ~$3.8*10^{10}$neutrinos/cm^2*s [2]. Experimental data for neutrinos signals from this high sensitivity torsion balance are presented and commented [3].

1. Introduction

The solar neutrinos flux may produce a transfer of momentum to the two crystals of sapphire and lead. These two crystals are the two weights of a torsion balance. They have different Debye temperature: 1000K for sapphire and 100K for lead.

In our case a torsion balance is involved to check the Eötvös experiment. Two equal masses with different concentrates, of sapphire and lead are positioned on a plate which is hanged on a wolfram wire of 1meter, in a vacuum room. Neutrinos will interact with sapphire crystal and torsion will appear in the balance. Gravitational and inertial forces will not give any torque. The vacuum inside the balance body has a value of 10^{-5} torr. This value of vacuum will assure no disturbance to any torque. Our balance has a sensitivity of $2.6\text{x}10^{-7}$dyn.

2. Experimental device

The kinematical scheme of our torsion balance is presented in Figure 1 and it consists from a wolfram wire, a sapphire crystal and an optical port. An autocollimator will measure the deviations of the balance.

In Figure 2 we can see the inside view of the torsion balance body. It is evidenced a plate with sapphire and lead weights. A mirror is in the middle to reflect the light from the autocollimator.

IOP Conf. Series: Journal of Physics: Conf. Series **940** (2017) 012036 doi:10.1088/1742-6596/940/1/012036

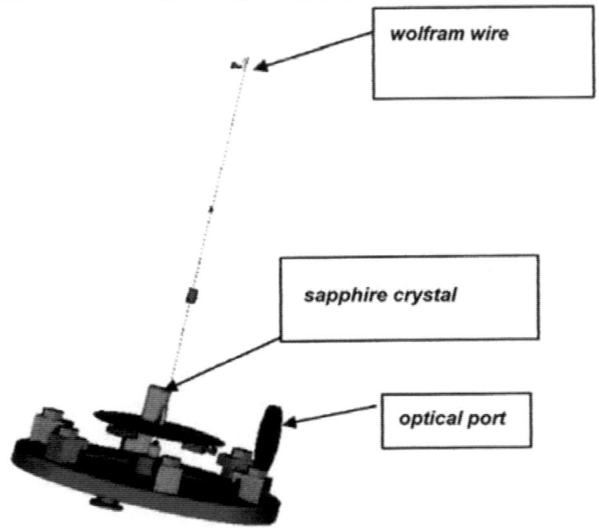

Figure 1. Kinematical scheme of the torsion balance.

Figure 2. Inside view of the torsion balance: two weights of lead and sapphire placed on an aluminum tray.

A coherent inelastic scattering is happening between a neutrino and the infinitely stiff crystals and this means that a large cross section is emphasized. Weber, in his experiments, calculated a total cross section which is proportional with N^2:

$$\sigma = \frac{4E_{\upsilon}^2 G_W^2}{\pi\hbar^4 c^4} N^2 \qquad (1)$$

where E_{υ} is neutrino energy, G_W is the weak-interaction coupling constant, N- is the number of neutrons.

A diurnal effect is presumed to appear. The neutrinos produced in the Sun will touch the Earth without passing any matter. If we measure solar neutrinos in the night, we have to account the fact that neutrinos will pass through the Earth.

The υ_e survival probability during the day is given by formula:

$$P_{ee}^D = cos^4\theta_{13}\left(\frac{1}{2} + \frac{1}{2}cos2\theta_S \bullet cos2\theta_{12}\right) + sin^4\theta_{13}. \qquad (2)$$

where θ_S is the mixing angle at the production point inside the Sun[4], θ_{12} and θ_{13} are the mixing angles between the first neutrino mass eigenstate and the two other active mass eigenstates.

The υ_e survival probability at night [4] when solar neutrinos pass through the Earth:

$$P_{ee}^N = P_{ee}^D - cos2\theta_s cos^2\theta_{13} < f_{reg} >_{zenith}, \qquad (3)$$

where f_{reg} means the regeneration effect in the Earth.

A maximum has to appear in the diurnal effect when the Sun is at zenith[3].

3. References

[1] Weber J 1988 *Phys. Rev.* D 38 32

[2] Nicolescu G 2013 *J. Phys.* G: *Nucl.Part.Phys.* 40 055201

[3] Cruceru M, Nicolescu G, Duliu O G and Cruceru I 2011 *Int.J.of Mod. Phys.*A 26 2773

[4] Antonelli V, Miramonti L,Pena Garay C and Serenelli A 2012, *Adv. in High En. Phys.* 2013 351926, Hindawi Publishing Corporation

Structure of ^{10}Be and ^{16}C nuclei via break-up reactions studied with the 4π Chimera array

D. Dell'Aquila[4], L. Acosta[11], F. Amorini[2], R. Andolina[3], L. Auditore[5], I. Berceanu[8], G. Cardella[1], M.B. Chatterjiee[9], E. De Filippo[1], L. Francalanza[4], B. Gnoffo[1], A. Grzeszczuk[10], G. Lanzalone[2,6], I. Lombardo[4], N. Martorana[2], T. Minniti[5], A. Pagano[1], E.V. Pagano[2,3], M. Papa[1], S. Pirrone[1], G. Politi[1,3], A. Pop[8], F. Porto[2], L. Quattrocchi[5], F. Rizzo[2,3], E. Rosato[4,†], P. Russotto[1], A. Trifirò[5], M. Trimarchi[5], G. Verde[1,7] and M. Vigilante[4].

[1] INFN-Sezione di Catania, Catania, Italy
[2] INFN-Laboratori Nazionali del Sud, Catania, Italy
[3] Dipartimento di Fisica e Astronomia, Università di Catania, Catania, Italy
[4] Dipartimento di Fisica, Università di Napoli Federico II and INFN Sezione di Napoli, Italy
[5] INFN Gruppo Collegato di Messina and Dipartimento di Fisica e Scienze della Terra, Università di Messina, Italy
[6] Facoltà di Ingegneria e Architettura, Università Kore, Enna, Italy
[7] Institute de Physique Nucléaire d'Orsay, Orsay, France
[8] Institute for Physics and Nuclear Engineering, Bucharest, Romania
[9] Saha Institute for Nuclear Physics, Kolkata, India
[10] Institut of Physics, University of Silesia, Katowice, Poland
[11] National Autonomous University of Mexico, Mexico City, Mexico
[†] deceased;

E-mail: `dellaquila@na.infn.it`

Abstract. The study of cluster states in neutron-rich Be and C isotopes is a subject of interest in Nuclear Physics. These states should be characterized by high deformation where α-clusters are bounded by valence neutrons. We performed a spectroscopic study of ^{10}Be and ^{16}C isotopes via projectile break-up reactions, by using radioactive beams produced at the INFN-LNS FRIBs facility and the Chimera 4π array. The possible evidence of a new ^{10}Be state at 13.5 MeV excitation energy was found in the ^{6}He+^{4}He disintegration channel. The spectroscopy of ^{16}C was studied via the ^{6}He+^{10}Be break-up channel; in this case we found the indication of a possible new state at about 20.6 MeV.

1. Introduction

The investigation of cluster structures in nuclei could be a powerful tool to explore the behaviour of nuclear forces in few body interacting systems [1]. These studies are important also in the astrophysical context; for example, the existence of the Hoyle state (0^+, 7.654 MeV) in ^{12}C makes possible the so called 3α process, justifying the observed carbon abundance in Universe.

In presence of extra neutrons, clustering phenomena could assume very different features from the self-conjugated nuclei, originating peculiar configurations, such as nuclear molecules. In the last case the extra-neutrons could act as covalent particles between the α-clusters [2, 3], increasing the stability of the whole configuration [3]. A typical example is the ^{9}Be nucleus that is weakly bound, while the self-conjugated ^{8}Be is unbound. An interesting neutron-rich Be isotope is the ^{10}Be, for which the spectroscopy is not well known and the possible existence of rotational and molecular bands is not definitively clarified [4]. For this nucleus, near the ^{4}He+^{6}He energy threshold the existence of a 0^+ state was predicted by AMD calculations [5] and experimentally confirmed [6]. This state can be described as a α:2n:α linear dimer. The existence of the molecular 2^+ and 4^+ states at 7.54 and 10.15 MeV was recently reported [7], while the subsequent 6^+ member is still missing. Other interesting examples are represented by carbon isotopes. A special attention is given to the ^{16}C isotope, for which various linear chain and triangular configurations have been recently predicted [8]. Unfortunately, the spectroscopy of ^{16}C is presently very poorly known [9, 10]. In this paper we report preliminary results of a new experimental investigation of ^{10}Be and ^{16}C spectroscopy via break-up reactions, that allows to explore the structure of these nuclei above the ^{4}He and ^{6}He decay thresholds.

2. The Experimental Technique

The experiment was carried out at the INFN-LNS by using beams provided by the FRIBs facility. A primary ^{18}O 55 MeV/u beam, impinging on a ^{9}Be production target, was used to produce neutron-rich fragments via the In Flight technique. A tagging system [11] made by a Micro Channel Plate and a Double Side Silicon Strip detector was also used to identify particle by particle the various components of the cocktail beam. We observed typical intensities of ^{16}C at 49.5 MeV/u ($\approx 10^5$ particles per second), ^{13}B ($\approx 5\times10^4$ particles per second) and ^{10}Be at 56 MeV/u ($\approx 4\times10^4$). These beams bombarded a CH_2 target. The subsequent break-up fragments were identified and tracked via the forward three rings of the CHIMERA 4π multi-detector array [12, 13], constituted by 144 ΔE-E Si-CsI(Tl) telescopes. To reconstruct the excitation energy E_x of the decaying nuclear states, we analyzed kinematical correlations between couples of break-up fragments identified by means of the ΔE-E technique and we reconstructed their relative energy (E_{rel}). The energy of decaying states is given by $E_x = E_{rel} + E_{thr}$, where E_{thr} is the energy emission threshold for the selected break-up channel.

3. Data Analysis

In the present work we reconstructed the relative energy spectra of the impinging ^{10}Be and ^{16}C nuclei respectively via the ^{4}He+^{6}He and ^{10}Be+^{6}He cluster break-up channels, as presented in Figs. 1 and 2.

3.1. ^{10}Be→^{4}He+^{6}He channel*

The relative energy ($E_{rel} + E_{thr}$) spectrum for the ^{4}He+^{6}He break-up channel is shown on Fig. 1 with the green histogram. The vertical arrows indicate the energies of ^{10}Be excited states known from literature. Despite the low statistics, it is very interesting to observe the appearance of a bump at $E_x \simeq 13.5$ MeV that could be the suggestion of a possible new state in ^{10}Be. To check the nature of this bump we evaluated the efficiency of our detection device via a MonteCarlo simulation. The behaviour of the estimated efficiency, respectively for the case of inelastic excitation on proton or carbon of the target, is represented by the blue (peaking at 26%) and purple (peaking at 6%) dashed lines. The efficiency curves are very smooth and differences between them can be attributed to the different kinematics. Also the calculated combinatorial background due to spurious coincidences does not show peaks in the region of interest where the 13.5 MeV signal is observed. For these reasons we suggest to attribute the 13.5 MeV bump to the decay from an excited state in ^{10}Be.

IOP Conf. Series: Journal of Physics: Conf. Series **940** (2017) 012037 doi:10.1088/1742-6596/940/1/012037

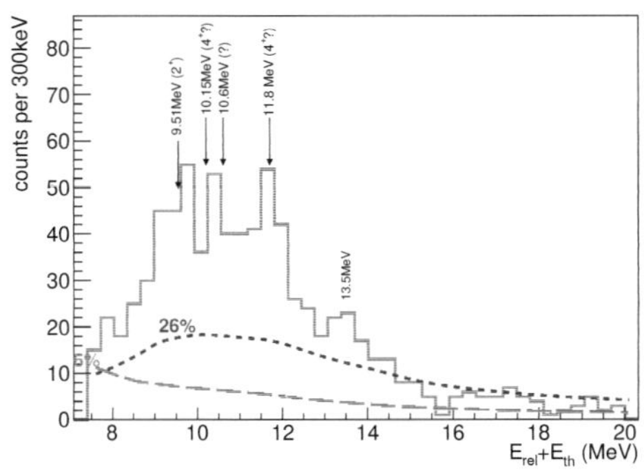

Figure 1. (color online) ^{10}Be relative energy spectrum ($E_{rel}+E_{thr}$) for the ^{4}He+^{6}He break-up channel. The blue and the purple dashed lines represent, respectively, the simulated detection efficiency for inelastic scattering on proton or carbon.

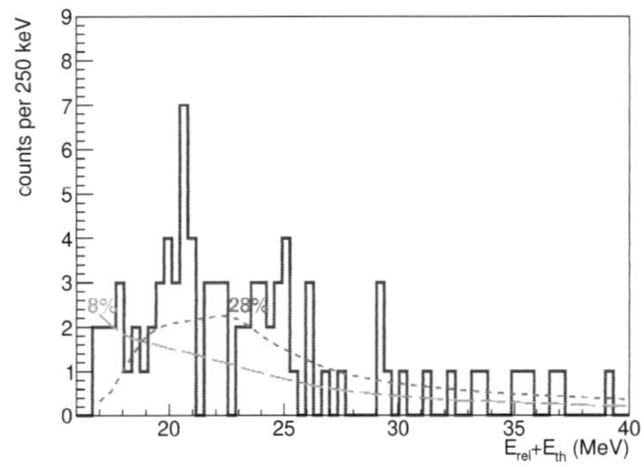

Figure 2. (color online) ^{16}C relative energy spectrum ($E_{rel}+E_{thr}$) for the ^{10}Be+^{6}He break-up channel. The red and the orange dashed lines represent, respectively, the simulated detection efficiency for inelastic scattering on proton or carbon.

3.2. $^{16}C^{*}\rightarrow{}^{10}He+{}^{6}He$ *channel*

The spectroscopy of ^{16}C was studied by using the ^{10}Be-^{6}He kinematical correlations. The corresponding relative energy ($E_{rel}+E_{thr}$) spectrum is shown, as the blue histogram, in Fig. 2. In this case the accumulated statistic is extremely low; anyway, we can find an yield enhancement at ≈20.6 MeV excitation energy. As in the case of ^{10}Be, it seems that this peak should not be due to an efficiency effect (red and orange lines). Another interesting point is that also previous works [9, 10] show a yield enhancement at about 21 MeV of ^{16}C excitation energy. This evidence, even with poor statistics, could be an indication of the possible existence of exotic ^{16}C shape configurations, theoretically predicted by [8] in this energy region; but a new experiment with improved statistic and energy resolution is needed to definitely confirm this hypothesis.

References

[1] Oertzen W V, Freer M and Kanada-En'yo Y 2006 *Phys. Rep.* **432** 43–113
[2] Beck C 2013 *Clusters in Nuclei* vol 1,2,3 (Heidelberg: Springer)
[3] Oertzen W V 1997 *Zeit. Phys.* **A 357** 355
[4] Fortune H and Sherr R 2011 *Phys. Rev.* **C 84** 024304
[5] Kanada-En'yo Y 1998 *J. Phys. G* **24** 1499
[6] Tilley D *et al.* 2004 *Nucl. Phys.* **A 745** 155
[7] Freer M *et al.* 2006 *Phys. Rev. Lett.* **96** 042501
[8] T Baba Y C and Kimura M 2014 *Phys. Rev.* **C 90** 064319
[9] Ashwood N I *et al.* 2004 *Phys. Rev.* **C 70** 0644607
[10] Leask P 2001 *Jour. Phys. G: Nucl. Part. Phys.* **27** B9–B14
[11] Lombardo I *et al.* 2011 *Nuc. Phys.* **B 215** 272–274
[12] Pagano A *et al.* 2004 *Nucl. Phys.* **A 734** 504
[13] Pagano A *et al.* 2012 *Nucl. Phys. News* **22** 25

IOP Conf. Series: Journal of Physics: Conf. Series **940** (2017) 012038 doi:10.1088/1742-6596/940/1/012038

Impact of rotation and magnetic fields in low mass AGB stars

J W den Hartogh[1,8], R Hirschi[1,2,8], C Georgy[1], 3, P Eggenberger[3], F Herwig[4,5,8], M Pignatari[6,8], U Battino[7,8], Jared Keown[4], C Ritter[4,5,8]

[1] Astrophysics Group, Lennard Jones Building, Keele University, Staffordshire ST5 5BG, UK
[2] Kavli Institute for the Physics and Mathematics of the Universe (WPI), The University of Tokyo, Kashiwa, Chiba 277-8583, Japan
[3] Observatoire de Genève, Université de Genève, Chemin des Maillettes 51, 1290 Versoix, Switzerland
[4] Department of Physics and Astronomy, University of Victoria, Victoria, BC V8W 3P6, Canada
[5] Joint Institute for Nuclear Astrophysics, University of Notre Dame, Notre Dame, IN 46556, USA
[6] Konkoly Observatory of the Hungarian Academy of Sciences, Konkoly Thege Miklós út 15-17, H-1121 Budapest, Hungary
[7] Department of Physics, University of Basel, Klingelbergstrasse 82, CH-4056 Basel, Switzerland
[8] NuGrid collaboration, http://www.nugridstars.org

E-mail: `j.den.hartogh@keele.ac.uk`

Abstract. After core helium burning in low and intermediate-mass stars, starts the AGB phase. In this phase, the s process takes place, which is believed to be at the origin of half of all elements heavier than iron. The role of rotation and magnetic fields on the AGB phase is still debated and uncertain. We have calculated stellar evolution models with MESA for stars with an initial mass of 1.5 and 3.0 solar masses. Our models include both rotation and the Taylor-Spruit (TS) dynamo. We show how these physical processes contribute to the total diffusion coefficient and how it will effect the transport of angular momentum and the s-process nucleosynthesis.

Our preliminary results confirm previous results that inclusion of rotation and the TS dynamo, compared to inclusion of rotation alone, results in an improvement of the predicted rotational period of white dwarfs. Inclusion of the TS dynamo reduces the rotationally induced mixing. The impact on the s-process nucleosynthesis is underway and will be presented in a forthcoming publication.

1. AGB stars and white dwarfs

After core helium burning is over an unstable phase of hydrogen shell burning and helium shell flash events (thermal pulses or TPs) starts [1]. The TPs are thin-shell instabilities in the helium shell, caused by a lack of energy production in the shell, leading to its contraction and compression. The thin-shell instability will then trigger a thermonuclear runaway or helium-shell flash. This flash leads to the extinction of the hydrogen shell and an expansion of the helium shell, which eventually decreases its temperature. The thermonuclear runaway is now over and helium burning continues in a stable manner. The reduced luminosity leads to a contraction of

the intershell region until the hydrogen layer is hot enough to ignite again. The slow-neutron-capture-process or s process occurs in between the two shells, after enough neutrons have been released by $^{13}C(\alpha, n)^{16}O$ [2]. Under these conditions heavier nuclei are synthesized by neutron capture, with iron being the seed. The production of heavy elements can reach Pb.
During the AGB phase the envelope is gradually pealed off by increasing mass loss, caused by pulsation-driven shocks in the atmosphere and dust-driven winds [3]. At the end of the phase a degenerate CO core is left, surrounded by a planetary nebula. The CO core then continues to cool down as a white dwarf.

2. Grid of models

To check the effect of rotation and the TS-dynamo, we ran stellar evolution models with MESA for two different masses (1.5 $M_\odot$ and 3.0 $M_\odot$) and two metallicities ($Z = 0.001$ and $Z = 0.02$). For each combination of mass and metallicity, we ran three models: one without rotation, one with rotation and without the TS-dynamo (R-B) and one with rotation and the TS-dynamo (R+B). We include convective boundary mixing. Rotating models have, on the main-sequence a lower temperature, because of a lower effective gravity, and become more luminous, because rotationally induced mixing brings extra fuel to the core. The core is therefore more massive and the Main-Sequence lifetime is longer in rotating models than in non-rotating models.

3. Results

The size and shape of the ^{13}C pockets is strongly affected by rotation, see Figure 1. Including rotation leads to a stretched out ^{13}C pocket and larger diffusion coefficients. This larger partial mixing zone is the subject of our future analysis. Including the TS dynamo decreases the rotationally induced diffusion coefficients, because those depend on the angular velocity as: $D \propto \Omega \frac{d\Omega}{dr}$. The TS dynamo decreases both terms by increasing the coupling between core and envelope.

In Figure 2 the averaged specific angular momentum of the core is plotted, for the initial ZAMS model and the final models of [4]. The rotating model without the TS dynamo have far too high final angular momentum. The inclusion of the TS dynamo results in a improvement of the predicted rotational period of white dwarf (see also [5] and [6]). The impact of the TS dynamo on the s-process nucleosynthesis is underway and will be presented in a forthcoming publication.

Acknowledgements

NuGrid acknowledges significant support from NSF grants PHY 02-16783 and PHY 09-22648 (Joint Institute for Nuclear Astro- physics, JINA). RH acknowledges the support from Eurocore project Eurogenesis and ERC Starting Grant No. 306901. RH acknowledges support from the World Premier International Research Center Initiative (WPI Initiative), MEXT, Japan. FH acknowledges support through an NSERC Discovery Grant (Canada). MP thanks the support from the "Lendület-2014" Programme of the Hungarian Academy of Sciences (Hungary) and from SNF (Switzerland). This work used the DiRAC Complexity system, operated by the University of Leicester IT Services, which forms part of the STFC DiRAC HPC Facility (www.dirac.ac.uk). This equipment is funded by BIS National E-Infrastructure capital grant ST/K000373/1 and STFC DiRAC Operations grant ST/K0003259/1. DiRAC is part of the National E-Infrastructure.

Bibliography

[1] Herwig, F, ARAA, 43:435–479
[2] Käppeler, F, Gallino, R, Bisterzo, S and Aoki, W, RvMP, 83:157-194
[3] Bloecker, T, A&A, 297:727
[4] Suijs, M P L, Langer, N, Poelarends, A-J, Yoon, S-C, Heger, A and Herwig, F, A&A, 481:L87–L90.
[5] Cantiello, M, Mankovich, C, Bildsten, L, Christensen-Dalsgaard, J and Paxton, B, ApJ, 788:93
[6] Eggenberger, P, Montalbán, J and Miglio, A, A&A, 544:L4
[7] Piersanti, L, Cristallo, S and Straniero, O, ApJ, 774:98

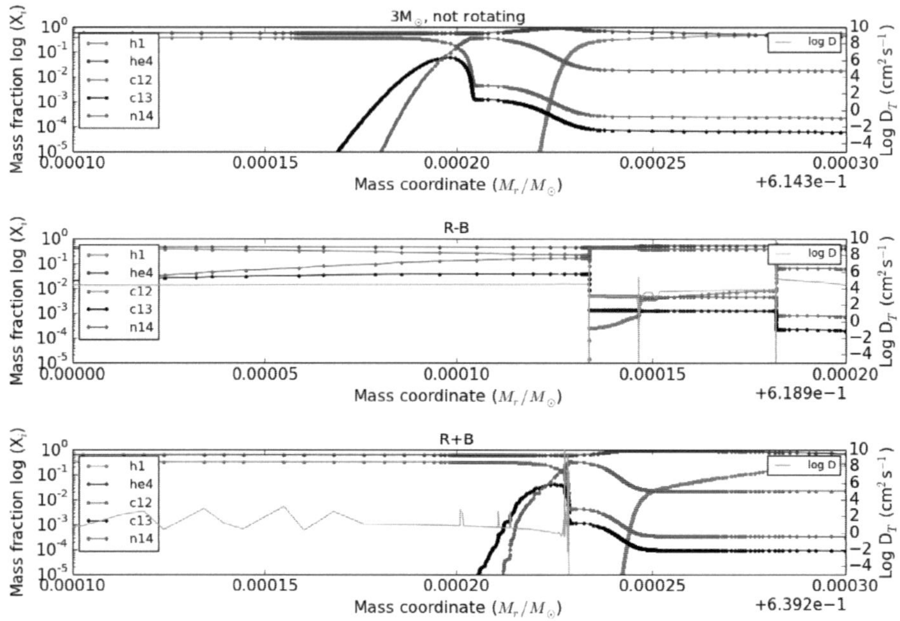

Figure 1: Abundances of important chemical elements within the ^{13}C pocket are plotted with the total diffusion coefficient, for the 3M$_\odot$, $Z = 0.02$ models. These figures confirm previous results of [1] and [7] that rotation has important consequences for s-process production in AGB stars.

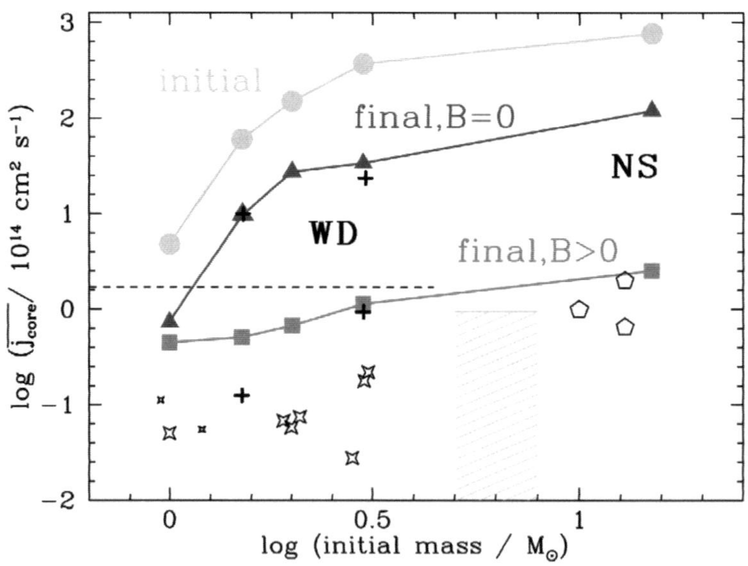

Figure 2: The two + signs at log(initial mass/M$_\odot$) $\simeq$ 0.5 and $\simeq$0.18 are the final values of our rotating models with $Z = 0.02$ and the full drawn lines are the results of [4]. The dashed line is the spectroscopic upper limit of DA WDs, the green area is populated by magnetic WDs, the stars are asteroseismic measurements of ZZ Ceti stars and the pentagons correspond to neutron stars. See [4] for the details on these observations. Adapted from Figure 5 of [4].

IOP Conf. Series: Journal of Physics: Conf. Series **940** (2017) 012039 doi:10.1088/1742-6596/940/1/012039

Explosion dynamics of parametrized spherically symmetric core-collapse supernova simulations

K Ebinger[1], A Perego[2], M Hempel[1], C Fröhlich[3], M Eichler[1], J Casanova[3], M Liebendörfer[1] and F-K Thielemann[1]

[1]Departement für Physik, Universität Basel, CH-4056 Basel, Switzerland
[2]Institut für Kernphysik, Technische Universität Darmstadt, D-64289 Darmstadt, Germany
[3]Department of Physics, North Carolina State University, Raleigh NC 27695, USA

E-mail: kevin.ebinger@unibas.ch

Abstract. We analyze the explosion properties and dynamics of spherically symmetric, parametrized core-collapse supernovae derived with the new PUSH method. To this end, we explore the progenitor mass range between 18 and 21 $M_{\odot}$. We find that it is possible to suggest a distinction between low and high compactness progenitors. We discuss the differences on the example of two reference runs.

1. Introduction

We discuss explosion dynamics of spherically symmetric core-collapse supernova simulations obtained with the recently presented PUSH method ([1], Fröhlich et al., and Eichler et al., this volume). The PUSH method provides a framework to study many important aspects of core-collapse supernovae: the effects of the shock passage through the star, explosive supernova nucleosynthesis and the progenitor-remnant connection. To trigger explosions in the otherwise non-exploding simulations, the PUSH method relies on the neutrino-driven mechanism. It taps the energy reservoir provided by the heavy neutrino flavours to locally increase the energy deposition in the gain region, mimicking the net effects of large multi-dimensional instabilities in spherically symmetric framework. This extra energy deposition is achieved by introducing a local heating term, which is only active where electron-neutrinos are heating and where neutrino-driven convection can occur. For the neutrino transport, we employ IDSA [2] for the electron flavour neutrinos, and ASL [3] for the heavy-lepton flavour neutrinos. We analyze the progenitor mass range between 18 and 21 $M_{\odot}$ [4], which corresponds to typical values for the progenitor of SN 1987A [5].

2. Explosion Dynamics

Contrary to traditional artificial methods such as pistons (e.g., [6],[7]) or thermal bombs (e.g., [8], [9]), now with PUSH, it is possible to analyze the feedback of the evolution of the system on the neutrino heating and hence the explosion dynamics. Furthermore, in our simulations the mass cut emerges naturally. In our models, the explosion energy is mostly generated by the energy deposition of neutrinos in the eventually ejected layers, especially within the first second after bounce. Neutrinos are required to deposit a cumulative energy (E_{idsa}+E_{push}) much larger than the actual explosion energy E_{expl} (see Table 1) to revive the shock and to generate a successful explosion that matches the expected energy range.This is mainly due to the fact that the advection timescale is much shorter than the explosion timescale. A large fraction of the energy deposited in the gain region is advected onto the PNS surface by the accreting mass before an explosion sets in.

By investigating typical values of explosion energies and explosion times obtained with PUSH, as well as the response of the simulations to variations of the PUSH parameters, it is possible to suggest a distinction between low compactness (LC) and high compactness (HC) progenitors. This was also suggested by the work of Nakamura et al. in two-dimensional supernova simulations [10]. We evaluate the compactness parameter [11] at an enclosed mass of 1.75 $M_\odot$ and we employ a value of 0.4 to discriminate between HC and LC models. Below, we compare the 19.2 and 20.0 $M_\odot$ ZAMS progenitor models [4] as representatives for HC and LC models (see Table 1 for the explosion properties of the two reference runs).

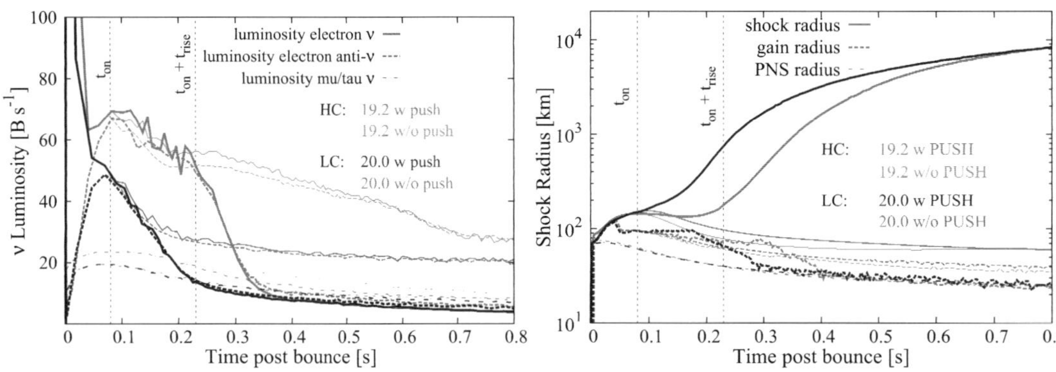

(a) Temporal evolution of the neutrino luminosities.

(b) Temporal evolution of the shock, the gain, and the PNS radii. t_{on} sets the time post bounce when PUSH starts to act. At $t = t_{on} + t_{rise}$ the temporal tuning prefactor of PUSH reaches its maximum.

Figure 1. Temporal evolution of the neutrino luminosities, the shock, the gain, and the PNS radii of the HC (red lines) and the LC (blue lines) progenitor with (thick lines) and without PUSH (thin lines).

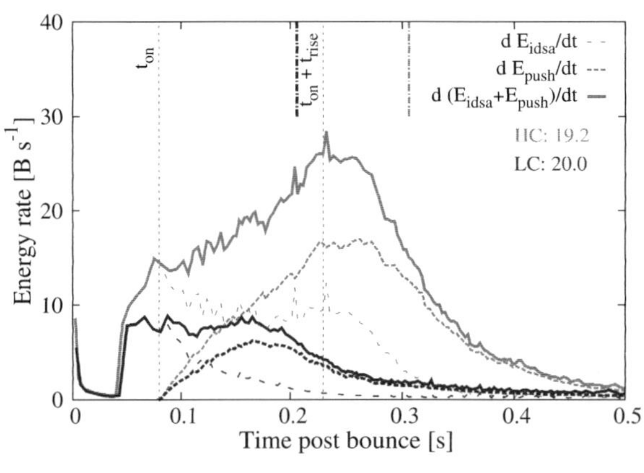

Figure 2. Temporal evolution of the neutrino energy deposition inside the gain region. The short colored vertical lines mark the times of explosion.

Quantity		HC	LC
ZAMS	($M_\odot$)	19.2	20.0
$\xi_{1.75}$	(-)	0.637	0.283
$t_{\rm expl}$	(ms)	307	206
$M_{\rm remn}$	($M_\odot$)	1.713	1.469
$E_{\rm expl}$ ($t_{\rm final}$)	(B)	1.36	0.57
$E_{\rm push}$ ($t_{\rm off} + t_{\rm rise}$)	(B)	3.51	1.08
$E_{\rm idsa}$ ($t_{\rm off} + t_{\rm rise}$)	(B)	2.76	1.01

Table 1. Explosion properties for the reference HC and LC runs. The compactness parameter evaluated at at an enclosed mass of 1.75 $M_\odot$ is denoted by $\xi_{1.75}$. Here, the explosion time $t_{\rm expl}$ is defined as the time when the shock reaches 500km, measured with respect to core bounce. Our final simulation time $t_{\rm final} \gtrsim 4.6$ s is much larger than the explosion time and allows the explosion energy $E_{\rm expl}$ to saturate. The value $E_{\rm push}$ denotes the cumulative energy deposited by PUSH and $E_{\rm idsa}$ represents the cumulative energy deposited by IDSA. Thereby, $t_{\rm off}$ stands for the time after bounce when PUSH is turned off ($t_{\rm off} = 1$ s).

Initially, the infalling velocities of unshocked matter are almost identical. However, the different density profiles of HC and LC models affect the evolution of the shock (see Figure 1b). For instance, $\rho_{19.2}/\rho_{20.0} \gtrsim 1.2$ outside the shock up to a radius of 2×10^8cm. The mass accretion rates start to differ between the two models around 30ms after the bounce. Immediate consequences of the difference in the accretion rates are that the neutrino luminosities are higher (Figure 1a), and that the shock stalling happens earlier and at a smaller radius (Figure 1b, visible in the case without PUSH) in the HC case. In the LC case, due to a lower accretion rate, the relatively small energy deposition rate of PUSH in the gain region (see Figure 2) is able to revive the shock a few milliseconds after PUSH is initiated (Figure 1b). As a result, the LC case explodes faster and with a lower explosion energy than the HC case (see Table 1).

3. Summary & Outlook

We have discussed the qualitative difference between a high and a low compactness explosion scenario in the progenitor mass range between 18 and 21 $M_\odot$ ZAMS that was investigated in [1] in order to reproduce observed properties of SN 1987A. In a next step we plan to conduct a broad progenitor study in the mass range between 15 and 75 $M_\odot$.

4. References

[1] Perego, A., Hempel, M., Fröhlich, C., Ebinger, K., Eichler, M., Casanova, J., Liebendörfer, M., Thielemann, F.-K., 2015, ApJ, 806, 275.
[2] Liebendörfer, M., Whitehouse, S. C., Fischer, T., 2009, ApJ, 698, 1174
[3] Perego, A., 2012, PhD thesis, University of Basel, Switzerland
[4] Woosley, S. E., Heger, A., Weaver, T. A., 2002 Reviews of Modern Physics,74, 1015
[5] Shigeyama, T., & Nomoto, K., 1990, ApJ, 360, 242
[6] Woosley, S. E., & Weaver, T. A., 1995, ApJS, 101, 181
[7] Limongi, M., & Chieffi, A., 2006, ApJ, 647, 483
[8] Thielemann, F.-K., Nomoto, K., Hashimoto, M. A., 1996, ApJ, 460, 408
[9] Umeda, H., & Nomoto, K., 2008, ApJ, 673, 1014
[10] Nakamura, K., Takiwaki, T., Kuroda, T., Kotake, K., 2014, arXiv:1406.2415v3
[11] O'Connor, E., Ott, C. D., 2011, ApJ, 730, 70

IOP Conf. Series: Journal of Physics: Conf. Series **940** (2017) 012040 doi:10.1088/1742-6596/940/1/012040

Explosive nucleosynthesis in core-collapse supernovae and the titanium problem

M Eichler[1], A Perego[2], M Hempel[1], C Fröhlich[3], K Ebinger[1], J Casanova[3], M Liebendörfer[1] and F-K Thielemann[1]

[1]Department für Physik, Universität Basel, CH-4056 Basel, Switzerland
[2]Institut für Kernphysik, Technische Universität Darmstadt, D-64289 Darmstadt, Germany
[3]Department of Physics, North Carolina State University, Raleigh NC 27695, USA

E-mail: marius.eichler@unibas.ch

Abstract. Based on a new mechanism to drive spherically symmetric core-collapse supernovae, PUSH, we perform full network nucleosynthesis calculations for different progenitors. While the $^{56-58}$Ni yields match the observational data very well for certain progenitors, the ejected titanium masses in our calculations are lower than the values inferred from observations. We demonstrate the dependence of ejecta composition on the progenitor structure and the mass cut. Furthermore, we discuss possible solutions to the well-known problem of titanium underproduction.

1. Introduction

Recently, the PUSH method for triggering core-collapse supernovae in 1D has been presented ([1], Fröhlich et al., and Ebinger et al., this volume). In this approach, a fraction of the energy of the heavy flavour neutrinos is deposited in the gain region to artificially increase the neutrino heating efficiency in otherwise non-exploding spherically symmetric models. Apart from studying explosion characteristics for a broad range of progenitors, this method provides a good framework for nucleosynthesis studies in core-collapse supernovae.

One of the long-standing problems associated with explosive nucleosynthesis in core-collapse supernovae is the discrepancy in the amount of ^{44}Ti ejected between observations and simulations, the latter usually being significantly lower than the former. While observations of SN 1987A have reported values between $(0.5 - 4) \times 10^{-4}$ $M_{\odot}$ [2, 3, 4, 5, 6, 7, 8], simulations (e.g., [9, 10]) typically predict around one order of magnitude less.

We give a short overview of our nucleosynthesis network and compare the composition of the ejecta from two different progenitors in our calculations with the observed values from SN 1987A in section 2. In section 3 we summarize the titanium underproduction problem and discuss possible solutions.

2. Composition of the ejecta

The PUSH method is calibrated on observables (explosion energy, $^{56-58}$Ni, and ^{44}Ti yields) of SN 1987A. To that end, we explore the progenitor space between 18 $M_{\odot}$ and 21 $M_{\odot}$, as it is known that the progenitor of SN 1987A is in that mass bracket. After applying a fallback of 0.1 $M_{\odot}$, many progenitors in that mass range are found to reproduce the ^{56}Ni yield from

SN 1987A well enough, however only about half of them also produce ^{57}Ni and ^{58}Ni to an amount that is needed to match the observational data. The reason can be found in the initial Y_e profiles of the progenitors and the position of the cutoff mass in the supernova simulations. The ejecta composition is plotted against mass coordinates in Figure 1 for two of the progenitors we consider (18.0 $M_\odot$ and 20.6 $M_\odot$) [11]. For the 18.0 $M_\odot$ model, the cutoff mass is 1.56 $M_\odot$ and a large part of the silicon shell is ejected. In this shell, the initial matter composition is slightly neutron-rich (due to a small contribution from ^{56}Fe) with $Y_e \simeq 0.498$ and the conditions for the production of ^{57}Ni and ^{58}Ni are favourable. The transition from silicon shell to oxygen shell for the 20.6 $M_\odot$ model happens around 1.74 $M_\odot$, i.e., inside the mass cut. Therefore, the matter ejected by this model is less neutron-rich and contains less ^{57}Ni and ^{58}Ni (see also [12]).

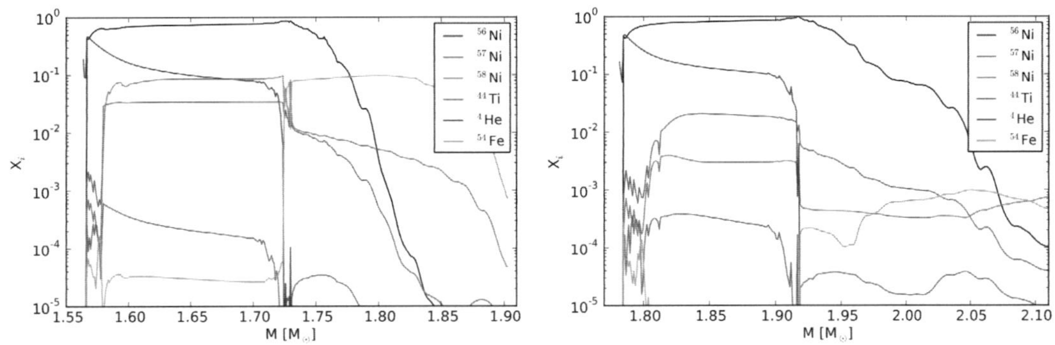

Figure 1. Ejecta composition of the 18.0 $M_\odot$ (left) and 20.6 $M_\odot$ (right) progenitors vs mass coordinates.

3. Titanium-44

With a fallback of 0.1 $M_\odot$ onto the proto-neutron star, the amount of $^{56-58}$Ni in the ejecta and the explosion energies agree very well with observations of SN 1987A for two of the progenitors we considered (18.0 $M_\odot$ and 19.4 $M_\odot$). However, the inclusion of fallback drastically reduces the ejected amount of ^{44}Ti, as it is produced in the innermost part of the ejecta. In Figure 2 the ejected amounts of ^{44}Ti (a) without and (b) with fallback are compared for four different progenitors and five different PUSH parameter sets, plotted against the respective explosion energies. The error box represents the observational values from Seitenzahl et al. (2014) [8].

Too low ^{44}Ti yield predictions have been a long-standing problem in supernova nucleosynthesis calculations (e.g., [9, 10]). Several uncertainties are connected to the production, destruction, and ejection of ^{44}Ti. First, ejecta in a supernova may be subject to convective overturn, which cannot be simulated in our 1D approach. To account for this, we can assume homogeneous mixing up to the outer boundary of the silicon shell before cutting off the fallback material (e.g., [13]). As a consequence, the ejected ^{44}Ti mass for the 18.0 $M_\odot$ progenitor increases to 2.70×10^{-5} $M_\odot$ compared to the previous yield of 1.04×10^{-5} $M_\odot$. Moreover, there are uncertainties in the main nuclear reactions that are responsible for the production and destruction of ^{44}Ti. The final amount of produced ^{44}Ti depends mainly on two reactions: ^{40}Ca$(\alpha,\gamma)^{44}$Ti and ^{44}Ti$(\alpha,p)^{47}$V. Recent measurements of the ^{44}Ti$(\alpha,p)^{47}$V reaction rate concluded that it may be considerably smaller than previous theoretical predictions [14]. Using this smaller cross section, our yield of ejected ^{44}Ti for the 18.0 $M_\odot$ progenitor rises to 1.49×10^{-5} $M_\odot$ with fallback and 5.65×10^{-5} $M_\odot$ without fallback (see Figure 3). If we include both the new cross section and homogeneous mixing (corresponding to the red line in Figure 3), the amount of ^{44}Ti in the

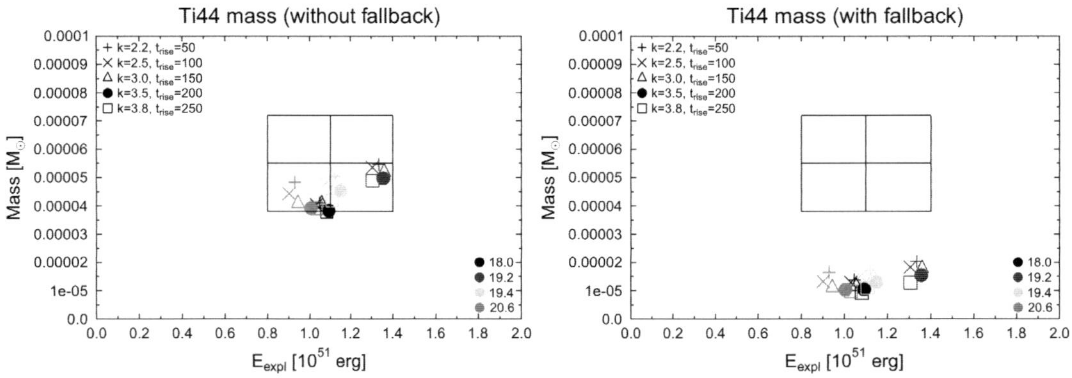

Figure 2. Ejected ^{44}Ti masses for different progenitors and PUSH parameter sets (a) without and (b) with the inclusion of 0.1 $M_{\odot}$ fallback.

ejecta is 3.99×10^{-5} $M_{\odot}$ including fallback, just within the error range of the observational value reported in Ref. [8].

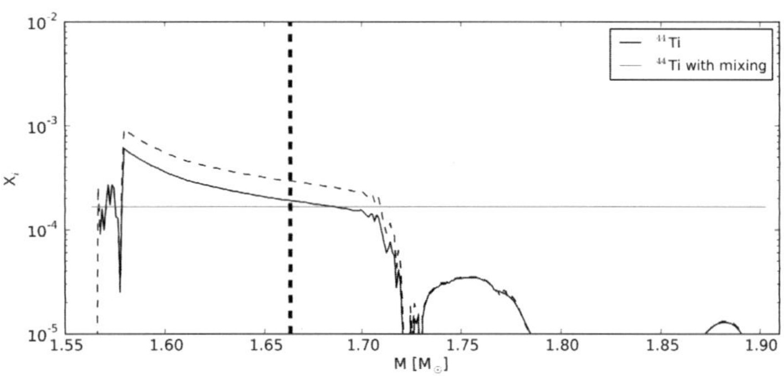

Figure 3. ^{44}Ti yields of the ejecta for the 18 $M_{\odot}$ star as in Figure 1 (black solid line), and with an updated ^{44}Ti$(\alpha,p)^{47}$V rate (black dashed line). The red line represents the distribution when in addition to the updated rate homogeneous mixing is assumed. The dashed vertical line indicates the mass cut if 0.1 $M_{\odot}$ fallback is applied.

References

[1] Perego, A., Hempel, M., Fröhlich, C., Ebinger, E., Eichler, M., Casanova, J., Liebendörfer, M., & Thielemann, F.-K. 2015, ApJ, 806, 275.
[2] Chugai, N. N., Chevalier, R. A., Kirshner, R. P., & Challis, P. M. 1997, ApJ, 483, 925.
[3] Fransson, C., & Kozma, C. 2002, New Astronomy Review, 46, 487.
[4] Jerkstrand, A., Fransson, C., & Kozma, C. 2011, A&A, 530, A45.
[5] Larsson, J., Fransson, C., Ostlin, G., et al. 2011, Nature, 474, 484.
[6] Grebenev, S. A., Lutovinov, A. A., Tsygankov, S. S., & Winkler, C. 2012, Nature, 490, 373.
[7] Grefenstette, B. W., Harrison, F. A., Boggs, S. E., et al. 2014, Nature, 506, 339.
[8] Seitenzahl, I. R., Timmes, F. X., & Magkotsios, G. 2014, ApJ, 792, 10.
[9] Thielemann, F.-K., Nomoto, K., & Hashimoto, M.-A. 1996, ApJ, 460, 408.
[10] Woosley, S. E., & Weaver, T. A. 1995, ApJS, 101, 181.
[11] Woosley, S. E., Heger, A., & Weaver, T. A. 2002, Reviews of Modern Physics, 74, 1015.
[12] Thielemann, F.-K., Hashimoto, M.-A., & Nomoto, K. 1990, ApJ, 349, 222.
[13] Umeda, H., & Nomoto, K. 2002, ApJ, 565, 385.
[14] Margerin, V., Murphy, A. S. J., Davinson, T., et al. 2014, Physics Letters B, 731, 358.

IOP Conf. Series: Journal of Physics: Conf. Series **940** (2017) 012041 doi:10.1088/1742-6596/940/1/012041

Measurement of the strengths of the resonances at 417, 458, 611, 632 and 1222 keV in the $^{22}Ne(p,\gamma)^{23}Na$ reaction

Federico Ferraro

Dipartimento di Fisica, Università degli Studi di Genova and INFN, Sezione di Genova
Via Dodecaneso 33, 16146 Genova, Italy

E-mail: federico.ferraro@ge.infn.it

Abstract. The $^{22}Ne(p,\gamma)^{23}Na$ reaction is part of the NeNa cycle of hydrogen burning. This cycle plays a key role in the nucleosynthesis of the elements between ^{20}Ne and ^{27}Al in red giant stars, asymptotic giant stars and classical nova explosions. The strengths of the resonances at proton energies above 400 keV are still affected by high uncertainty. In order to reduce this uncertainty, a precision study of some of the most intense resonances between 400 keV and 1250 keV has been performed at the HZDR 3 MV Tandetron. The target, made of ^{22}Ne implanted in a 0.22 mm thick Ta backing, has been characterized using the 1222 keV and 458 keV resonances, well known in literature. Subsequently, the strengths of the resonances at 417, 458, 611, 632 and 1222 keV were determined. Two HPGe detectors equipped with active anti-Compton shielding have been used.

1. The $^{22}Ne(p,\gamma)^{23}Na$ reaction and the NeNa cycle of hydrogen burning.

The $^{22}Ne(p,\gamma)^{23}Na$ reaction is part of the NeNa cycle of hydrogen burning (see Figure 1), active in red giant branch (RGB) stars, asymptotic giant branch (AGB) stars, classical novae (CN) and supernovae Ia (SN Ia) explosions. This cycle greatly influences the nucleosynthesis of the elements between ^{20}Ne and ^{27}Al because of its link to the MgAl cycle [1, 2]. At temperatures of about 0.4 GK (for CN) up to 1.5 GK (for SN Ia) and above, the greatest contribution to the reaction rate comes from the resonances between 400 keV and 1250 keV (see Figure 2), some of which are affected by high uncertainty [3]. Because of the lack of knowledge on some of these resonances, the $^{22}Ne(p,\gamma)^{23}Na$ reaction rate is very uncertain, asking for a deeper investigation of the existing resonances and a better determination of their strengths [4].

2. Experiment

The measurements have been performed at the 3 MV Tandetron accelerator of Helmholtz-Zentrum Dresden-Rossendorf in Dresden, Germany. A solid target has been used, made of ^{22}Ne implanted on a 0.22 mm thick Ta backing. The typical proton current provided by the accelerator was 10 μA. The target was water-cooled and a copper tube in front of the target (see Figure 3 for a section drawing of the setup), connected to a voltage of -100 V, provided the secondary electrons suppression, so that the collected charge has been measured at 1% uncertainty.
Two high purity germanium (HPGe) detectors equipped with BGO active shielding and

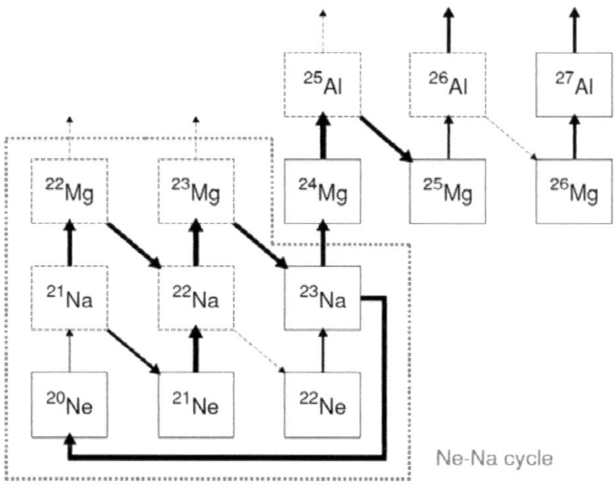

Figure 1. The NeNa cycle and its link to the MgAl cycle.

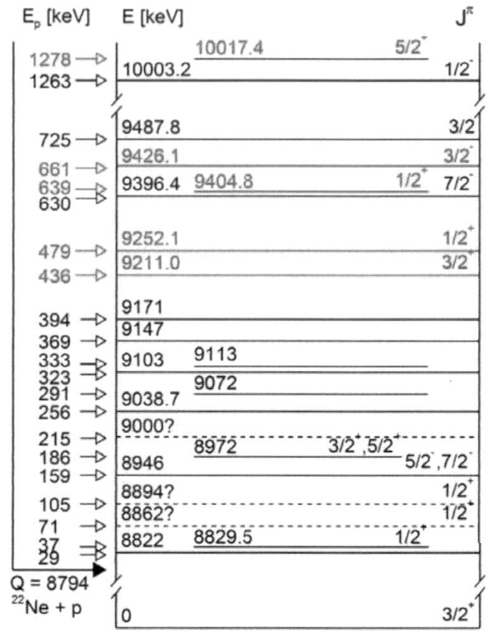

Figure 2. ^{23}Na excited states. In red the measured resonances (E_p in the laboratory frame of reference).

surrounded by lead have been positioned at 55° and 90° (see Figure 4 for a picture of the setup). A lead collimator has been put in front of each detector in order to shield the BGO and better define the solid angle for the HPGe detectors. The active BGO shielding, together with the anti-coincidence trigger, provided a background reduction by a factor of 3 without any significant reduction of the peak efficiency.

The yield curve of the well known narrow resonance at 1222 keV permitted to monitor the target stability over time and to measure the implantation profile. The energy and efficiency calibrations have been carried out owing to some standard radioactive sources (^{137}Cs, ^{60}Co and ^{88}Y) together with the well known reaction ^{27}Al(p, γ)^{28}Si, which permitted the extension of the range of these measurements up to 10763 keV. A target scan has been done for each resonance and the yield curve has been studied. A long run has been acquired on the top of each resonance yield curve to determine the resonance strength.

3. Analysis

The ^{23}Na characteristic peaks have been recognized in the spectra and the Ta/^{22}Ne ratio has been determined normalizing our results to the 1222 keV and 458 keV resonance strengths, thus improving the determination of the stoichiometry, which turned out to be Ta/^{22}Ne $= 7.92 \pm 0.04_{\text{stat}} \pm 0.34_{\text{syst}} \pm 0.56_{\omega\gamma}$ as shown in [7].

The primary transition peaks have been clearly identified for each resonance and the background subtraction has been done. The total yield has been calculated using

$$Y_{tot} = \frac{e}{Q} \sum_i \frac{N_{\gamma_i}}{W_{\gamma_i} \eta_{\gamma_i}} \tag{1}$$

where $\frac{e}{Q}$ is the reciprocal of the total number of incident particles, N_{γ_i}, W_{γ_i} and η_{γ_i} are respectively the number of counts, the angular correlation coefficient [6, 7] and the efficiency of

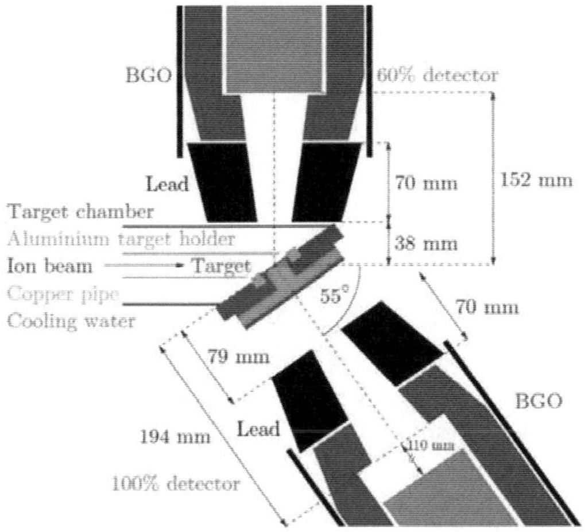

Figure 3. Setup drawing. BGO detectors are used as anti-Compton veto.

Figure 4. Setup picture. On the left side the target chamber is visible.

the ith primary transition photon. The stopping power in ^{22}Ne and Ta has been evaluated with SRIM and the resonance strengths were determined using the equation

$$Y_{max} = \frac{\lambda^2}{2}\omega\gamma\frac{M+m}{M}\frac{1}{\epsilon} \tag{2}$$

where λ is the De Broglie wavelength, $\omega\gamma$ is the resonance strength, M and m are respectively the masses of the target nucleus and the projectile and ϵ is the effective stopping power in the Ta - ^{22}Ne compound.

4. Results

In conclusion, we measured ^{22}Ne(p, γ)^{23}Na resonance strengths for E_{CM}=417, 458, 611, 632, 1222 keV. The obtained strengths and their impact on the reaction rate are reported in [7], together with a more complete description of the experiment.
These measurements are complementary to the resonance strengths measured by LUNA at lower energies [4, 8, 10] and will improve our knowledge of the thermonuclear reaction rate.

Acknowledgments

This work has been supported by the European Union (SPIRIT EC contract no. 227012) and by NAVI (HGF VH-VI-417).

References

[1] Marion J and Fowler W 1957 *Astrophys. J.* **125** 221-32
[2] Iliadis C, Champagne A, José J, Starrfield S and Tupper P 2002 *Astrophys. J. Suppl. Ser.* **142** 105-37
[3] Iliadis C, Longland R, Champagne A E and Coc A 2010 *Nuclear Physics* A **841** 251-322
[4] Cavanna F *et al.* 2014 *Eur. Phys. J.* A **50** 179
[5] Longland R *et al.* 2010 *Phys. Rev.* C **81** 055804
[6] Viitasalo M, Piiparinen M and Anttila A 1972 *Z. Physik* **250** 387-394
[7] Depalo R *et al.* 2015 *Phys. Rev.* C **92** 045807
[8] Cavanna F *et al.* 2015 *Phys. Rev. Lett.* **115** 252501
[9] Ferraro F *et al.* *PoS NIC XIII* 093
[10] Depalo R and Cavanna F *PoS NIC XIII* 087

IOP Conf. Series: Journal of Physics: Conf. Series **940** (2017) 012042 doi:10.1088/1742-6596/940/1/012042

Neutron capture cross section of ^{85}Kr

S Fiebiger[1], Z Slavkovská[1], U Giesen[2], T Heftrich[1], M Krtička[3], R Reifarth[1], S Schmidt[1], B Thomas[1], M Weigand[1], A Zadeh[1]
[1]Goethe University Frankfurt, Germany
[2]Physikalisch-Technische Bundesanstalt Braunschweig, Germany
[3]Charles University Praque, Czech Republic

Abstract. Neutron capture and β^--decay are competing branches of the s-process nucleosynthesis path at ^{85}Kr, which makes it an important branching point. The knowledge of its neutron capture cross section is therefore an essential tool to constrain stellar models of nucleosynthesis. The goal is the measurement of the ^{85}Kr(n,γ) cross section via the time-of-flight method. For this, several methods for the production of ^{85}Kr will be investigated. One of these is the irradiation of a ^{82}Se target with an α-beam. Here, the produced ^{85}Kr stays trapped inside the crystalline structure of the selenium. Due to technical difficulties and low yields, an alternative method is the use of reactor produced ^{85}Kr. For future measurements of the neutron capture cross section of ^{85}Kr at FRANZ (Frankfurter Neutronenquelle am Stern-Gerlach-Zentrum), the goal is the use of a target with a high isotopic purity to reduce the background from ^{83}Kr.

1. Motivation

A part of the s-process path and a matter of great interest over the years has been the isotope ^{85}Kr [1]. It represents a branching point in the s process [2, 3], because the β-decay rate and the neutron capture rate compete. Therefore the mass flow during the s process depends on the stellar conditions during the production, like temperature and neutron density.

Besides its importance for the s process, ^{85}Kr is of interest in other applications as well. It could for example, help to constrain the broad distribution of ^{86}Kr/^{82}Kr ratios observed in presolar grains, small SiC bonds, that are believed to grow in the outer regions of Red Giants. Trapped inside the SiC crystal matrix are isotopes that can than be examined further. At the moment, the understanding of the ^{86}Kr distribution is only hindered by the poorly known neutron capture cross section of ^{85}Kr.

Furthermore, ^{85}Kr serves as an important part in the Rb/Sr cosmochronometer as it influences the production of ^{87}Rb, which in turn feeds ^{87}Sr. This impact on the ratio of the stable isotopes ^{86}Sr/^{87}Sr could be used to constrain cosmological models of the universe.

2. Target Production via ^{82}Se(α,n)

2.1. Sample preparation

For the measurement of the reaction ^{85}Kr(n,γ) a sufficiently pure sample has to be produced. An anticipated challenge is the presence of other isotopes, which will cause background. For the experiment at Physikalisch-Technische Bundesanstalt (PTB) the α-beam was entirely stopped in the target material, a thick target layer of Se was produced by melting Se onto an aluminum backing. The selenium powder was placed in the gap in the center of the backing and heated in an oven to its melting point of 221°C. To achieve a homogeneous layer of Se, which was essential for an activation experiment, several steps had to be undertaken. First, the droplets were spread mechanically using a spattle after reducing the oven temperature to approximately 100 °C. Afterwards more selenium powder was put in the gap as a part of it remained on the spattle. The heating procedure was repeated until a smooth glassy black layer of Se was formed. The resulting thicknesses of the Se layers were between 240 μm and 400 μm based on the weight of the samples. This was sufficient for the planned experiment as the range of α-particles of 15 MeV in Se is only 0.1 mm [4].

2.2. Experiment at Physikalisch-Technische Bundesanstalt

In February 2014, the activation experiment ^{nat}Se(α,n) was performed at PTB Braunschweig. The goal was to measure unknown thick target yields and α-induced production cross sections for several different reactions for the first time. In particular ^{82}Se(α,n)^{85}Kr. Several optimizations of the setup would have to be made to produce ^{85}Kr in an amount sufficient for a time-of-flight experiment [5]. The most important is a better cooling to counter the small thermal conductivity of Se, which was the reason that the α-current had to be held at a low level. The spectroscopic analysis of the reaction product ^{85m}Kr was conducted using a High-Purity Germanium (HPGe) detector at PTB. Because of the long half life of the ^{85}Kr ground state and the small intensity of its strongest γ-emission line at 514 keV, it was not possible to use the HPGe detector at PTB for its spectroscopic analysis. Therefore it was undertaken at Goethe University Frankfurt using a Low Energy Photon Spectrometer (LEPS) detector, which has the advantage of a very good energy resolution. This allows the separation of the 514 keV γ-line following the decay of ^{85}Kr^{GS} from the 511 keV background. From the measured thick target yields the preliminary cross sections of ^{85m}Kr and ^{85}Kr have been calculated to σ_{85mKr} = 37.46 mb and σ_{85Kr} = 124.61 mb respectively (Figure 1).

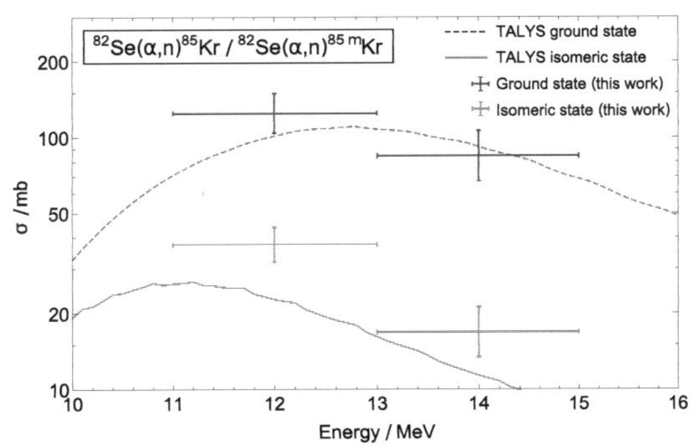

Figure 1. Comparison of determined α-induced production cross section of ^{85}Kr to the theoretically determined from TALYS. The TALYS values are in good agreement with the measured values. The energy error of ± 1 MeV originates from the subtraction of two thick target yields.

3. Target production using reactor produced ^{85}Kr

^{85}Kr is also a product of the fission reaction ^{235}U(n,f) in a nuclear reactor and is with an isotopic abundance of about 12%. Though easier to obtain, reactor produced ^{85}Kr is contaminated with other Kr isotopes, which will cause background during (n,γ) measurements. The biggest problem is ^{83}Kr, which has a higher Q-value than ^{85}Kr. Such a mixed sample, with an isotope ratio of approximately 1.8:3.7:10.6:1:6.2 (^{82}Kr:^{83}Kr:^{84}Kr:^{85}Kr:^{86}Kr), is already available. An additional

IOP Conf. Series: Journal of Physics: Conf. Series **940** (2017) 012042 doi:10.1088/1742-6596/940/1/012042

measurement of pure ^{83}Kr will be necessary as it will produce a significant background during measurement. Except for ^{83}Kr, which has a higher Q-value (10.52 MeV) than ^{85}Kr (9.86 MeV), all other Q-values are smaller than for ^{85}Kr. Therefore capture events on other isotopes can be discriminated applying a Q-value cut [7, 8], provided the measurement is performed using a 4π-γ-detector. To verify that this sample is viable for an ^{85}Kr(n,γ) experiment, simulations with DICEBOX [9] were performed to obtain γ-cascades of the (n,γ) reactions of ^{83}Kr and ^{85}Kr (Figure 2). With these, further simulations with GEANT3 [10] are possible to find a way of distinguishing captures on the two isotopes in a calorimetric measurement.

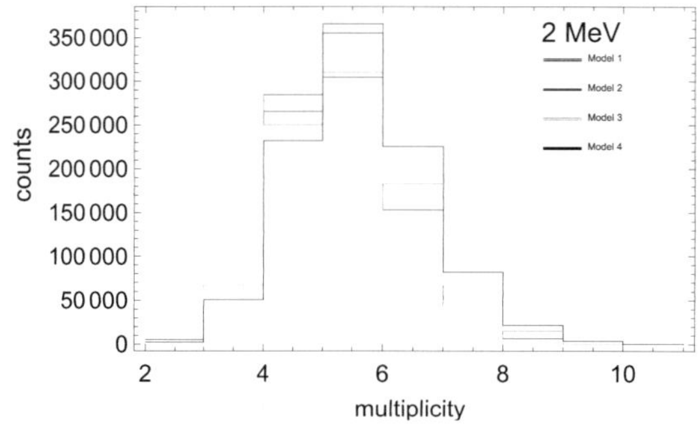

Figure 2. An example result of the DICEBOX simulations of ^{83}Kr. Shown are the multiplicities of all 2 MeV γ's of a cascade.

4. Time of flight experiment

With a suitable sample [5], it will be possible to perform an experiment to investigate the reaction ^{85}Kr(n,γ) at FRANZ (Frankfurter Neutronenquelle am Stern-Gerlach Zentrum) in Frankfurt. Neutrons at FRANZ will be produced via the reaction ^{7}Li(p,n) [3]. The high neutron flux at the sample position will be increased by reducing the neutron flightpath to about 8 cm [11].

Acknowledgments

This project was supported by the European Research Council under the European Union's Seventh Framework Programme (FP/2007-2013)/ERC Grant Agreement n. 615126.

References

[1] Raut R, Tonchev A P, Rusev G, Tornow W, Iliadis C, Lugaro M, Buntain J, Goriely S, Kelley J H, Schwengner R, Banu A and Tsoneva N 2013 *Physical Review Letters* **111** 112501 (*Preprint* `1309.4159`)

[2] Abia C, Busso M, Gallino R, Dominguez I and Straniero Oand Isern J 2001 *Ap. J.* **559** 1117

[3] Reifarth R, Lederer C and Käppeler F 2014 *Journal of Physics G Nuclear Physics* **41** 053101

[4] Ziegler J 1980 *Handbook of Stopping Cross-Sections for Energetic Ions in all Elements* vol 5 (New York: Pergamon Press) this is a full BOOK entry

[5] Couture A and Reifarth R 2007 *Atomic Data and Nuclear Data Tables* **93** 807

[6] Crouch E A C 1977 *Atomic Data Nuclear Data Tables* **19** 417–532

[7] Wisshak K, Voss F, Arlandini C, Bečvář F, Straniero O, Gallino R, Heil M, Käppeler F, Krtička M, Masera S, Reifarth R and Travaglio C 2001 *Physical Review Letters* **87** 251102 (pages 4) URL `http://link.aps.org/abstract/PRL/v87/e251102`

[8] Reifarth R, Bredeweg T A, Alpizar-Vicente A, Browne J C, Esch E I, Greife U, Haight R C, Hatarik R, Kronenberg A, O'Donnell J M, Rundberg R S, Ullmann J L, Vieira D J, Wilhelmy J B and Wouters J M 2004 *Nucl. Instr. Meth. A* **531** 530

[9] Bečvář F 1998 *Nucl. Instr. Meth. A* **417** 434

[10] Apostolakis J 1993 Cern program library long writeup, w5013 Tech. rep. CERN, GEANT library http://wwwinfo.cern.ch/asd/geant/

[11] Reifarth R, Haight R C, Heil M, Käppeler F and Vieira D J 2004 *Nucl. Instr. Meth. A* **524** 215

Rotational excitation of the Hoyle state in ^{12}C

R Garg,[1,*] C Barton,[1] C Aa Diget,[1] S Courtin,[2] G Fruet,[2] H O U Fynbo,[3] A Howard,[3] A Illana,[4,†] D G Jenkins,[1] I Marroquin,[4] O S Kirsebom,[3] M V Lund,[3] I Moore,[5] A Perea,[4] J Refsgaard,[3] J E Riley,[1] S Rinta-Antila,[5] L Sinclair,[1] O Tengblad[4] and The IGISOL Collaboration[5]

[1] Department of Physics, University of York, YO10 5DD, UK
[2] Institut Pluridisciplinaire Hubert Curien, 67037 Strasbourg Cedex 2, France
[3] Department of Physics and Astronomy, University of Aarhus, 8000 Arhus C, Denmark
[4] Instituto Estructura de la Materia, CSIC, Serrano 113bis, E-28006, Madrid, Spain
[5] Department of Physics, University of Jyväskylä, FIN-40351 Jyväskylä, Finland
[†] Present address: IKS, Leuven, Belgium

E-mail: *rg847@york.ac.uk

Abstract. ^{12}C is synthesised in stars by fusion of three α particles. This process occurs through a resonance in the ^{12}C nucleus, famously known as the Hoyle state. In this state, the ^{12}C nucleus exists as a cluster of α particles. The state is the band-head for a rotational band with the 2^+ rotational excitation predicted in the energy region 9 - 11 MeV. This rotational excitation can affect the triple-α process reaction rate by more than an order of magnitude at high temperatures (10^9 K). Depending on the energy of the resonance, the knowledge of the state can also help determine the structure of the Hoyle state. In the work presented here, the state of interest is populated by beta decay of radioactive ^{12}N ion beam delivered by the IGISOL facility at JYFL, Jyväskylä.

1. Introduction

The ^{12}C nucleus has a cluster structure in the Hoyle state, which is a 0^+ resonance just above the triple-α threshold. With the non-spherical structure of the state, ^{12}C has a rotational band built upon this state. A recent measurement by Zimmerman *et al.* [1], along with the other observations [2–5] suggest that the first excitation of this band, the 2^+ state, is to be found in 9 – 11 MeV energy region. The second and higher lying member of this rotational band, 4^+ has been found experimentally [6], but the 2^+ excitation has proven difficult to study because of the the presence of a broad 0^+ resonance at 10.3 MeV [4, 7].

At higher temperature in stars (10^9 K), this rotational excitation state becomes relevant to the triple-α process, affecting the reaction rate. In addition to its importance in determining the ^{12}C production rate, the state is also important when considering the structure. With the energy of this first rotational excitation known, the determined moment of inertia can settle the debate on the structure of the ^{12}C nucleus in the Hoyle state as different theoretical models predict different arrangements of the cluster in the Hoyle state [8].

The experiment was performed at JYFL, Jyväskylä, Finland. The IGISOL (Ion-Guide Isotope Separator OnLine) facility delivered a beam of the radioisotope ^{12}N which β decays to ^{12}C, populating the states of interest. Due to the selection rules of β decay, this method gives some selectivity over the spin-parity of states populated. From the parent nucleus, ^{12}N, with 1^+ spin-parity ground state, only 0^+, 1^+ and 2^+ states are populated in the daughter nucleus, ^{12}C. This prevents the strong states, such as the 3^- at 9.6 MeV to affect the spectra.

2. ^{12}N beam production and detector arrangement

The IGISOL facility at JYFL uses the ion-guide method to produce radioactice ion beams. In this technique, the ions are produced via a nuclear reaction of an energetic beam on a target foil. The ions recoil out of the target and are transported to the extraction electrode via a buffer gas through a skimmer. After the extraction, they are mass separated using a magnetic field.

The beam is stopped at a thin carbon foil in the experimental chamber. In total, 5.9% of the decays of ^{12}N populate states above the triple-α threshold [9]. Such decays are typically followed by the breakup of ^{12}C into three α particles which were detected using an array of detectors. Figure 1 shows the cubic arrangement of six DSSDs (Double-sided Silicon Strip Detectors) around the foil. These detectors are made from 50mm$\times$50mm silicon wafers, segmented into 16 strips, both on front and back. Five of these (except the one on the outside), are thin detectors to detect α-particles. The sixth is a thick detector for detecting β-particles. One HPGe (High Purity Germanium) detector was also used for the detection of γ-particles.

3. Analysis and Results

3.1. Energy calibrations

Energy calibrations were carried out using the radioactive sources, ^{239}Pu, ^{241}Am and ^{244}Cm, which emit α-particles of known energies, 5157, 5486 and 5805 keV respectively. For calibrating the thick backing detector and Ge detector, the electron source ^{207}Bi and the γ-ray source ^{60}Co were used respectively. In order to extrapolate these calibrations to the low energy region, two main corrections are to applied. They are explained in the following.

3.1.1. Calibration offsets and non-linearity for beta particle detection: The Compton scattered events between the thick DSSD and the HPGe detector have been used to check the linearity of amplification gains as a function of energy. For this, it is required that the ^{60}Co source is placed such that the radiation has to pass through the Si detector to reach the Ge detector. γ particles that are scattered from Si at a low angle before reaching the Ge detector deposit part of their energy in both detectors. Such events can be seen as two diagonal lines in Figure 2, which shows the energy deposited in Si detector plotted against the energy deposited in HPGe (similar to work in Ref [10]).The intercepts made by these lines on x axis are 5 -10 keV off from 1173 and 1332 keV (^{60}Co γ rays energies), indicating offset in calibrations of Si detector towards the lower energy. The pulser data also indicates slight non-linearity in the calibrations. The correction will be done using the pulser data and cross-checked against the ^{60}Co data as in Figure 2.

3.1.2. Detector dead-layer and source thickness effects: The dead layer on the surface of the DSSDs vary from 50 - 500 nm depending upon the detector. α particles lose

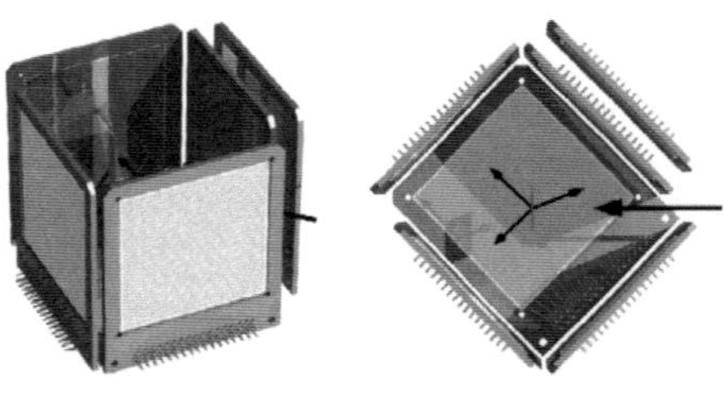

Figure 1: 3D model of DSSD arrangement. (Side view and top view respectively)

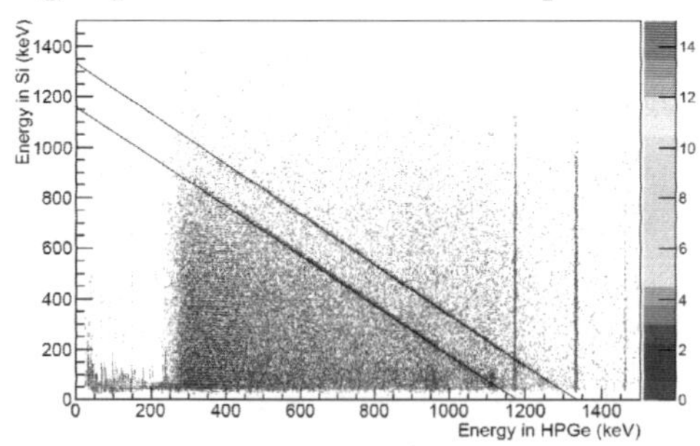

Figure 2: γ-ray energy deposited in DSSD vs HPGe.

20 - 160 keV energy in the dead-layer before reaching the active material of the detector, with the loss being a nonlinear function of energy. Similarly, the α particles lose energy while exiting the source. Corrections of these effects have not yet been fully applied in Figure 3, which explains the energy offset of 200 keV for the 12.7 MeV state.

3.2. β-triple-α coincidences

To begin with the analysis on data, we identify the ^{12}C state's energy and constrain the spin-parity. The excitation energy is reconstructed by adding the energies of all three α-particles to the triple-α threshold energy (7.275 MeV). To constrain the spin-parity, the events are gated on the breakup channel. ^{12}C breakup can either be sequential, or it can be a three-body breakup. The sequential breakup involves the ^{8}Be nucleus via either the 0^+ ground state or the 2^+ excited state. 0^+ and 2^+ states predominantly breakup through the ^{8}Be ground state. In such breakups, the first α-particle takes approximately two-thirds of the excitation energy above the threshold. In Figure 3, the diagonal line with the slope 2/3 corresponds to the first α-particle from such events. This channel is prohibited for the 1^+ state. The selection on the ^{8}Be (g.s.) channel therefore cleans the data of 1^+ component.

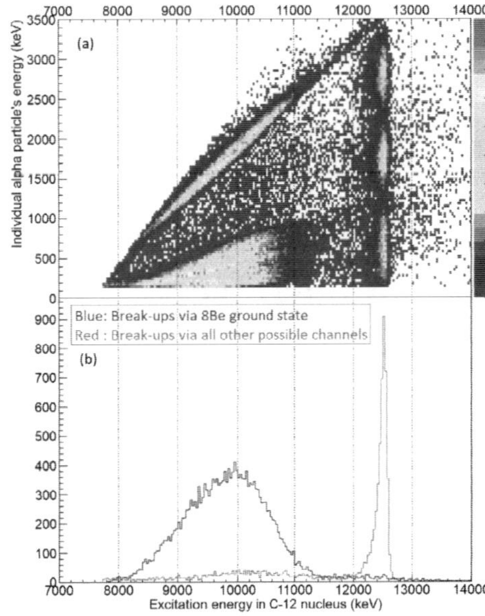

Figure 3: (a) Individual α-particle energy against excitation energy. (b) Projection of the above, gated on breakup channels.

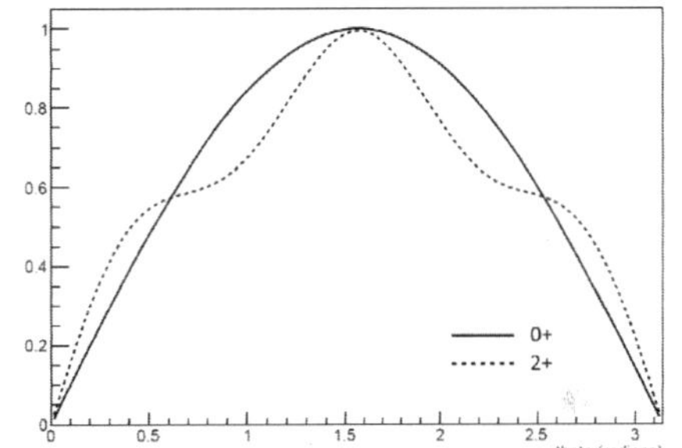

Figure 4: Predicted angular correlation of first α-particle and the β-particle.

3.3. Future work on β-α angular correlation

To discriminate between the 2^+ and 0^+ component, we will use of the angular correlation between the β-particle and the first α-particle from the breakup. If the emitted β-particle in a decay is relativistic, the daughter nucleus is left in a polarised state along the axis defined by the β-particle emission direction. This results in an angular correlation between subsequent radiation and the β-particle. With the high Q-value (16.316 MeV [2]) of β-decay of ^{12}N, the relativistic assumption is applicable here. Figure 4 shows the difference in angular correlations for 0^+ and 2^+ states. Work on the detailed correlation analysis is ongoing. The data shown in Figure 3 is β-triple-α coincidence data, and will be used in correlation study.

References

[1] Zimmerman W R *et al.* 2013 *Phys. Rev. Lett.* **110** 0152502
[2] Fynbo H O U and Diget C A 2014 *Hyperfine interactions* **223** 1–3
[3] Hyldegaard S *et al.* 2010 *Phys. Rev. C* **81** 024303
[4] Itoh M *et al.* 2011 *Phys. Rev. C* **84** 054308
[5] Freer M *et al.* 2009 *Phys. Rev. C* **80** 041303
[6] Ogloblin A *et al.* 2014 *EPJ Web of Conferences* **66** 02074
[7] Ajzenberg-Selove F 1990 *Nucl. Phys. A.* **506**
[8] Epelbaum E *et al.* 2012 *Phys. Rev. Lett.* **109** 252501
[9] Hyldegaard S *et al.* 2009 *Phys. Rev. C.* **80** 044304
[10] Bildstein V *et al.* 2012 *Eur. Phys. J. A* **48** 85

IOP Conf. Series: Journal of Physics: Conf. Series **940** (2017) 012044 doi:10.1088/1742-6596/940/1/012044

Maxwellian Neutron Spectrum generation and Stellar Cross-Section measurements: measurement of the $^{197}Au(n,\gamma)$ MACS.

P Jiménez-Bonilla[1,3,*], J Praena[2,3], J M Quesada[1]

[1]Atomic, Molecular and Nuclear Physics Department. University of Seville, Spain.
[2]Atomic, Molecular and Nuclear Physics Department. University of Granada, Spain.
[3]Centro Nacional de Aceleradores (US-JA-CSIC), Spain.

E-mail: jimenezbpablo@gmail.com

Abstract. Maxwellian-averaged cross-sections (MACS) are needed as an input for the models of stellar s- and r-processes nucleosynthesis. MACS can be obtained from activation measurements, irradiating a sample with the neutron field generated by the $^{7}Li(p,n)^{7}Be$ reaction at 1912 keV proton energy. At this energy, the neutron energy spectrum is close ($R^2 \leq 0.9$) to a Maxwellian one of kT=25 keV. However, it was shown that shaping the energy of the incident proton beam is possible to generate a neutron field with an energy spectrum much closer to a real Maxwellian ($R^2 > 0.995$), therefore avoiding or minimizing corrections in the MACS calculation. We show a preliminary result of an experiment performed at JRC-IRMM (Geel) to confirm our method. We have measured the $MACS_{30}$ (kT=30 keV) of the $^{197}Au(n,\gamma)$ reaction, at CNA (Seville). We obtained 612 mb, in good agreement with the latest measurements.

1. Introduction

Neutron capture is the main process responsible for the nucleosynthesis of the major part of the elements beyond iron [1]. In the helium-burning phase in red giants stars, neutrons are produced via (α,n) reactions. These neutrons are quickly thermalized through elastic scattering, and their velocities are represented by a Maxwell-Boltzmann distribution (or Maxwellian in unit of energy). A temperature of $T=3.48\cdot10^6$ K, typical of He-burning in red giants stars, corresponds to a Maxwellian distribution with kT=30 keV. The Maxwellian-averaged cross-section (MACS) or stellar cross-section of the involved isotopes is a key parameter for modeling the stellar sites. The MACS is defined as the ratio between $\langle\sigma v\rangle$ (reaction rate per particle pair) and v_T (particle most probably thermal velocity for a temperature T). Its expression is showed in eq. 1.

$$MACS \equiv \langle\sigma\rangle = {\langle\sigma v\rangle}/{v_T} = \frac{2}{\sqrt{\pi}}\cdot\frac{1}{(kT)^2}\cdot\int_0^\infty \sigma(E)\cdot E\cdot e^{-\frac{E}{kT}}dE \quad . \qquad \text{(Eq. 1)}$$

MACS can be obtained by time-of-flight (TOF) and activation techniques. In case of activation, a neutron flux with an energy spectrum close to a Maxwellian at 25 keV ($MACS_{25}$) ($\approx$ 90% area

---

* To whom any correspondence should be addressed.

overlapping) can be generated using the $^{7}Li(p,n)^{7}Be$ near the threshold [2,3]. To obtain the MACS, the experimentally measured cross-section value must be corrected for spectrum difference. In addition to this, if the $MACS_{30}$ is needed an extrapolation from the $MACS_{25}$ is necessary. For both, correction and extrapolation, the $\sigma(E)$ knowledge or assumption is mandatory (including resonances). However, it was shown [4] that it's possible to generate a more realistic (>99% area overlapping) Maxwellian, avoiding $\sigma(E)$ assumption and minimizing the corrections.

2. Maxwellian Neutron Spectrum generation method: experiment and simulations.

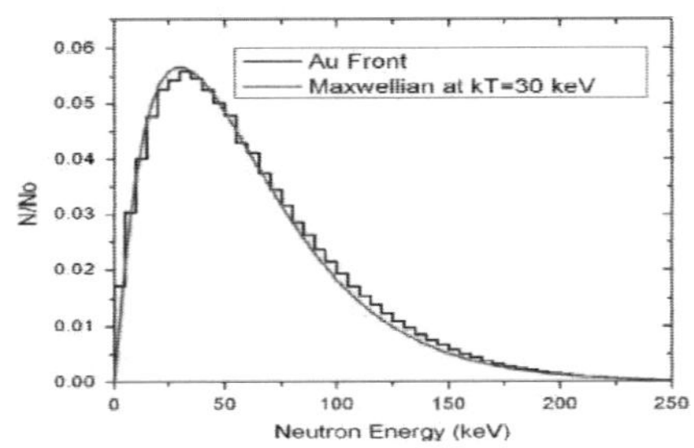

Figure 1. Simulated neutron spectrum for a GPD (Ec=1851 keV, FWHM=154 keV) on Li target.

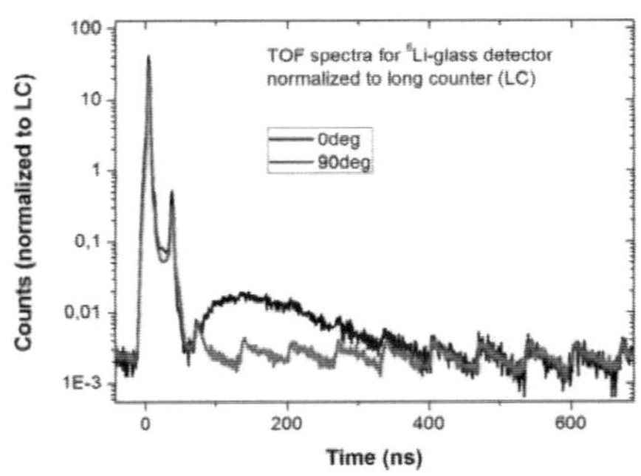

Figure 2. Normalized TOF spectra at 0° (black) and 90° (red).

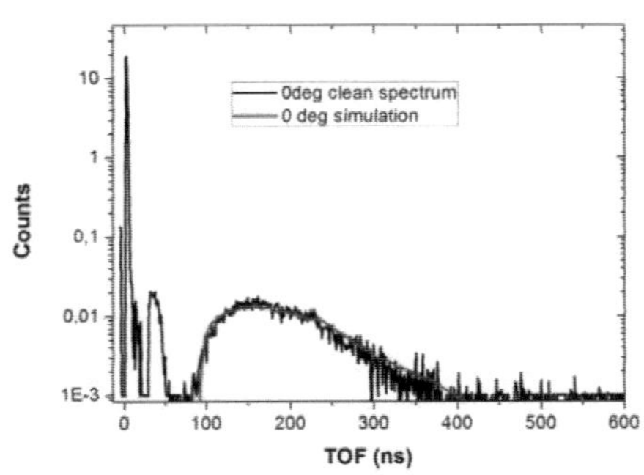

Figure 3. 0° TOF spectrum compared to SIMPRO procedure.

Our method for the production of Maxwellian Neutron Spectra is based on the shaping of the proton beam to a Gaussian-like distribution (GPD), introducing a foil degrader upstream of the lithium target [4]. The method was checked in [5,6]. Different combinations of proton energy and degrader material and thickness are suitable for generating Maxwellian-30. The GPDs were measured [5] and neutron spectra were obtained by means of analytical descriptions of the differential neutron yield in angle and energy of the $^{7}Li(p,n)$ reaction and Monte Carlo (MC) simulations of the neutron transport through the experimental setup (SIMPRO procedure) [5,6]. Fig.1 shows the neutron spectrum integrated over the Au sample (see Section 3) obtained with SIMPRO and compared to Maxwellian-30.

For an experimental confirmation of our method, we carried out an experiment at Van de Graaff at JRC-IRMM, briefly described here; more details will be given in a forthcoming paper. The proton beam energy was 3.7 MeV and the Al degrader 75 μm in thickness. We used the usual setup at IRMM for neutron spectrometry by time-of-flight (TOF) technique in this energy range that consisted of 625 kHz and 1-2 ns pulsed proton beam, with ^{6}Li-glass detectors located at 52 cm, more details of the electronic chain, detectors and materials can be found for similar experiments at IRMM [7].

Our experiment suffered from a contamination due to the proton beam halo at the accelerator source. At source, the beam was chopped by means of deflection plates sweeping it across a small aperture. When the proton beam was not passing across the aperture, a small part of the halo passed through it and was bunched. This effect is clearly shown in Fig.2, where the black line corresponds to TOF spectrum at 0° and red line at 90° (same detector). The highest peak in both spectra (t=0-50 ns) corresponds to prompt γ rays (so-called γ-flash) from reactions in the Al degrader, LiF (thick) target and Cu backing. Then, for higher times, neutrons are detected at 0° (from ≈80-400 ns) and not at 90°. The lack of separation between the γ-peak and the neutron "panettone" is due to the beam halo, producing a larger γ-peak in time than usual proton beam with no halo, and the high energy neutrons (up to 250 keV or 76 ns). Since the system was based on the pulse-height information of the detector, it was not possible to distinguish the γ-n contamination by combined pulse-shape analysis and TOF information. However, the neutron contamination is three orders of magnitude lower than the γ one, the same ratio than γ-flash and neutron "panettone" in Fig.1, therefore a reasonable clean TOF spectrum can be obtained by the subtraction of the 90° to 0° spectrum, both normalized to a long counter used as neutron monitoring. Fig.3 shows a clean 0° TOF spectrum compared to a time spectrum obtained with our SIMPRO procedure. In this case, SIMPRO consists of the inclusion in a MCNP file of the analytical description of the differential neutron yield in angle and energy of the ^{7}Li(p,n) for GPD (Ec=1851 keV, FWHM=154 keV). The MCNP file contains a detailed description of the ^{6}Li-glass detector at 52 cm from LiF target and Cu backing. Then, the time spectrum is obtained by means of Tally Time Card and a Tally Multiplier Card for the ^{6}Li(n,t)^{4}He reaction on the cell which contained the ^{6}Li-glass. The good agreement between the clean TOF spectrum and the SIMPRO suggests a preliminary experimental confirmation of our method and to continue cleaning all the measured angles.

3. ^{197}Au(n,γ) Stellar Cross Section measurement at kT=30 keV.

Gold is not a standard in the astrophysical energy range. However, it is commonly used as reference for neutron-capture cross-section measurements with $MACS_{30}$=582±9 mb[8]. A large effort has been made to solve the $MACS_{30}$ discrepancies between different experiments and techniques [8], in particular between TOF ones and the activation of [3]. Recently, we carried out the experiment of [3], quasi-Maxwellian, with Au flat sample and we obtained a value slightly higher, 626±25 mb [9]. The main difference between $MACS_{30}$ value in [3] and [9] is associated to the use of MC simulations for the neutron scattering correction. In [8] a new recommend value is given, 613±7 mb. Here, we measure the $MACS_{30}$ with our method for Maxwellian-30 generation by means of GPD. Our measurement was carried out at the 3MV Tandem accelerator at CNA (Seville) with the same setup and procedure detailed in [5],[6] and [9]. Irradiation time was 210 min and measured time was 13h. A complete MCNPX simulation (gammas and neutron transport) of the setup and HPGe was carried out for the needed corrections [9]. Eq. 2, 3 and table 1 indicate the measured values and corrections. The obtained result for the ^{197}Au(n,γ) $MACS_{30}$ is 612 ± 20 mb, very similar to our previous measurement [9] and the new recommend value 613±7 mb [8].

Table 1. Measured values and corrections.

	Value	ε (%)	
C_{Au}	64100	0.7	Gamma counts.
C_{Be}	545800	0.3	Gamma counts.
$I_{\gamma Au}$	95.54	0.1	Gamma intensity
$I_{\gamma Be}$	10.45	0.04	Gamma intensity
N_{Au}	6.4E-4	1.5	Mass thickness.
f_{dAu}	0.1281	0.01	Decay correction.
f_{dB}	0.0070	0.1	Decay correction.
$\varepsilon_{Be/Au}$	0.80	1.9	Gamma efficiency.
K_n	1.75	1.3	Neutron flux corr.
K_{macs}	1.01	1.6	Maxwellian correct.

$$MACS = \frac{2}{\sqrt{\pi}} \cdot K_{macs} \cdot \sigma_{exp} \quad \text{(Eq. 2)}$$

$$\sigma_{exp} = \frac{1}{N_{Au}} \cdot \frac{A_{Au}}{\Phi} = \frac{1}{N_{Au}} \cdot \frac{C_{Au}}{C_{Be}} \cdot \frac{[f_d \cdot I_\gamma \cdot \varepsilon]_{Be}}{[f_d \cdot I_\gamma \cdot \varepsilon]_{Au}} \cdot \frac{1}{K_n} \quad \text{(Eq.3)}$$

References

[1] E.M.Burbidge, G.R.Burbidge, W.A.Fowler, F.Hoyle, Rev.Mod.Phys.,v 29, n 4, 547-650 (1957).
[2] H. Beer, F. Käppeler. Phys. Rev. C, vol. 21, nº 2, p.534–544 (1980).
[3] W. Ratynski, F. Käppeler, Phys. Rev. C, 37 (1988) 595.
[4] P. F. Mastinu, G. Martín-Hernández and J. Praena. Nucl. Inst. and Meth. A 601 (2009) 333-338.
[5] J. Praena, P.F. Mastinu, M. Pignatari *et al.,* Nucl. Inst. and Meth. A, 727 1-6 (2013).
[6] J. Praena, P.F. Mastinu, R. Capote *et al.,* Nuc. Data Sheets 120 (2014) 205–207.
[7] G. Feinberg, M. Friedman, A. Krása , *et al.,* Phys. Rev. C, 85, 055810 (2012).
[8] I. Dillmann, R.Plag, F.Käppeler, A.Mengoni, C.Heinz, M. Pignatari. "The new KADoNiS v1.0 and its influence on the weak s-process nucleosynthesis". PoS (NIC XIII Conf.) 057 (2014)
[9] P. Jiménez-Bonilla and J. Praena. Proceedings of Science PoS (NIC XIII) 102 (2014).

Nucleosynthesis of intermediate mass stars: inferences from the observed abundances in photoionized nebulae of the Local Group

W. J. Maciel[1], **R. D. D. Costa**[1], **O. Cavichia**[2]

[1] University of São Paulo, Rua do Matão 1226, São Paulo, Brazil
[2] UNIFEI, Av. BPS, 1303, Itajubá, Brazil

E-mail: wjmaciel@iag.usp.br

Abstract. Photoionized nebulae, comprising HII regions and planetary nebulae, are excellent laboratories to investigate the nucleosynthesis and chemical evolution of several elements in the Galaxy and other galaxies of the Local Group. Our purpose in this investigation is threefold: (i) to compare the abundances of HII regions and planetary nebulae in each system in order to investigate the differences derived from the age and origin of these objects, (ii) to compare the chemical evolution in different systems, such as the Milky Way, the Magellanic Clouds, and other galaxies of the Local Group, and (iii) to investigate to what extent the nucleosynthesis contributions from the progenitor stars affect the observed abundances in planetary nebulae, especially for oxygen and neon, which places constraints on the amount of these elements that can be produced by intermediate mass stars.

1. Introduction

Planetary nebulae (PN) have strong emission lines of H, He, O, N, Ne, S, Ar, etc., including forbidden lines and recombination lines. The analysis of these lines gives abundances accurate to about 0.2 dex or better. The abundances include elements that are probably not significantly produced by the progenitor stars (O, S, Ne, Ar), and therefore contribute to the study of the chemical evolution of the host galaxies. They also provide accurate abundances of elements that are produced by the progenitor stars (He, N, C), so that they are also useful to study nucleosynthesis processes in intermediate mass stars. Some of these elements are difficult to study in stars, and better observed in photoionized nebulae, which comprise PN and HII regions. Blue Compact Galaxies (BCG) and Emission Line Galaxies (ELG) can also be included as low metallicity HII regions. In this work, we compare the observed abundances of these elements in PN and HII regions in several galaxies in the Local Group.

2. The data

The data used in this investigation include a large sample of PN and HII regions in the following galaxies: The Milky Way (MW), the Large Magellanic Cloud (LMC), the Small Magellanic Cloud (SMC), M31, M32, M33, M51, M81, M101, NGC 185, NGC 105, NGC 300, NGC 628, NGC 3109, NGC 5194, and the Sextans galaxy. The total PN sample includes about 1200 objects, while the HII region sample has about 500 objects, so that a reasonably large sample of Local Group objects is considered. The data sources are given by Maciel et al. (2014) [1], with recent data by Stanghellini et al. (2014) [2], Reyes et al. (2015) [3], Berg et al. (2015) [4], and Croxall et al. (2015) [5].

3. Results

3.1 Elements produced by the PN progenitor stars

In this work, we have stressed the elements that are not expected to be substantially produced by the PN progenitor stars, so that we will present a general outline of the elements that are affected by the progenitor star evolution, such as He and N. The abundances of these elements are particularly changed by the dredge-up processes that occur in the intermediate mass stars.

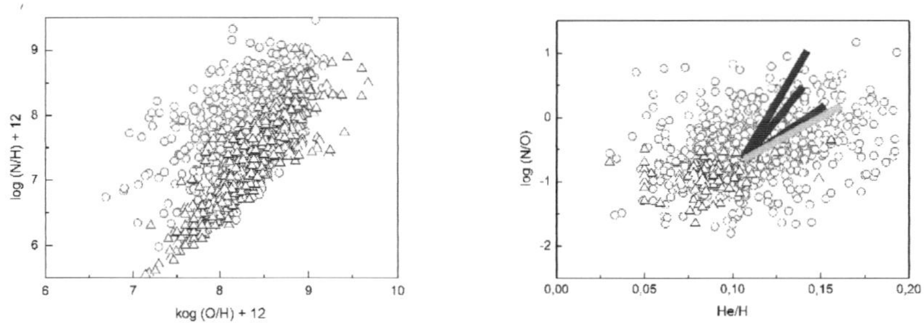

Figure 1. Nitrogen and helium abundances

Figure 1 shows the N/H and N/O ratios as functions of O/H and He/H, respectively, where the empty red circles refer to PN, and the blue triangles are HII regions in the Local Group. As expected, PN show an increase in both N and He compared to most HII regions in the sample. It can be seen that for these elements some dispersion is observed even for HII regions, so that part of the nitrogen is probably secondary. The excess nitrogen in PN is essentially produced by their progenitor stars, in a strong contrast with the remaining elements studied here, as we will see in the next section. The green and blue lines in the second plot are from theoretical models by Marigo et al. (2003) [6] with $Z = 0.019$, and Karakas (2010) [7], with $Z = 0.02$, 0.004, 0.008, respectively. These are synthetic evolutionary models for thermally-pulsing AGB stars with initial masses of 1 to 6 solar masses, in which up to three dredge-up episodes occur, apart from hot-bottom processes (HBB) for the most massive objects. According to these models, progenitors having 0.9 to 4 solar masses and solar composition can explain the "normal" abundances, $He/H < 0.15$, while for objects with higher enhancements ($He/H > 0.15$), masses of 4 to 5 solar masses are needed, plus an efficient HBB. For intermediate mass stars, agreement with theoretical models is fair, but abundance determinations should be improved and expanded.

3.2 Elements not produced by the PN progenitor stars

The abundances of the elements O, Ne, S, and Ar are probably not significantly affected by the evolution of the PN progenitor stars, and the observed correlations are much better determined. The measured abundances therefore reflect the interstellar abundances at the time the progenitor stars were born. Oxygen can be used as a metallicity proxy, and an accurate relation between O and Fe can be determined (see for example Ramirez et al. 2013) [8]. This is important, since in photoionized nebulae Fe is mostly locked up in grains. The average slope of the Fe – O correlation is $\delta = 1.11$ (thin disk) and $\delta = 1.31$ (thick disk). Here we concentrate on distance-independent correlations, to avoid distance determination problems, which affect processes such as the abundance gradients observed in the Milky Way and other spirals.

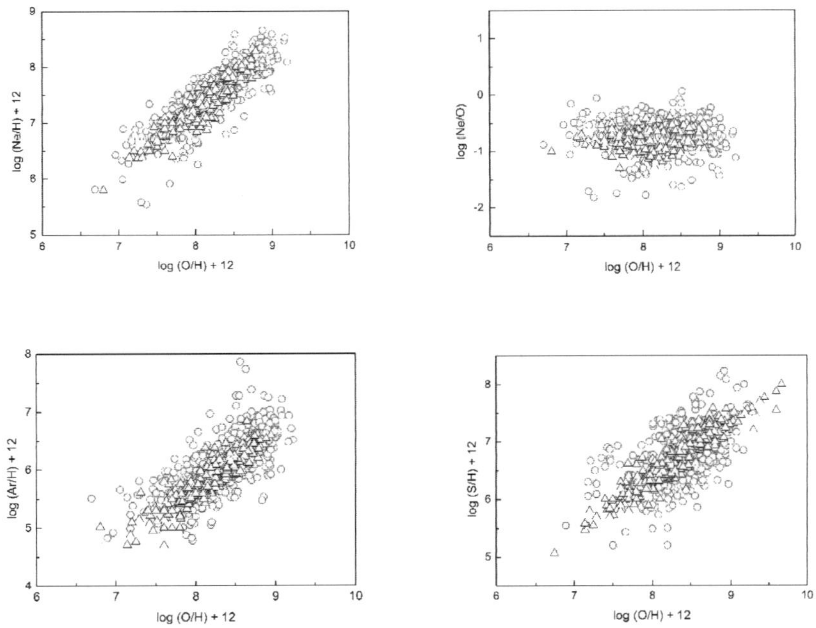

Figure 2. Abundances of Ne, Ar, and S

The main results are shown in Figure 2 for Neon, Argon, and Sulphur. Again, the empty red circles are Local Group PN, while the blue triangles represent HII regions and BCG. PN abundances of O, Ne, S, and Ar show good correlations, indicating that Ne, S, and Ar vary in lockstep with O. The most accurate result is for neon (figure 2a), which does not suggest any important contribution from the progenitor stars, as the PN data dispersion is essentially constant at all metallicities. The abundances relative to oxygen are essentially constant, which can be seen in figure 2b for neon. Argon presents similar results as neon, with a somewhat higher dispersion, probably due to the weakness of the Ar lines (figure 2c). Sulphur abundances in PN may present the "sulphur anomaly", that is, lower abundances than expected at a given metallicity. This is probably due to incorrect ICFs (ionization correction factors), but is not apparent in the bulk of Local Group objects, as can be seen from figure 2d, where a particularly extended metallicity range can be observed. All correlations for HII regions show smaller dispersions, as expected, a result that is not affected by the inclusion of either BCG or ELG. The same trends are observed in both types of photoionized nebulae. The dispersion observed in PN is probably real, and reflects the different ages of the progenitor stars, but it should be noted that part of the dispersion may be due to the different abundance sources considered.

Acknowledgements. This work was partially supported by CNPq and FAPESP.

References

[1] Maciel W J, Costa R D D and Cavichia O 2014 *PoS* (NIC XIII) 030
[2] Stanghellini L, Magrini L, Cassola V and Villaver E 2014 *A&A* **567** A88
[3] Reyes, R E C *et al.* 2015 *RMAA* **51** 133
[4] Berg D A *et al.* 2015 *ApJ* **806** 16
[5] Croxall K V *et al.* 2015 *ApJ* **808** 42
[6] Marigo P *et al* 2003 *A&A* **409** 619
[7] Karakas A 2010 *MNRAS* **403** 1413
[8] Ramirez I, Allende-Prieto C and Lambert D L 2013 *ApJ* **764** 78

IOP Conf. Series: Journal of Physics: Conf. Series **940** (2017) 012046 doi:10.1088/1742-6596/940/1/012046

Effectiveness of using a magnetic spectrograph with the Trojan Horse method

S. Manwell[1], A. Parikh[2], A. A. Chen[1], N. de Séréville[3], P. Adsley[4], D. Irvine[1], F. Hammache[3], I. Stefan[3], R. F. Longland[5], J. Tomlinson[6], P. Morfuace[3] and B. Le Crom[3]

[1] Department of Physics and Astronomy, McMaster University, Hamilton L8S 4M1, CA
[2] Departament de Física i Enginyeria Nuclear, Universitat Politècnica de Catalunya, Barcelona 08034, ES
[3] Institut de Physique Nucléaire d'Orsay, Orsay 91400, FR
[4] Department of Physics, Stellenbosch University, Stellenbosch 7602, ZA
[5] Department of Physics, North Carolina State University, Raleigh 27695-8202, US
[6] Department of Physics, University of York, York YO10 5DD, UK

E-mail: S. Manwell: spencer.manwell@gmail.com, A. Parikh: xrayburst@gmail.com, A. A. Chen: chenal@mcmaster.ca, N. de Séréville: deserevi@ipno.in2p3.fr

Abstract. The Trojan Horse method relies on performing reactions in a specific kinematic phase space that maximizes contributions of a quasi-free reaction mechanism. The hallmark of this method is that the incident particle can be accelerated to high enough energies to overcome the Coulomb barrier of the target, but once inside the target nucleus the relative motion of the clustered nuclei allows the reaction of interest to proceed at energies below this Coulomb Barrier. This method allows the experimentalist to probe reactions that have significance in astrophysics at low reaction energies that would otherwise be impossible due to the vanishing cross section. Traditionally the Trojan Horse method has been applied with the use of silicon detectors to observe the reaction products. In this study we apply the Trojan Horse method to a well studied reaction to examine the potential benefits of using a splitpole magnetic spectrograph to detect one of the reaction products. We have measure the three body ^{7}Li(d,αn)α reaction to constrain the energy ^{7}Li(p,α)α cross section. Measurements were first made using two silicon detectors, and then by replacing one detector with the magnetic spectrograph. The experimental design, limitations, and early results are discussed.

1. Introduction

The Trojan Horse method (THM) is an indirect approach to estimating reaction cross sections at energies below the Coulomb Barrier. This technique has been the topic of many publications covering the method's theoretical and experimental attributes, e.g. Refs. [1-5]. Here we apply the THM to a well studied reaction that can be used to estimate the low energy cross section of the astrophysically important ^{7}Li(p,α)α reaction [1]. We are interested in the quasi-free (QF) ^{7}Li(d,αn)α reaction that will be used to determine a relative cross section the 2-body ^{7}Li(p,α)α reaction. The THM usually uses of silicon detectors to measure reaction ejectiles in coincidence. Our aim is to evaluate potential benefits of measuring the emerging nuclei with a combination of a magnetic spectrograph and one silicon detector, rather than exclusively using silicon detectors.

2. Experimental Setup

In this study, performed at the IPN-Orsay Tandem Accelerator, we bombarded a 200 $\mu g/cm^2$ CD_2 target with a 20 MeV ^7Li beam with currents between 10 and 150 enA. We measured outgoing α-nuclei under strict kinematic conditions that make the QF reaction mechanism favourable [1]. Initially we used two silicon detectors and then later replaced one with the Splitpole magnetic spectrograph. The two silicon detectors were placed at $(45.3 \pm 0.7)°$ and $(44 \pm 1)°$. The latter was replaced by the magnetic spectrograph which was placed at $(45.55 \pm 0.5)°$; note that all angles are relative to the beam axis. A set of collimators were applied to the silicon detectors during both arrangements that were used to constrain the acceptance angle of each. Energy calibration of the silicon detectors was performed using a triple-α source and two additional α energy lines retrieved by the position sensitive detector within the spectrograph after gating on α-nuclei. This latter approach served to provide the energy calibration of the spectrograph itself.

3. Kinematic Data Selection

QF yield is maximized by constraining the relative angle (θ_{12}) of the emerging alpha particles, α_1 and α_2, to be 90°. This angular separation was used to determine the detector placements during both phases of the measurement. Constraining θ_{12} also helps ensure that the energy given to the spectator (neutron) in the reaction will be minimized. Despite the careful selection of the QF reaction mechanism the sequential decay (SD) of $^8Be^*$ is expected to have a strong contribution in this kinematic phase space. Kinematic calculations revealed that low spectator energy corresponds to α-α coincidences in an arc with α energies between 14 and 21 MeV. Our measurement agrees with kinematic calculations which can be seen in Figures 1 and 2. The data that were selected for further analysis, i.e. those corresponding to the QF mechanism, are shown in red.

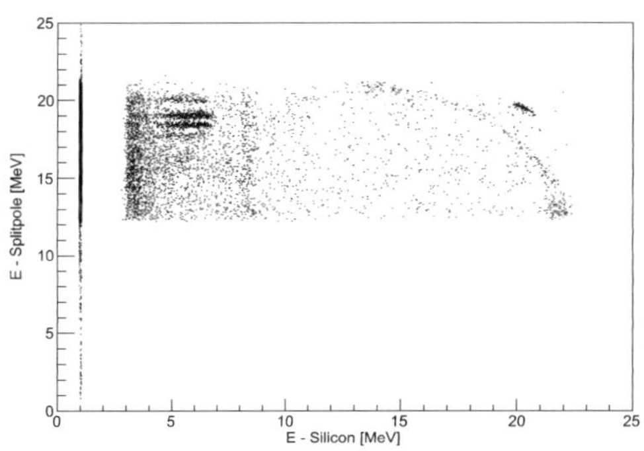

Figure 1. Two dimensional energy spectrum of α_1-α_2 coincidences in the silicon-spectrograph setup.

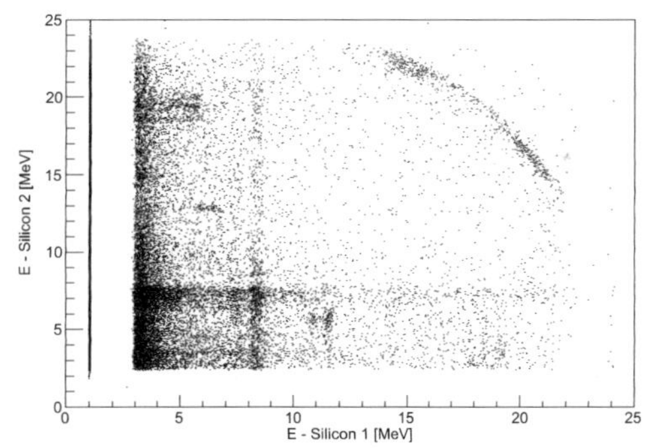

Figure 2. Two dimensional energy spectrum of α_1-α_2 coincidences in the silicon-silicon setup.

According to the results of Ref. [1], the QF mechanism is expected to dominate the reaction yield for spectator momenta less that 40 MeV/c, whereas SD events become preferred when that limit is exceeded. Figures 3 and 4 show the spectator momenta for the two setups. The relative energy in the QF entrance channel by: $E_{7Lip} = E_{\alpha1\alpha2} - Q_2$ [7] where $E_{\alpha1\alpha2}$ is the relative energy of the emerging α nuclei and Q_2 is the Q-value of the two-body reaction of interest. In this arrangement δE_{7Lip} has a strong dependence on the uncertainty in the relative angle of the emerging α nuclei. In the silicon-silicon and silicon-spectrograph setups the relative angles were $\theta_{12} = (89 \pm 1)°$ and $(90.9 \pm 0.9)°$, respectively. The angular positions of the detectors were limited by uncertainties related to collimator size, direction of the beam, and readability of the ruled angular positions of the silicon detectors within the reaction chamber.

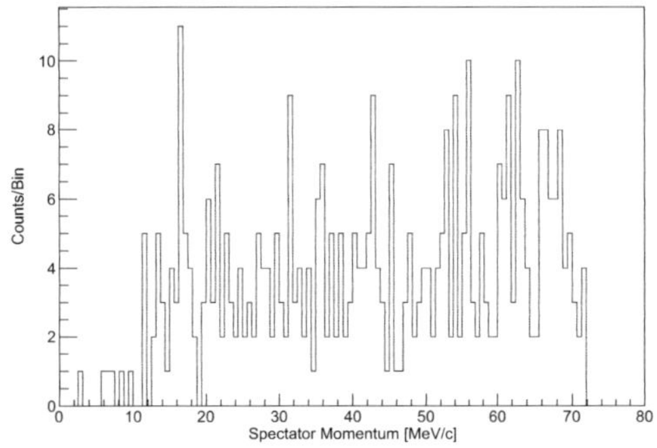

Figure 3. Spectator momentum spectrum from silicon-spectrograph setup.

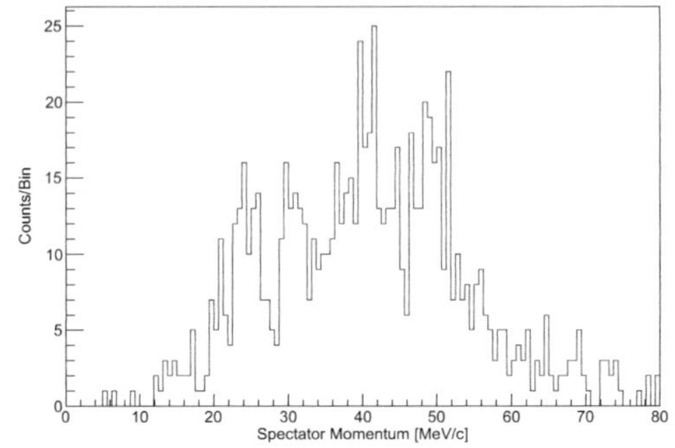

Figure 4. Spectator momentum spectrum from silicon-silicon setup.

4. Conclusion and Outlook

The aim of this work is to evaluate the possible advantage of using a magnetic spectrograph in Trojan Horse studies with respect to traditional silicon detectors. The well-known ^{7}Li(p,α)α reaction studied by the THM in the past [1] has been used as a benchmark for our work. In this work we have used a the Splitpole magnetic spectrograph at the IPN-Orsay Tandem Accelerator facility in Orsay, France. Our analysis is in progress but our early findings are shared here.

Our data show strong kinematic similarities with the data presented in Ref. [1], with the loci shown in our 2-D coincidence spectra and the range of associated spectator momenta both in reasonable agreement with expectations. Work is underway to select QF yields more carefully based on calculated SD contributions. This will allow a more rigorous comparison to the data of [1]. We have found that δE_{cm} has strong dependence on θ_{12}. Although the Splitpole is expected to reduce uncertainty in the energy measurements of emerging nuclei, further measures are needed to constrain $\theta_1 + \theta_2$ to determine the relative angle between the detectors to higher precision. Looking forward, cross section estimates of the ^{7}Li(p,α)α reaction will be determined and normalized to directly measured results [8, 9]. Afterwards, a comparison will be made to the results of reference [1] to gauge any possible advantage to incorporating a magnetic spectrograph in Trojan Horse experiments.

Acknowledgments

We acknowledge the financial support from National Sciences and Engineering Research Council of Canada (NSERC).

References

[1] Spitaleri C, et al., 1999 *Phys. Rev.* C **60** 055802
[2] La Cognata M, et al., 2007 *Phys. Rev.* C **76** 065804
[3] La Cognata M, et al., 2008 *Phys. Rev. Lett.* **101** 152501
[4] Romano S, et al., 2008 *J. Phys.* G: *Nucl. Part. Phys.* **35** 014008
[5] Wen Q -G, et al., 2008 *Phys. Rev.* C **78** 035805
[6] Spitaleri C, et al., 2010 *Nucl. Phys.* **A834** 639
[7] Typel S and Baur G 2003 *Ann. Phys.* **305** 228
[8] Rolfs C and Kavanagh R W 1986 *Nucl. Phys.* **A455** 179
[9] Mani G S, et al., 1964 *Nucl. Phys.* **60** 588

Synthesis of Heavy Elements in the Ejecta of Neutron Star Mergers

Dirk Martin[1,*], **Albino Perego**[1], **Almudena Arcones**[1,2], **Friedrich-Karl Thielemann**[3], **Oleg Korobkin**[4], **Stephan Rosswog**[4]

[1]Institut für Kernphysik, Technische Universität Darmstadt, Schlossgartenstraße 2, Darmstadt, D-64289, Germany

[2]GSI Helmholtzzentrum für Schwerionenforschung GmbH, Planckstr. 1, Darmstadt D-64291, Germany

[3]Department of Physics, University of Basel, Klingelbergstraße 82, 4056, Basel, Switzerland

[4]The Oskar Klein Centre, Department of Astronomy, AlbaNova, Stockholm University, SE-106 91 Stockholm, Sweden

E-mail: `dirk.martin@physik.tu-darmstadt.de`

Abstract. We present a nucleosynthesis study on the neutrino-driven wind after neutron star merger. Post-processing tracers from a hydrodynamical simulation, we determine nucleosynthesis yields that depend on both life time of the massive neutron star and polar angle. In comparison with the dynamic ejecta we find a complementary nucleosynthesis, which can explain the r-process pattern beyond the first r-process peak. Additionally, we predict possible forms of light curves for the detection of a kilonova.

1. Introduction

The merger of two neutron stars has been recognized as a possible site for the rapid neutron capture process (r-process) for a long time [1, 2]. Being a both neutron-rich and explosive environment, even the heaviest elements up to uranium can be produced. Further evidence for the r-process in mergers is provided by the detection of a near-infrared excess in the afterglow several days after a short gamma-ray burst (sGRB) [3, 4], which has been interpreted as a "kilonova" [5]. Since here substantial mass fractions of r-nuclei decay radioactively, the resulting heating rate powers this electromagnetic transient. It can serve as a valuable counterpart [6] in coincidence with detecting gravitational waves (GWs), emitted by such systems.

At least three different kinds of ejected matter can be distinguished [7]: dynamic ejecta due to gravitational torques and hydrodynamical processes, neutrino-driven winds, and viscously driven material by the late-time disk dissolution. In this proceeding, we focus on the neutrino-driven wind and its interplay with the dynamic ejecta.

2. Method

Our nucleosynthesis study is based on the first three-dimensional, long-term ($t_{\mathrm{sim}} \approx 190$ ms), Newtonian simulation of a central massive neutron star (MNS) surrounded by a thick accretion

*DM gratefully acknowledges support by the Nuclear Astrophysics Virtual Institute and the Wilhelm and Else Heraeus Foundation to participate in Nuclear Physics in Astrophysics VII.

disk with an initial mass of $M_{\text{disk}} \approx 0.19 M_\odot$ (for further details see [8]). Neutrino-matter interactions are modeled with a spectral neutrino leakage scheme, taking into account neutrino absorption. To calculate nucleosynthesis yields, we use tracer particles that are advected passively. The tracers are distributed proportional to the density in the beginning. Throughout the simulation, every tracer records the temporal profile of ejected material, in particular density, internal energy, electron fraction and velocity. These data are post-processed with a full nuclear reaction network [9] including neutrino absorption on nucleons. We determine heating rates by computing the energy generation by nuclear reactions, which essentially originates from beta-decays. Over 17000 unbound tracers are used to compute total masses, defined as the sum over the final mass fraction distributions times the mass of all tracers [10].

3. Results

Both amount and composition of matter ejected in the neutrino-driven wind depend on the evolution of the accretion disk and the life time of the MNS. This dependence is explored in Fig. 1, showing the unbound mass as a function of the electron fraction and time.

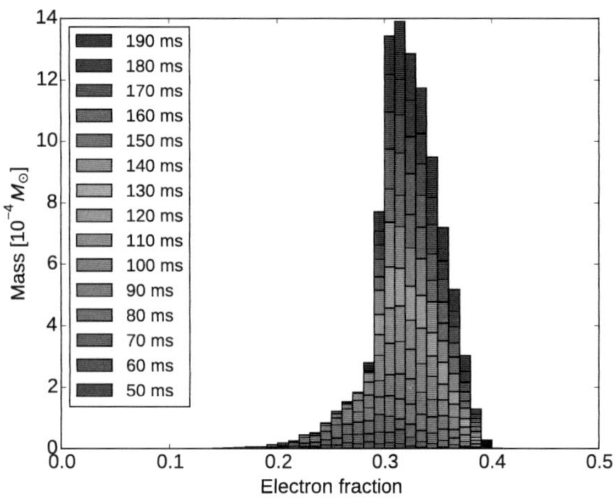

Figure 1. Electron fraction distribution of the ejected mass at different simulation times. Y_e covers values from 0.2 to 0.4 with a final average of $Y_e \sim 0.33$. Very few material with $Y_e < 0.3$ is ejected within the first 100 ms. From 110 ms on, the growth rate of unbound mass becomes roughly constant, as the substantial mass accretion rate inside the disk leads to quasi steady-state conditions with neutrinos as major cooling source. Therefore, neutrino interaction rates tend to approach equilibrium. In total, material in excess of up to $9 \cdot 10^{-3} M_\odot$ ($\sim 5\%$ of M_{disk}) becomes unbound until 190 ms after the merger.

In the neutrino-driven wind mainly nuclei with $A < 130$ are produced, i.e. lighter heavy elements. This distribution is a direct consequence of the key parameter Y_e. The left panel of Fig. 2 shows the total masses for different simulation times (colored lines). Heavy nuclei beyond the second r-process peak ($A \sim 130$) form in nearly negligible amounts at early times only, as it is apparent by the overlap of the three curves. Similar to [11], we find a threshold of $Y_e \approx 0.25$ beyond which no more r-process elements are created. Considering complete mixing with the dynamic ejecta [12] that feature a robust strong r-process due to fission cycling (black line in the right part of Fig. 2), the whole r-process mass fractions are formed from the first peak ($A \sim 80$) to the third peak ($A \sim 195$). The combined pattern is compatible with the solar r-process abundances [13] as well as the abundances in the oldest r-process enriched stars [14].

While full mixing occurs, e.g., on a time scale of star formation, these two types of ejecta are well separated in the direct aftermath of a merger. This initial separation has important implications for the possible detection of a kilonova. Using a semi-analytical model [15], we calculate the resulting combined light curves in four different angular bins (see right panel of Fig. 2). The light curves peak in all directions after ~ 4 h, caused by the decay of nuclei with $A \sim 80$. Furthermore, there is a second peak $3 - 4$ d later in the last two angular bins closest to the disk, originating from larger mass fractions of the heavy r-process ejecta. Due to their relatively high opacities of roughly 10 cm^2/g [16] the medium becomes delayed transparent when the temperature has decreased (while rapidly shifting the light curve to the infrared wavelengths).

IOP Conf. Series: Journal of Physics: Conf. Series **940** (2017) 012047 doi:10.1088/1742-6596/940/1/012047

The shapes of the combined light curves significantly depend on the observation angle. This anisotropy needs to be taken into account when assessing the detectability of optical counterparts for neutron star mergers.

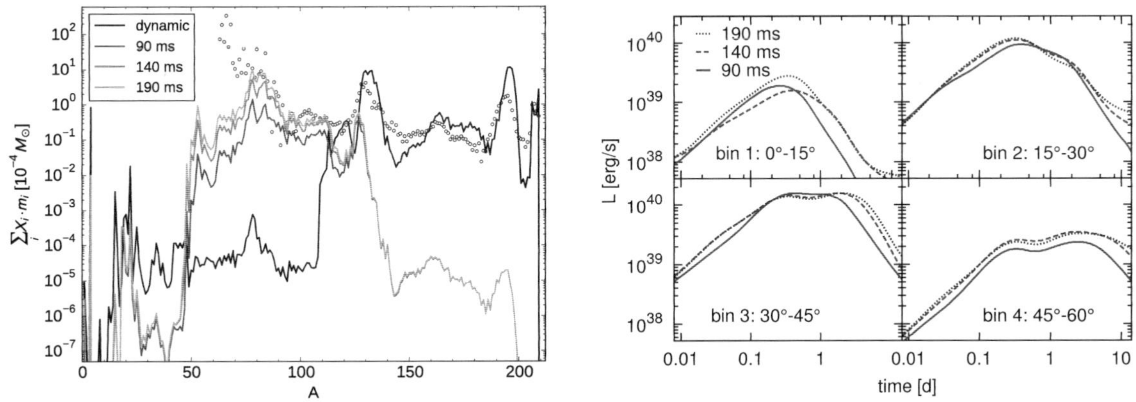

Figure 2. Left: comparison of the nucleosynthesis yields for neutrino-driven wind at three different simulation times and dynamic ejecta. Solar abundances are shown with dots for comparison. Right: contributions of the combined ejecta to the luminosity in the angular bins.

4. Conclusions

We have shown that neutrino-driven wind ejecta play a relevant role in the nucleosynthesis of neutron star mergers and it is crucial to disentangle the interplay of the different ejecta. Further detailed simulations (e.g. in general relativity [17, 18]), observations that constrain r-process abundances, and nuclear physics experiments advancing toward the most neutron-rich nuclei will help to pin down the role of mergers for the origin of the heavy elements in the universe.

References

[1] Lattimer J M and Schramm D N 1974 *ApJ* **192** L145
[2] Eichler D, Livio M, Piran T and Schramm D N 1978 *Nature* **340** 126
[3] Berger E, Fong W and Chornock R 2013 *ApJL* **774** L23
[4] Tanvir N R, Levan A J, Fruchter A S, Hjorth J, Hounsell R A, Wiersema K and Tunnicliffe R L 2013 *Nature* **500** 547
[5] Metzger B D, Martínez-Pinedo G, Darbha S, Quataert E, Arcones A, Kasen D, Thomas R, Nugent P, Panov I V and Zinner N T 2010 *MNRAS* **406** 2650
[6] Nissanke S, Kasliwal M and Georgieva A 2013 *ApJ* **767** 124
[7] Rosswog S 2015 *International Journal of Modern Physics* D **24** 30012
[8] Perego A, Rosswog S, Cabezón R M, Korobkin O, Käppeli R, Arcones A and Liebendörfer M 2014 *MNRAS* **443** 3134
[9] Winteler C, Käppeli R, Perego A, Arcones A, Vasset N, Nishimura N, Liebendörfer M and Thielemann F-K 2012 *ApJ* **750** L22
[10] Martin D, Perego A, Arcones A, Thielemann F-K, Korobkin O and Rosswog S 2015, accepted for publication in *ApJ* (*Preprint* arXiv:1506.05048)
[11] Kasen D, Fernández R and Metzger B D 2015 *MNRAS* **450** 1777
[12] Korobkin O, Rosswog S, Arcones A and Winteler C 2012 *MNRAS* **426** 1940
[13] Lodders K 2003 *ApJ* **591** 1220
[14] Sneden C, Cowan J J and Gallino R 2008 *ARA&A* **46** 241
[15] Grossman D, Korobkin O, Rosswog S and Piran T 2014 *MNRAS* **439** 757
[16] Kasen D, Badnell N R and Barnes J 2013 *ApJ* **774** 25
[17] Wanajo S, Sekiguchi Y, Nishimura N, Kiuchi K, Kyutoku K, Shibata M 2014, *ApJL* **789** L39
[18] Sekiguchi Y, Kiuchi K, Kyutoku K and Shibata M 2015 *Phys. Rev.* D **91** 064059

IOP Conf. Series: Journal of Physics: Conf. Series **940** (2017) 012048 doi:10.1088/1742-6596/940/1/012048

TACTIC: the TRIUMF Annular Chamber for Tracking and Identification of Charged particles

L Martin[1], G Ruprecht[1(3)], U Hager[1(4)], P-A Amaudruz[1], L Buchmann[1], S P Fox[2], A M Laird[2], E Martin[2]

[1]TRIUMF, Vancouver, Canada,
[2]University of York, UK,
[3]Institut für Festkörper-Kernphysik, Berlin, Germany,
[4]Michigan State University, East Lansing, USA
Affiliations at time of research, differing current affiliations given in parentheses

E-mail: LarsMartin1978@gmail.com

Abstract. An in-depth characterization of the TACTIC detector was performed using data from a ^{148}Gd alpha source and some test runs with a stable ion beam. The detector is an active target time-projection chamber with a blind central region for maximizing beam tolerance and GEM-based electron amplification, equipped with a modern digitizing data acquisition system allowing the recording of full signals. The system was developed to study the reaction ^{8}Li$(\alpha, n)^{11}$B, which is important for bridging the mass 8 gap in scenarios of low ^{4}He density like Inhomogeneous Big Bang Nucleosynthesis and the production of r-process seeds in supernovae. Both energy resolution and tracking accuracy were found to agree with theoretical predictions and Geant4 simulations. The ^{8}Li beam rate capability of the system is predicted to be of the order of 10^5s^{-1}, several orders of magnitude higher than most previous measurements of the same reaction, while still maintaining a high detection efficiency of 70 % to 80 %.

1. Motivation

The reaction ^{8}Li$(\alpha, n)^{11}$B is considered to be important for bridging the mass 8 gap in scenarios like inhomogeneous Big Bang nucleosynthesis [1] and r-process seed formation in supernovae [2].

While several experiments have measured a cross section for this reaction, there is significant disagreement between the results (see [3] and references therein), and they all suffer from severe rate limitations. To address this, a new time projection chamber (TPC) was built to measure and identify the charged reaction product ^{11}B [4]. The design includes a blind central region to avoid rate saturation due to detected beam particles.

2. Setup

TACTIC is an active target time-projection chamber with radial field geometry [4–6]. A schematic showing the main parts of the detector is given in Figure 1.

In a TPC ion tracks are reconstructed from ionization electrons collected on a segmented anode, one spatial coordinate being calculated from the electron drift time. In TACTIC the anode is segmented into 8 azimuthal sectors of 60 pads each along the beam direction z. A gas electron multiplier (GEM, [7]) amplifies the charge before it enters the amplifier and data acquisition system (DAQ). In order to not overwhelm the DAQ with irrelevant beam-induced

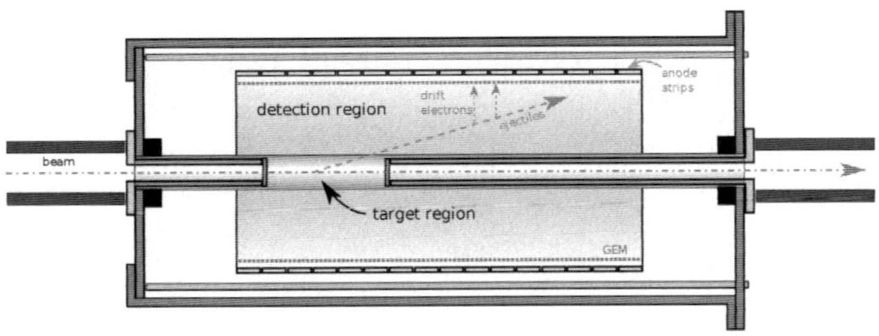

Figure 1. Schematic of basic setup: The beam enters the gas-filled target volume via a mylar window and unscattered beam particles leave the detector again via an additional window downstream. Drift electrons are collected on the segmented anode.

signals, the target region is separated from the detection region by a cathode wire cage. This means only ion tracks at a significant angle to the beam axis produce signals, while the detector remains blind to "unscattered" beam particles.

The DAQ comprises 11 VF48 48-channel sampling ADCs [8], which can record and output the full pulses from each pad or perform some limited on-board data reduction. Tracks affecting more than one VF48 module have to be assembled in the analysis using recorded time stamps.

The main fundamental difficulty of the setup lies in the cylindrical field. While most TPCs use a homogeneous field, leading to a linear relationship between drift time and position, in a radial field the drift velocity is radius-dependent, making an analytical solution impractical. To address this a radius formula was derived by simulation with GARFIELD [9]. Tracking is further complicated by the fact that TACTIC is self-triggering, resulting in the drift times only being known relative to each other. To calculate the radius coordinate the assumption must be made that the last detected pulse originates from a track point just outside the cathode cage.

3. Test results

Several tests were performed with a ^{148}Gd alpha source and stable beam to characterize the tracking capabilities and resolutions as well as efficiency and rate limitations. Any serious problem with the radius calculation would most likely lead to an apparent overall curvature of analyzed tracks. Figure 2 shows the coefficient of a quadratic fit to 1000 tracks as a function of the number of points per track. It is clear that the curvature only strays from zero for tracks with very few points, more consistent with a noise effect than a problem with the radius calculation.

A working tracking algorithm should reproduce the isotropic angular distribution of the alpha source. Figure 3 shows that this is indeed the case for small to medium angles in most sectors, with major discrepancies at large angles, which are explainable due to the small number of measured points and track angle effects on the pulse shape. It is important to keep in mind that large angles are not needed for the detection of ^{11}B in the reaction of interest, however. Fitting the shown expected isotropic angular distribution to the histogram (fit range $20° < \theta < 40°$) for each sector results in a measured source activity of 3278.6(10) Bq, compared with the independently measured activity of 3206(24) Bq. The apparent discrepancy is due to the fact that only the statistical counting error is reported, disregarding other uncertainties.

The cylindrical symmetry gives TACTIC a high geometric efficiency of about 69.5 %, which could be increased to 81.8 % with a relatively simple modification of the cathode wire assembly in the future. The rate capability of TACTIC is limited mainly by the DAQ. Beam tests found the system saturates at an event rate of $2800\,s^{-1}$ or a data rate of $19\,MB\,s^{-1}$. Assuming the on-board data reduction can be utilized fully, the data limit should not be reached in normal operation, leaving only the count rate limit. For elastic scattering in typical experimental conditions[1] for ^{8}Li$(\alpha, n)^{11}$B, Geant simulations show this to be reached at a beam rate of $10^4 s^{-1}$ to $10^5 s^{-1}$, this is compared to other measurements of ^{8}Li(α, n) in Table 1.

[1] e.g. 500 mbar He/CO_2 gas, 10 MeV ^{8}Li beam

IOP Conf. Series: Journal of Physics: Conf. Series **940** (2017) 012048 doi:10.1088/1742-6596/940/1/012048

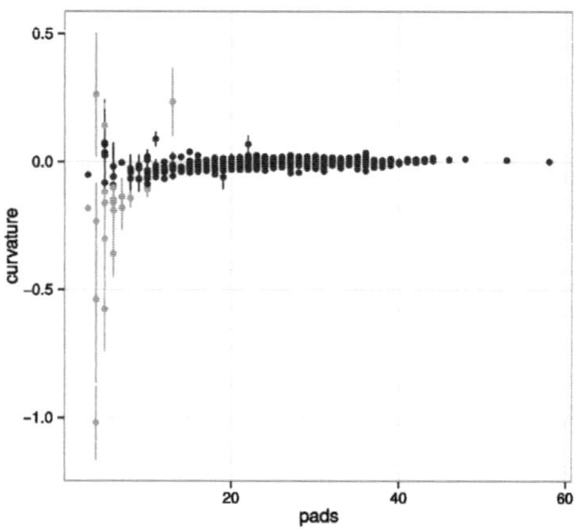

Figure 2. Coefficient of a quadratic fit to 1000 tracks. A significant curvature is only seen for tracks involving very few points.

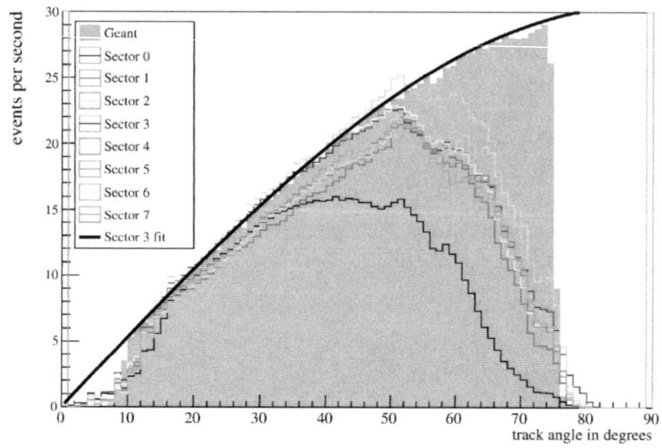

Figure 3. Angular distribution of alpha tracks measured in all 8 sectors compared to Geant4 simulation. Only scaled by measurement time, ignoring differences in geometric efficiency.

Experiment	MUSIC	Polycube	RIKEN [10]	TACTIC
Max. Rate (s^{-1})	23 to 2400	$\approx 10^2$	$\approx 10^5$	10^4 to 10^5
Efficiency (%)	≈ 100	10 to 20	2 to 12.56	70

Table 1. Rate capabilities and efficiency for $^8Li(\alpha, n)^{11}B$ measurements (from citations in [3]).

Of course superior rate capabilities are only useful if particle identification can be achieved. While a qualitative understanding of the energy response of the detector exists, GEM effects and pulse-shape-induced angle dependencies of the gain mean energy alone cannot be used for particle identification in TACTIC yet. Geant simulations do however look promising in terms of utilizing the ion stopping length in the gas as an indicator of particle identity.

4. Conclusion

TACTIC is a functioning tracking detector, albeit with a poorly understood energy response. More work is needed to determine to what extent this impedes the particle identification required for the intended reaction $^8Li(\alpha, n)^{11}B$. If particle identification can be achieved to a sufficient accuracy, TACTIC should be capable to improve on previous statistics.

References

[1] Kajino T and Boyd R N 1990 *Astrophys. J.* **359** 267–276
[2] Terasawa M, Sumiyoshi K, Kajino T, Mathews G J and Tanihata I 2001 *Astrophys. J.* **562** 470
[3] Ishiyama H *et al.* 2006 *Phys. Lett. B* **640** 82 – 85
[4] Laird A *et al.* 2007 *Nucl. Instrum. Methods Phys. Res., Sect. A* **573** 306 – 309
[5] Martin L *et al.* 2010 *J. Phys. Conf. Ser.* **202** 012040
[6] Fox S P *et al.* 2011 *J. Phys. Conf. Ser.* **312** 052007
[7] Bouclier R, Capeans M, Dominik W, Hoch M, Labbe J C, Million G, Ropelewski L, Sauli F and Sharma A 1997 *IEEE T. Nucl. Sci.* **44** 646 –650
[8] Martin J P and Amaudruz P A 2006 *IEEE T. Nucl. Sci.* **53** 715 – 719
[9] Veenhof R GARFIELD website at CERN `http://garfield.web.cern.ch`
[10] Ishiyama H *et al.* 2012 *Journal of Instrumentation* **7** C03036

IOP Conf. Series: Journal of Physics: Conf. Series **940** (2017) 012049 doi:10.1088/1742-6596/940/1/012049

Manganese Abundances in the Stars with Metallicities -1 <[Fe/H]< +0.3

T Mishenina[1], T Gorbaneva[1], M Pignatari[2,3,4], F-K Thielemann[3] S Korotin[1]

[1] Astronomical Observatory of Odessa, National University Shevchenko Park 65014 Odessa Ukraine
[2] Konkoly Observatory, Hungarian Academy of Sciences, Konkoly Thege Miklos ut 15-17 H-1121 Budapest Hungary
[3] Department of Physics, University of Basel, Klingelbergstrabe 82 4056 Basel Switzerland
[4] The NuGrid collaboration, http://www.nugridstars.org

E-mail: tmishenina@ukr.net

Abstract. We estimate the Mn abundances in the atmospheres of 247 F-G-K-type dwarf stars belonging to the thin and thick disk populations in the metallicity range -1 < [Fe/H] < +0.3. The observations were conducted using the 1.93 m telescope at Observatoire de Haute-Provence (OHP, France) equipped with the echelle type spectrographs ELODIE and SOPHIE. The abundances were derived under the LTE approximation; the synthetic spectrum for the Mn lines was computed accounting for the hyperfine structure. Starting from the results obtained, we discuss the evolution of the [Mn/Fe] ratio with respect to [Fe/H] in the galactic disk.

1. Introduction

Mn (Z=25) is an element of the iron group. Previous works (e.g., [1] and references therein) showed that in metal-poor stars Mn has a different behavior with respect to Fe compared to other iron-peak elements. Concerning the galactic disk, the uncertainties affecting the nucleosynthesis in thermonuclear supernovae (SNe Ia) [2] are a major limitation to our understanding of the evolution of Mn. For instance, while it is clear that SNe Ia produced most of the Mn in the solar system (e.g., [3]), the reproduction of the observed [Mn/Fe] trend in the galactic disk by theoretical galactic chemical evolution models requires robust SNIa abundance yields and a defined picture of the contribution from different types of SNe Ia (e.g., [4]). In this work we provide: 1) the measurement for Mn abundances for 247 F-G-K-type dwarf stars belonging to the different galactic disk populations and 2) the discussion on the origin of the Mn production in the galactic disk.

2. Observation, atmospheric parameters, and abundance determination

The spectra of the high-resolution echelle spectrographs SOPHIE and ELODIE on board 1.93m telescope of OHP (France) with resolving power R = 75000 and 42000 for the wavelengths range 4400 – 6800 Å were used. The models by [5] were applied to determine Mn abundance. The atmospheric parameters of stars were taken from [6]. The Mn abundance were obtained with the LTE STARSP software package [7] and the VALD atomic data [8] using the lines of Mn I 4783,

4823, 5432, 6013 and 6021 Å with taken into account the HFS. As [9] found the LTE [Mn/Fe] ratio might underestimate the real Mn abundances up to 0.5-0.6 dex for the metal-poor stars, and about 0.1 dex for stars of solar like metallicities. The difference in the NLTE corrections varies for lines of different multiplets are often exceeds 0.10 dex. The LTE Mn abundances obtained from the lines of different multiplets must show the systematic difference between them. As we can see from figures 1, 2 there is no systematic trend observed. Furthermore, the development of an adequate model of the Mn atom is complicated by the absence of detailed computations of atomic data. Therefore, the LTE determinations of the Mn abundance are correct, and the total measurement error does not exceed 0.10 dex accounting for the uncertainty in the stellar parameters and the accuracy of the synthetic spectrum fitting.

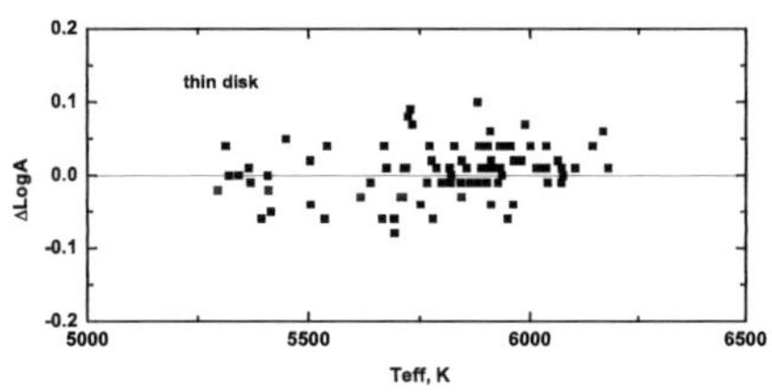

Figure 1. The difference between Mn abundances determined with the lines 4783 and 6013 Å on Teff for thin disk stars.

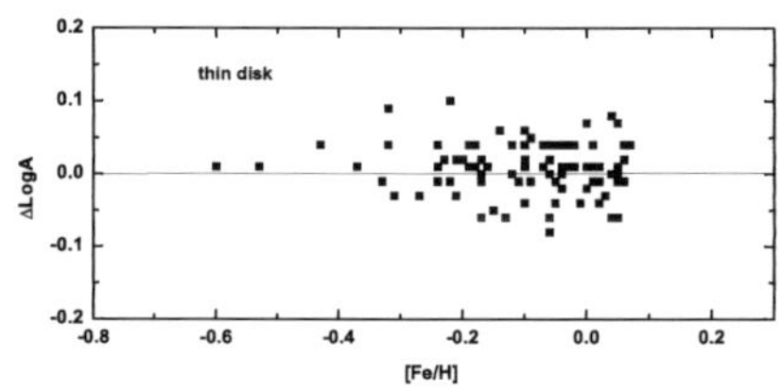

Figure 2. The diffrence between Mn abundances determined with the lines 4783 and 6013 Å on [Fe/H] for thin disk stars.

3. Results and discussion

In the figure 3 the dependence of [Mn/Fe] vs. [Fe/H] is shown for our stellar sample: thin disk stars are marked as open symbols, thick disk stars as full symbols, the Hercules stream stars and unclassified stars are marked as black asterisks and points, respectively.

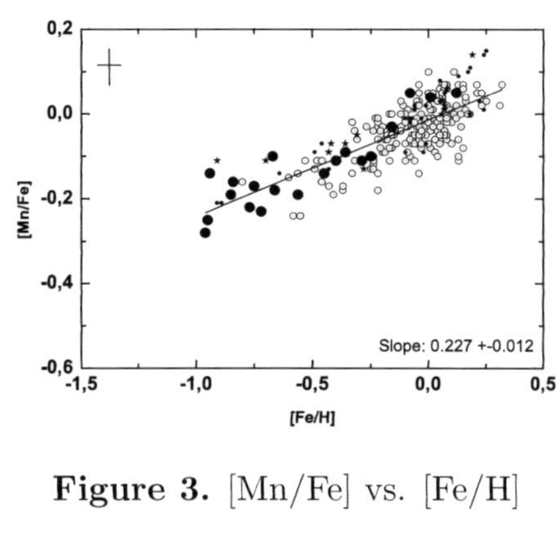

Figure 3. [Mn/Fe] vs. [Fe/H]

For stars with [Fe/H] < -1, the [Mn/Fe] ratio is dominated by core-collapse supernovae (CCSNe, e.g., [3]). In massive stars the only stable isotope of Mn (^{55}Mn) is formed mostly by the SN explosion in incomplete explosive Si-burning conditions as ^{55}Co (e.g., [10], [11], [12]). As odd-Z element, in principle Mn production is increasing with the increasing of the initial metallicity (e.g., [11], [13]). Most of Fe is produced as radiogenic ^{56}Fe from radioactive ^{56}Ni. This isotope is mainly made under complete explosive Si-burning conditions as primary product (i.e., independently from the initial metallicity of the star). Present uncertainties related to the

CCSN engine, to the relevance of asymmetries before and after the CCSN, and the SN-shock propagation through the massive star progenitor (e.g., [14], [15], [16], [17]) affect the final Mn/Fe ratio in the SN ejecta, possibly explaining the large [Mn/Fe] spread observed in the early Galaxy. Given that CCSNe produced about 30-50 % of the Fe observed in the solar system, this means that 12-20 % of the solar Mn is made by CCSNe. For [Fe/H] > -1, the [Mn/Fe] ratio in disk stars is increasing up to the solar ratio (figure 3), due to the contribution from thermonuclear supernovae (SNIa ,e.g., [2]) which on average is larger for Mn than for Fe.

The rising [Mn/Fe] trend has been proposed as a evidence of: 1) a gradual enrichment by SNIa ejecta of the ISM [18]; 2) the increasing Mn/Fe ratio with the increasing in the SNIa yields of the progenitor metallicity [19]; 3) the overlapping contribution of different types of SNIa. Thus, this trend may be used for indirect diagnostic of the relative contribution from different types of SNIa [4], [20].

4. Conclusions

– Within the metallicity range of our target stars, we confirm that the [Mn/Fe] ratio is increasing with the increase of metallicity. This is due to the contribution to the Galactic chemical evolution of Mn from SNe Ia, which is increasing compared to Fe with the increasing of the metallicity of the supernova progenitor.

– The possibility to use the [Mn/Fe] as a diagnostic of the relative contribution from different types of SNe Ia is appealing, but the impact of the different uncertainties in the stellar yields and from galactical chemical evolution simulations need to be fully considered.

Acknowledgments

TM, MP, FKT and SK thank for the kind support of the Swiss National Science Foundation under the project SCOPES No. IZ73Z0152485. MP acknowledges signicant NuGrid support by NSF grants PHY 02-16783 and PHY 09-22648 (Joint Institute for Nuclear Astrophysics, JINA), NSF grant PHY-1430152 (JINA Center for the Evolution of the Elements), and EU MIRG-CT-2006-046520. We are grateful for the support of the Lendulet-2014 Programme of the Hungarian Academy of Sciences (MP, Hungary) and SNF (MP and FKT, Switzerland).

References

[1] Yong D, Norris J E, Bessell M S *et al* 2013 *Astrophys. J.* **762** 26
[2] Hillebrandt W, Kromer M, Ropke F K and Ruiter A J 2013 *Frontiers of Physics* **8** 116
[3] Kobayashi C, Karakas A and Umeda H 2011 *Mon.Not.Royal.Astron.Soc.* **414** 3231
[4] Seitenzahl I R, Cescutti G, Ropke F K, Ruiter A and Pakmor R 2013 *Astron. Astrophys.* **559** L5
[5] Castelli F and Kurucz R L 2004 *arXiv:astro-ph/0405087*
[6] Mishenina T V, Pignatari M, Korotin S A *et al* 2013 *Astron. Astrophys.* **552** 128
[7] Tsymbal V V 1996 *ASP Conf. Ser.* **108** 198
[8] Kupka F, Piskunov N E, Ryabchikova T A *et al* 1999 *Astron. Astrophys. Suppl. Ser.* **138** 119
[9] Bergemann M and Gehren T 2008 *Astron. Astrophys.* **492** 823
[10] Woosley S E and Weaver T A 1995 *Astrophys. J. Suppl. Ser.* **101** 181
[11] Thielemann F-K, Nomoto K and Hashimoto M-A 1996 *Astrophys. J.* **460** 408
[12] Woosley S E, Heger A and Weaver T A 2002 *Reviews of Modern Physics* **74** 1015
[13] Nomoto K, Kobayashi C and Tominaga N 2013 *Ann. Rev. Astron. Astrophys.* **51** 457
[14] Thielemann F-K, Hirschi R, Liebendorfer M and Diehl R 2011 *Lecture Notes in Physics* vol 812, ed R Diehl, D H Hartmann, and N Prantzos (Berlin: Springer) p 153
[15] Hix W R *et al* 2014 *AIP Advances* **4** 041013
[16] Wongwathanarat A, Mueller E and Janka H-T 2014 *ArXiv e-prints*
[17] Smith N and Arnett W D 2014 *Astrophys. J.* **785** 82
[18] Kobayashi C and Nomoto K 2009 *Astrophys. J.* **707** 1466
[19] Cescutti G, Matteucci F, Lanfranchi G A and McWilliam A 2008 *Astron. Astrophys.* **491** 401
[20] Kobayashi C, Nomoto K and Hachisu I 2015 *Astrophys. J.* **804** 24

Role of nuclear reactions on stellar evolution of intermediate-mass stars

H Möller[1], S Jones[2], T Fischer[3] and G Martínez-Pinedo[1]

[1] Technische Universität Darmstadt, Inst. für Kernphysik, Schlossgartenstr. 2, Darmstadt, DE
[2] Heidelberger Institut für Theoretische Studien, Schloss-Wolfsbrunnenweg 35, Heidelberg, DE
[3] Uniwersytet Wrocławski, Institute of Theoretical Physics, Plac Maksa Borna 9, Wrocław, PL

E-mail: hmoeller@theorie.ikp.physik.tu-darmstadt.de

Abstract. The evolution of intermediate-mass stars (8 - 12 solar masses) represents one of the most challenging subjects in nuclear astrophysics. Their final fate is highly uncertain and strongly model dependent. They can become white dwarfs, they can undergo electron-capture or core-collapse supernovae or they might even proceed towards explosive oxygen burning and a subsequent thermonuclear explosion. We believe that an accurate description of nuclear reactions is crucial for the determination of the pre-supernova structure of these stars. We argue that due to the possible development of an oxygen-deflagration, a hydrodynamic description has to be used. We implement a nuclear reaction network with ~200 nuclear species into the implicit hydrodynamic code AGILE. The reaction network considers all relevant nuclear electron captures and beta-decays. For selected relevant nuclear species, we include a set of updated reaction rates , for which we discuss the role for the evolution of the stellar core, at the example of selected stellar models. We find that the final fate of these intermediate-mass stars depends sensitively on the density threshold for weak processes that deleptonize the core.

1. Introduction

Intermediate-mass stars from 8 to 12 $M_\odot$ denote the transition region of stars that can become either white dwarfs or undergo gravitational collapse. Starting from very similar initial masses, various final stages of stellar evolution are possible. These include ONe-white dwarfs, electron-capture or core-collapse supernovae and potentially even thermonuclear runaways [1, 2]. Despite their large abundance and their very interesting evolution, there is a distinct lack of stellar evolution models in that mass range. Seminal work goes back to [3, 4] while recent work has been published by [5, 6, 7]. Looking at the nuclear physics involved, these stars are especially interesting because unlike their massive counterparts ($\gtrsim$ 12 $M_\odot$), the same burning stages take place at higher densities. For example, neon burning operates at densities above 10^9 g/cm^3 compared to $\sim 10^6$ g/cm^3 in massive stars. From the recently published stellar evolution models by [5], it becomes obvious that starting from very similar initial masses, different models can lead to various final stages of stellar evolution. One particular reason for this is the strong sensitivity regarding the nuclear physics input. Here, especially the large density dependence of the weak processes involved should be mentioned. In [8], we point out the key nuclear reactions that we think should be considered in simulations of degenerate ONe-cores until possible core-collapse.

IOP Conf. Series: Journal of Physics: Conf. Series **940** (2017) 012050 doi:10.1088/1742-6596/940/1/012050

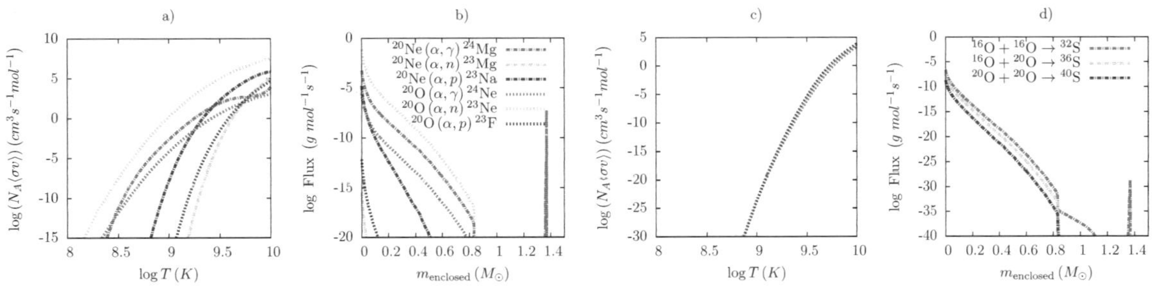

Figure 1. Panel a) shows $(\alpha, \{\gamma, \mathrm{n}, \mathrm{p}\})$-rates for ^{20}O and ^{20}Ne taken from the JINA Reaclib database [9]. Panel c) shows fusion rates for ^{16}O and ^{20}O (^{16}O-channel from [9], others from [8]). Panels b) and d) show the flux of the reactions displayed in a) and c) for conditions of a 8.75 $M_\odot$ model from [5] prior to neon burning. The flux $f_{\mathrm{AB}\to\mathrm{X}}$ is: $\rho N_{\mathrm{A}} \langle\sigma v\rangle_{\mathrm{AB}\to\mathrm{X}} Y_{\mathrm{A}} Y_{\mathrm{B}}$, with the density ρ, the abundances Y_{A} and Y_{B} and the rate $N_{\mathrm{A}} \langle\sigma v\rangle_{\mathrm{AB}\to\mathrm{X}}$.

2. Evolution of degenerate ONe-cores

It was shown in [5] that for a range of initial progenitor masses from 8 to 9 $M_\odot$, the star will form a degenerate ONe-core towards the end of its stellar life. Furthermore, it was found that in a narrow range between 8.7 and 8.8 $M_\odot$, Urca cooling occurs prior to the neon-burning phase by electron capture (EC) on odd-A sd-shell nuclei (^{23}Na, ^{25}Mg, ^{27}Al) above densities of 10^9 g/cm^3. Afterwards, heating due to EC on even-A sd-shell nuclei (first ^{24}Mg) occurs once densities of 2×10^9 g/cm^3 are reached. Above 6×10^9 g/cm^3, EC on ^{20}Ne set in and produce considerable amounts of ^{20}O via the chain $^{20}\mathrm{Ne}\,(e^-, \nu_e)\,^{20}\mathrm{F}\,(e^-, \nu_e)\,^{20}\mathrm{O}$. Hence, the stellar core consists mainly of ^{16}O and ^{20}Ne, ^{24}Ne and some fraction of ^{20}O (up to 4%, depending on the model). The situation is different compared to massive stars where no EC-phase occurs before neon burning. In this case, neon burning proceeds via the following reactions: $^{20}\mathrm{Ne}\,(\gamma, \alpha)\,^{16}\mathrm{O}$ and $^{20}\mathrm{Ne}\,(\alpha, \gamma)\,^{24}\mathrm{Mg}$. Due to the presence of ^{20}O however, additional reaction channels need to be taken into account: $^{20}\mathrm{O}\,(\alpha, \{\gamma, \mathrm{n}, \mathrm{p}\})\,\{^{24}\mathrm{Ne}, ^{23}\mathrm{Ne}, ^{23}\mathrm{F}\}$. As can be seen in panel a) of Figure 1, the reaction $^{20}\mathrm{O}\,(\alpha, n)\,^{23}\mathrm{Ne}$ is dominating in the relevant range of temperatures. Hence, we argue that the α-capture on ^{20}O can be a competitive process and should be considered. In panel b) of Figure 1, it can be seen that for the model which has up to 4% of ^{20}O, the reaction $^{20}\mathrm{O}\,(\alpha, n)\,^{23}\mathrm{Ne}$ is dominant in the stellar core during neon burning.

Once the star reaches temperatures in excess of 1.7 GK, oxygen fusion sets in. As there is also ^{20}O available, the fusion can now proceed via three channels: $^{16,20}\mathrm{O} + ^{16,20}\mathrm{O} \to ^{32,36,40}\mathrm{S}^*$. For details about the calculation of the neutron rich channels, the reader is referred to [8]. As can be seen in panel c) of Figure 1, the three reaction rates are comparable. For the reaction fluxes, we find that the contribution of the $^{16}\mathrm{O} + ^{20}\mathrm{O}$-channel reaches up to 15%, while the $^{20}\mathrm{O} + ^{20}\mathrm{O}$-channel is much less important. In addition to that, the fusion involving ^{20}O has Q-values of $\sim$ 30 MeV. This does not only increase the rate of energy release during the fusion phase of oxygen by up 30%, but it also allows for more exotic decay channels including multi-neutron emission.

3. Hydrodynamic modeling

In this study, our goal is to simulate the dynamical stellar evolution starting from neon burning until a possible core-collapse, using the models from [5]. For this, we use the implicit 1D GR-hydrodynamic code AGILE [10] and couple it to a nuclear reaction network with 200 nuclear species in an operator split scheme. Regarding the nuclear physics input, we use an updated set of rates for O/Ne-burning (c.f. Section 2) and EC-rates from [11]. The other rates are taken from the JINA Reaclib database [9].

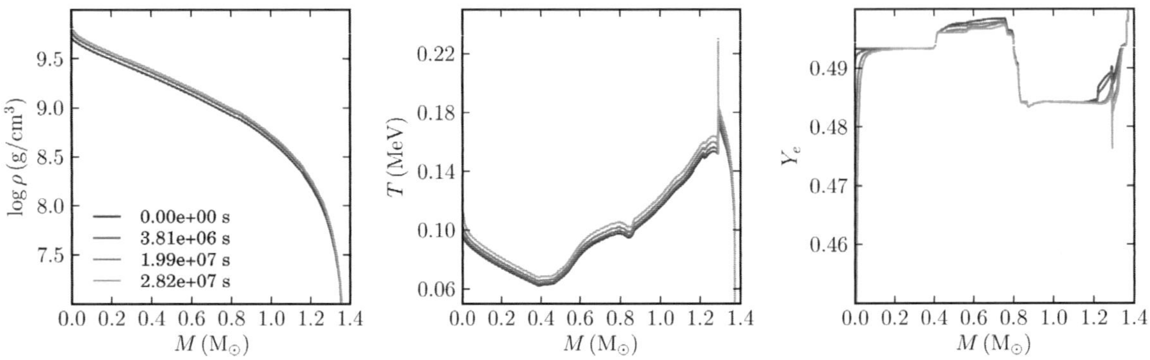

Figure 2. Hydrodynamic simulation of an 8.8 $M_\odot$ model from [5]. Panels a),b) and c) show density, temperature and Y_e profiles over enclosed mass for 4 different times until neon is ignited.

We simulate the first about 10^7 seconds of an 8.8 $M_\odot$ model, starting prior to the onset of neon burning until the neon/oxygen flame develops at the edge of the core. The results are displayed in Figure 2. Once the flame is ignited, its propagation is driven by heat transfer across flame front due to the conduction of degenerate electrons. However, estimating the width of the flame front suggests that the relevant length scales of flame propagation, which are on the order of centimeters, cannot be resolved in a hydrodynamical simulation which is resolving length scales of kilometers. In order to proceed the simulation, a parametric description of the flame has to be used that is capable of modeling the oxygen deflagration as it moves through the star.

4. Conclusion

Burning processed operate differently in intermediate-mass stars and their final stellar evolution is very sensitive to the nuclear physics input. The last stage of stellar evolution towards a possible core-collapse has to be modeled hydrodynamically. As our current simulations fail to resolve the propagation of the oxygen flame, our current work is aiming at finding an appropriate description that will allow us to evolve our models further.

Acknowledgments

H. M. is member of the HGS-HIRe Graduate School. T.F. acknowledges support from the Narodowe Centrum Nauki (NCN) within the "Maestro" program under contract No. DEC-2011/02/A/ST2/00306. G.M.P. is partly supported by the Helmholtz International Center for FAIR within the framework of the LOEWE program launched by the state of Hesse, and the Helmholtz Association through the Nuclear Astrophysics Virtual Institute (VH-VI-417).

References

[1] Isern J, Canal R and Labay J 1991 ApJ **372** L83–L86
[2] Gutierrez J, Garcia-Berro E, Iben Jr I, Isern J, Labay J and Canal R 1996 ApJ **459** 701
[3] Nomoto K 1984 ApJ **277** 791–805
[4] Nomoto K 1987 ApJ **322** 206–214
[5] Jones S *et al.* 2013 ApJ **772** 150 (*Preprint* 1306.2030)
[6] Takahashi K, Yoshida T and Umeda H 2013 ApJ **771** 28 (*Preprint* 1302.6402)
[7] Schwab J, Quataert E and Bildsten L 2015 MNRAS **453** 1910–1927 (*Preprint* 1504.05194)
[8] Möller H, Jones S, Fischer T and Martínez-Pinedo G 2014 Impact of Nuclear Reactions on the Fate of Intermediate-mass Stars *Nuclei in the Cosmos (NIC XIII)*
[9] Cyburt R H *et al.* 2010 ApJS **189** 240–252
[10] Liebendörfer M, Rosswog S and Thielemann F K 2002 ApJS **141** 229–246 (*Preprint* astro-ph/0106539)
[11] Langanke K and Martínez-Pinedo G 2000 *Nuclear Physics A* **673** 481–508 (*Preprint* nucl-th/0001018)

IOP Conf. Series: Journal of Physics: Conf. Series **940** (2017) 012051 doi:10.1088/1742-6596/940/1/012051

Sensitivity to neutron captures and β-decays of the enhanced s-process in rotating massive stars at low metallicities

N Nishimura[1], R Hirschi[1,2] and T Rauscher[3,4]
[1]Astrophysics, Faculty of Natural Sciences, Keele University, Keele ST5 5BG, UK
[2]WPI Kavli IPMU, The University of Tokyo, Kashiwa 277–8583, Japan
[3]Centre for Astrophysics Research, University of Hertfordshire, Hatfield AL10 9AB, UK
[4]Department of Physics, University of Basel, 4052 Basel, Switzerland

E-mail: n.nishimura@keele.ac.uk

Abstract. The s-process in massive stars, producing nuclei up to $A \approx 90$, has a different behaviour at low metallicity if stellar rotation is significant. This *enhanced s-process* is distinct from the s-process in massive stars around solar metallicity, and details of the nucleosynthesis are poorly known. We investigated nuclear physics uncertainties in the enhanced s-process in metal-poor stars within a Monte-Carlo framework. We applied temperature-dependent uncertainties of reaction rates, distinguishing contributions from the ground state and from excited states. We found that the final abundance of several isotopes shows uncertainties larger than a factor of 2, mostly due to the neutron capture uncertainties. A few nuclei around branching points are affected by uncertainties in the β-decay.

1. Enhancement of weak s-process by stellar rotation

The s-process in massive stars ($\gtrsim 10M_{\odot}$) is called “weak s-process”, because the major products are limited to lighter s-process elements up to $A \approx 90$ (heavier elements up to Pb and Bi are produced in low mass AGB stars, called *main s-process*, see [1] for a review). It takes place during the He-core and C-shell burning, and the main neutron source is an α-capture reaction ^{22}Ne(α, n)^{25}Mg following the reaction sequence $^{14}\mathrm{N}(\alpha,\gamma)^{15}\mathrm{F}(\beta^-)^{18}\mathrm{O}(\alpha,\gamma)^{22}\mathrm{Ne}$. For very metal-poor stars, the behaviour of this weak s-process drastically changes due to rotation, and heavy nuclei with $A \approx 138$, including Ba, are produced (see [2, 3]; Frischknecht et al., 2015, in preparation). Rotation-induced mixing between the helium-burning and hydrogen-burning convective zones increases the abundance of primary ^{14}N (and thus of ^{22}Ne). This increases the neutron captures and enhances the weak s-process (hereafter, we call it *enhanced s-process*, “e.s-process”).

The evolution of the e.s-process pattern is shown in Figure 1. This is based on a simplified evolutionary track from a $25M_{\odot}$ model, effectively accounting for rotation-induced mixing by increasing the primary ^{14}N, as introduced in a previous study [4]. The abundance pattern at $T = 219$ MK is similar to the weak s-process producing isotopes up to $A = 90$. As temperature increases ($T = 236$ MK), the distribution goes to the higher A region and finally reaches $A \approx 140$. The final abundances show production in the range $90 \leq A \leq 140$.

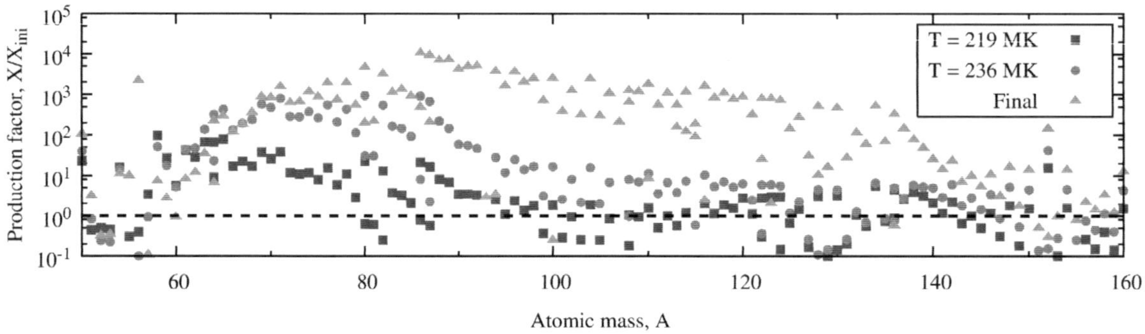

Figure 1. Evolution of abundances by the e.s-process by the production factors.

2. Uncertainty of reaction rates and Monte-Carlo simulation

Nucleosynthesis in the e.s-process is quantitatively different from the standard weak s-process. The impact of reactions relevant to stellar burning (α-captures) of lighter nuclei has been studied before, but with a focus on neutron-source and -poison reactions [4]. However, the impacts of uncertainties in (n,γ) and β-decay on the path of the e.s-process are poorly known. We investigated, therefore, the role of nuclear physics input on the path of the e.s-process. For this purpose, we performed Monte-Carlo (MC) simulations focusing on (n,γ) and β-decay reactions. We use a MC framework with a general reaction network which is applicable to a variety of nucleosynthesis processes [5, 6].

The input of nuclear physics uncertainty (i.e., of reaction rates) is crucial for studying impacts on the nucleosynthesis yields. In the present study, we assumed that reaction rates have a temperature-dependent uncertainty because the relative contributions by the ground state (g.s.) and excited states to the rate change with temperature and experimental cross sections, if available at all, only constrain g.s. contributions. Following the prescription in [7, 8], experimental uncertainties are used for the g.s. contributions to (n,γ) rates, whereas a factor 2 is used for excited state uncertainties. We simply apply a constant value 2 for theoretical rates. A similar approach is used for β-decay rates, based on partition functions to determine the importance of excited states. The uncertainty at lower temperatures ($T < 10^7$ K) corresponds to the one of measured decays, while the uncertainty becomes larger as the temperature increases (for details, see [8]). A uniform random distribution between the upper and lower limit of the

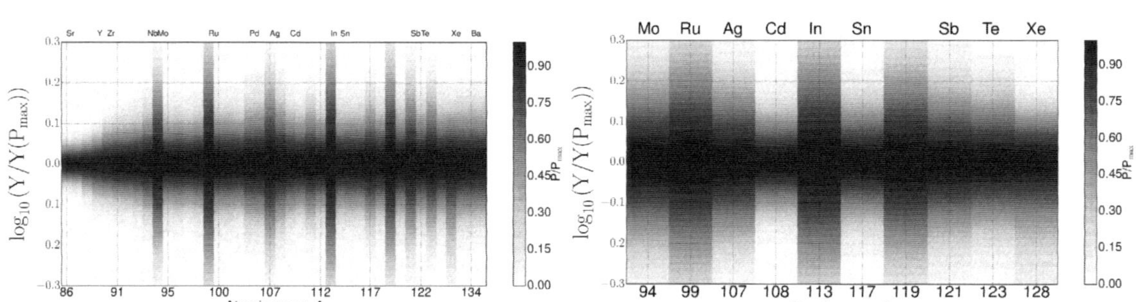

Figure 2. Uncertainties in final isotope production when all reactions are varied for stable isotopes that e.s-process produces (left) and selected isotopes with larger uncertainty (right). For each isotope, the normalized probability density distribution $Y/Y(P_{\max})$ is shown.

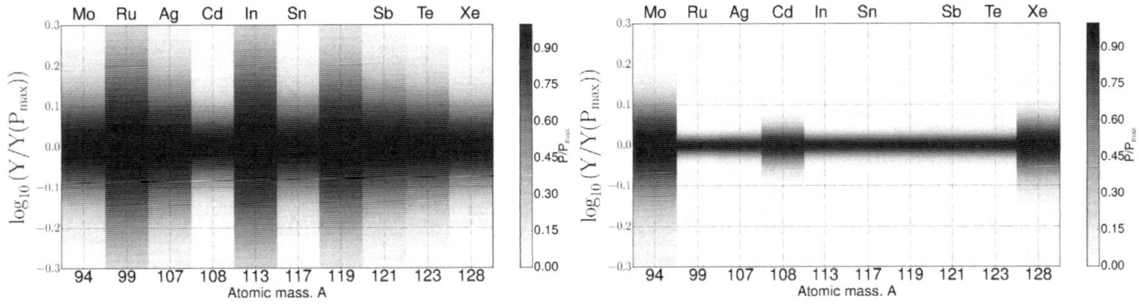

Figure 3. Same as Figure 2, where only (n,γ) (left) and only β-decay (right) are varied.

reaction rate at a given temperature was used for the MC variation factors.

Figure 2 shows the resulting production uncertainty for the cases where we varied all (n,γ) reactions and β-decays. We chose to show abundance uncertainties for stable s-process isotopes with $86 \leq A \leq 136$ in the left panel, which cover the main products of the e.s-process. We plot isotopes showing the largest uncertainties in the right panel. The colour distribution corresponds to the normalized probability density distribution of the uncertainty in the final abundance. Values of 0.114 and 0.301 for the probability density distribution correspond to 30 % and a factor of 2 uncertainty, respectively.

Some isotopes show larger uncertainties, while most others are within the ± 30 % range. In order to identify the role of the reactions for the final uncertainty, we varied neutron captures and β-decays separately. Figure 3 shows the results for the isotopes with the largest uncertainties, the left and right panels correspond to variation of all (n,γ) and all β-decays, respectively.

The results clearly indicate that (n,γ) reactions have a dominant impact on the nucleosynthesis uncertainty and that β-decays have a limited importance: (i) the uncertainties in (n,γ) leads to a general 30 % uncertainty in the final abundances, several isotopes having a higher value up to a factor of 2; (ii) β-decays only affect a few isotopes around the branching points at $A \approx 94$, 108, and 128.

3. Conclusion

In this study, we evaluated the impact on e.s-process nucleosynthesis of nuclear physics uncertainties using MC calculations. The method can identify the importance of reactions and we found that (n,γ) reactions dominate the total uncertainty, with a few important contributions from β-decays around branching points. Our method is a robust way to identify key reaction rates to support further investigations in nuclear astrophysics regarding the e.s-process.

We were supported by the BRIDGCE UK (www.astro.keele.ac.uk/bridgce), the ERC (EU-FP7-ERC-2012-St Grant 306901, EU-FP7 Adv Grant GA321263-FISH), and the UK STFC (ST/M000958/1), COSMOS (STFC DiRAC Facility) at DAMTP in University of Cambridge.

References

[1] Käppeler F, Gallino R, Bisterzo S and Aoki W 2011 *Reviews of Modern Physics* **83** 157–194 1012.5218
[2] Pignatari M, Gallino R, Meynet G, Hirschi R, Herwig F and Wiescher M 2008 *ApJL* **687** L95–L98 0810.0182
[3] Frischknecht U, Hirschi R and Thielemann F K 2012 *A&A* **538** L2 1112.5548
[4] Nishimura N, Hirschi R, Pignatari M et al 2014 *AIP Conference Series* **1594** *1 p.146–151*
[5] Nishimura N, Hirschi and Rauscher T 2015 *PoS (NIC XIII)* p 127
[6] Rauscher T, Travaglio C Gallino R, Nishimura N and Hirschi 2015 *PoS (NIC XIII)* p 141
[7] Rauscher T, Mohr P, Dillmann I and Plag R 2011 *ApJ* **738** 143 1106.1728
[8] Rauscher T 2012 *ApJS* **201** 26 1205.0685

IOP Conf. Series: Journal of Physics: Conf. Series **940** (2017) 012052 doi:10.1088/1742-6596/940/1/012052

Towards an activation cross section measurement of the ^{17}O(p,γ)^{18}F reaction in a wide energy range

A. Ornelas[1,2], Gy. Gyürky[1], Z. Halász[1], Cs. M. Oláh[1], Zs. Fülöp[1], I. Rajta[1] and I. Vajda[1]

[1] Atomki, H-4001 Debrecen, POB. 51, Hungary
[2] University of Debrecen, H-4028 Debrecen, Hungary

E-mail: ornelas@atomki.mta.hu

Abstract.
The CNO cycles are fusion processes in stars that convert hydrogen to helium. These hydrogen burning processes occur in several sites and stages of stellar evolution, such as red giants, asymptotic giant branch (AGB) stars, massive stars, and classical novae.

One of the important reactions in the CNO-III and CNO-IV cycles is ^{17}O(p,γ)^{18}F [1]. The only available total cross section measurement in a wide energy range for this reaction dates back to several decades ago [2] which makes the theoretical extrapolation to astrophysical energies more difficult and introduces uncertainty.

The aim of the present work is to provide precise total cross section data in the energy range between about 500 keV and the 2 MeV using the activation method. The experimental campaign at the new tandetron accelerator of Atomki is in progress.

1. Introduction

The ^{17}O(p,γ)^{18}F reaction competing with the ^{17}O(p,α)^{14}N process opens the way to the third and fourth CNO cycles of hydrogen burning which take place at elevated temperatures in several different stellar environments. This reaction plays therefore an important role in e.g. the nucleosynthesis of the rare oxygen isotopes and the important radionuclide ^{18}F [3].

In order to provide the precise reaction rate, the low energy cross section of the ^{17}O(p,γ)^{18}F reaction has been measured by several groups recently ([4], and references therein). The only available total cross section data in a wide energy range, however, dates back to several decades ago [2] and the validity of the low energy part of these data was already questioned [5]. The aim of the present work is therefore to measure the cross section of the ^{17}O(p,γ)^{18}F reaction in the energy interval between E_p= 500 keV and E_p= 2000 keV.

2. Experimental approach

The reaction cross section is determined using the activation method as the reaction product ^{18}F is radioactive, decays by positron emission which can be measured by detecting the 511 keV annihilation radiation. The activation method provides directly the astrophysically important total cross section independently from the various transitions in the residual nucleus.

The present experiment is the first scientific project to be performed at the new 2 MV tandem accelerator installation in the Atomki institute in Debrecen, Hungary. The proton beam energy of the new accelerator was thus calibrated first using some of the well known resonances in the ^{27}Al(p,γ)^{28}Si reaction.

The Ta_2O_5 targets were prepared by anodic oxidation of tantalum backings in isotopically enriched water [6]. To determine the target thickness and stoichiometry, the targets were studied with the RBS technique, using an alpha beam of 1.6 MeV at the microprobe facility of Atomki. The target thickness was then calculated using the SIMNRA code to fit the obtained spectra (see Fig. 1).

Target stability was periodically monitored with ^{17}O(p,γ) and ^{18}O(p,γ) resonance scans detecting the prompt gamma radiations.

For the cross section measurements the targets were irradiated by proton beams of several μA intensity for a few hours, after which they were transported to a calibrated offline low background detector for the measurement of the ^{18}F β^+ decay (Fig. 2).

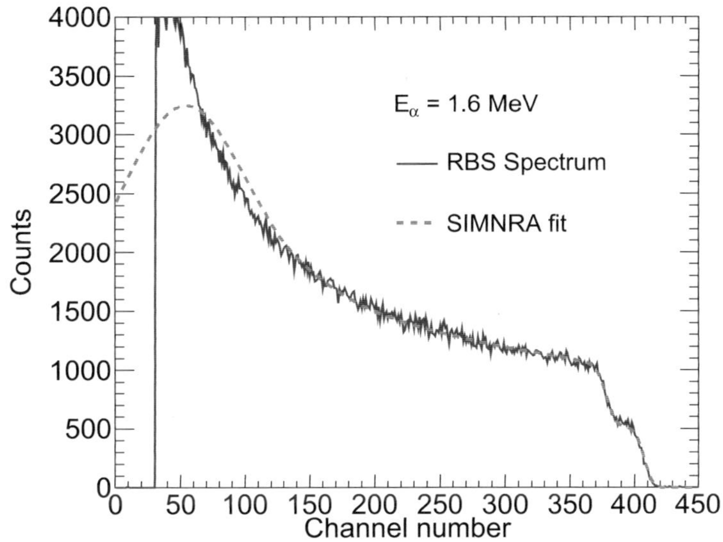

Figure 1. Measured and simulated RBS spectrum of a Ta_2O_5 target taken at $E_\alpha = 1.6$ MeV with the detector at 135°.

3. Analysis

The main decay mode of ^{18}F is the β^+ emission (96.7 %) with half-life of 109.77 minutes. To track this β^+ decay curve, several 10 - minute gamma - spectra were taken to measure the 511 keV annihilation radiation, allowing the calculation of the activity after the activation of the Ta_2O_5 target.

In Fig. 2 a typical ^{18}F β^+ decay curve is shown fitted with an exponential decay curve superimposed on a constant 511 keV background.

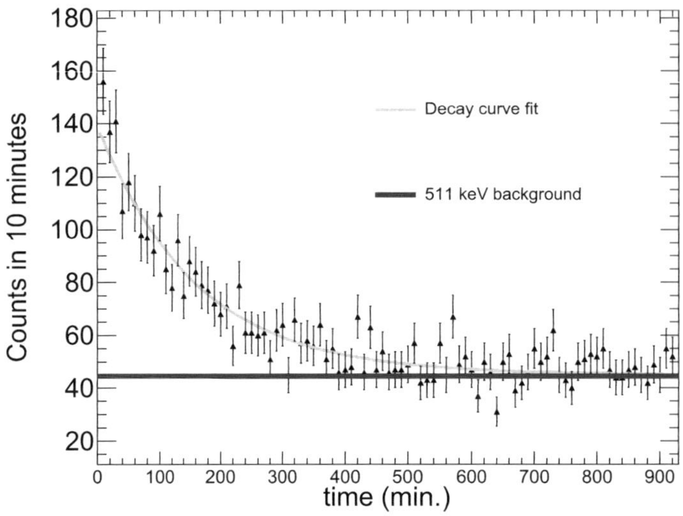

Figure 2. β^+ decay curve of ^{18}F after an activation with 1 MeV protons.

4. Status of the experiments and outlook

A few test irradiations were carried out between E_p = 580 keV and E_p = 1800 keV in order to study the feasibility of the measurement. It is found that the decay of the 511 keV γ-radiation can be well described by the half-life of ^{18}F and therefore no disturbing short-lived positron emitter is present in the target.

The analysis of the obtained data as well as the experimental campaign are still in progress and the final results will be given in a forthcoming publication.

Acknowledgments

We thank A. Formicola and the LUNA collaboration for giving us access to the Ta_2O_5 target preparation device. This work was supported by the SROP-4.2.2.B-15/1/KONV-2015-0001 project, by the European Union, co-financed by the European Social Fund and by OTKA grants No. K101328 and K108459.

References

[1] José J and Iliadis C 2011 *Reports on Progress in Physics* **74** 096901
[2] Rolfs C 1973 *Nuclear Physics A* **217** 29 – 70
[3] Wiescher M *et al.* 2010 *Annual Review of Nuclear and Particle Science* **60** 381–404
[4] Buckner M Q *et al.* 2015 *Phys. Rev. C* **91**(1) 015812
[5] Scott D A *et al.* 2012 *Phys. Rev. Lett.* **109**(20) 202501
[6] Caciolli A *et al.* 2012 *The European Physical Journal A* **48** 144

IOP Conf. Series: Journal of Physics: Conf. Series **940** (2017) 012053 doi:10.1088/1742-6596/940/1/012053

Beta-decay half-lives for short neutron rich nuclei involved into the r-process

I Panov[1,2,3], **Yu Lutostansky**[3], **F-K Thielemann**[1]

[1]Department of Physics, University of Basel, Basel, Switzerland
[2]Institute for Theoretical and Experimental Physics NRC Kurchatov Institute, Moscow, Russia
[3]NRC Kurchatov Institute, Moscow, Russia

E-mail: igor.panov@itep.ru

Abstract. The beta-strength function model based on Finite Fermi-Systems Theory is applied for calculations of the beta-decay half-lives for short neutron rich nuclei involved into the r-process. It is shown that the accuracy of beta-decay half-lives of short-lived neutron-rich nuclei is improving with increasing neutron excess and can be used for modeling of nucleosynthesis of heavy nuclei in the r-process.

1. The model

During r-process modeling, when the nuclei heavier than iron are formed, we have to solve systems of equations, in which tens of thousands of reactions between thousands of nuclei should be involved. The majority of these nuclei are short-lived and are not investigated experimentally. Beta-decays are the key reactions for the r-process, and the majority of these rates can be obtained only theoretically. Since the theory of weak interaction was developed by E. Fermi in 1934, a number of different beta-decay models were considered, which permit to calculate the beta-decay half-lives - $T_{1/2}$ of neutron-rich short-lived heavy nuclei, using the formalism of beta-strength functions $S_\beta(E)$. The structure of the beta-strength function has a resonance form, and giant the Gamow-Teller resonance (GTR) plays an important role [1]. Although it lies much higher than the energy window of beta-decay Q_β, the GTR has an indirect influence on $T_{1/2}$ values, when part of its strength is redistributed into low lying states according to sum rules which defines a "suppression factor" of allowed β-transitions [2]. The resonance structure of the beta-strength function $S_\beta(E)$ is considered since $T_{1/2}$ calculations obtained within the "Gross-theory" [3], in which the GTR only was taken into account with the energy of the known analogue resonance, which was possible only for heavy nuclei [4]. Different models of the strength function $S_\beta(E)$, based on Finite Fermi Systems Theory (FFST) [5], taking into account the GTR and low-lying states, were developed [2] for extended calculations of nuclear properties of neutron rich nuclei. Nowadays probably the most consistent approach [6] is derived on the basis of the density functional DF3 [7], but self-consistent calculations are done only for a small part of nuclei (~10%) participating in the r-process and the data set is insufficient yet for applying it to the r-process models. The widely used predictions of beta-decay rates [8, 9] did not reconsider this approach yet and application to the r-process models [10] demands more detailed examination of beta-decay parameters involved.

IOP Conf. Series: Journal of Physics: Conf. Series **940** (2017) 012053 doi:10.1088/1742-6596/940/1/012053

The $T_{1/2}$ values, calculated in the framework of the BETA-model [11, 12] have been used by now for r-process modelling, defining the importance of beta-delayed processes in the nucleosynthesis of heavy nuclei [10]. It is well known that the beta-decay strength function - $S_\beta(E)$ is defined by the characteristics of charge-exchange excitations of nucleus - so called isobaric states (IS). These states are located in the neighbor nucleus, but energetically connected with ground state of mother nucleus. For calculations of IS parameters in the Finite Fermi-systems Theory one has to solve the complicated system of equations for the effective nuclear field.

The integral equations of FFST with local nucleon-nucleon interaction in the form of Landau-Migdal after a number of averaging procedures were transformed to secular equations for which a number of averaged parameters of nucleus A(N,Z) were calculated [2,11]. The poles of the effective field of the equations define the energies of excited states, and the residues of the field in the poles define the squares of matrix elements of the β-transitions from these states to the ground state to the daughter nucleus as well as the contributions of particle-hole configurations into the isobaric state ω_k , and as a consequence, the degree of its collectivity. As it was shown earlier [2], all the solutions belong to 3 types of states: ω_+ (spin-flip transitions $l + 1/2 \rightarrow l - 1/2, \Delta j = +1$), ω_0 (core polarization states, $j - j$ transitions, $\Delta j = 0$) and ω_- (back spin-flip transitions $l - 1/2 \rightarrow l + 1/2$, Δj = -1). Among them the transition with Δn=0 and $V_{pn}(r) = Const$ are the most collective ones and have the large square matrix elements. In this model, we considered consistently only allowed β-transitions. With regard to forbidden β-transitions, first-forbidden ones were taken into account, using systematics of experimental data for transitions to the low-lying states with a "hindrance factor" as in [13, 14].

2. Results and Conclusions

On the basis of extended calculations of beta-decay half-lives $T_{1/2}$ for isotopes from krypton to uranium a comparison with the experimental values of $T_{1/2}$ as well as with other predictions [9] was done. The results for strontium isotopes are plotted and shown at figure 1. The present $T_{1/2}$ values are decreasing with neutron excess and follow the experimental dependence on A. A small decline occurs in the values of $T_{1/2}$ for heavier isotopes, as for example for Sr isotopes with A >105. This small decline is connected with existence of back-spin-flip states in the region close to the ground state of the daughter nucleus. For nuclei heavier than lead this effect is not seen so clearly, but the values of $T_{1/2}$ are usually smaller than those [9] used for astrophysical nucleosynthesis.

The comparison of the calculated values of beta-decay half-lives with experimental data in the range from Rb to Pb have shown that the difference can achieve 2 orders of magnitude for long-lived nuclei, but with decreasing of $T_{1/2}$ (and increasing of Q_β) the accuracy of calculations is increased predictably rather well for short-lived r-process nuclei (figure 2). This is the reason to apply our calculations for the r-process modelling. The present $T_{1/2}$ values for neutron rich isotopes usually are shorter than widely used RPA-predictions of $T_{1/2}$ values [9] and difference increased closer to neutron drip line, and for nuclei heavier Pb. Phenomenologically this effect was taken into account in recent calculations of the r-elements yields in different scenarios [10,15].

The beta-decay is one of the main properties of nuclei, needed for the r-process nucleosynthesis modelling, and effect strongly the abundances of heavy elements, in part the position and the form of peaks [10,12]. The extended calculations of beta-decay rates of short lived neutron-rich nuclei were compared with other predictions and measurements. The new revision of nuclear data for r-process nuclei was done on the basis of fitting model parameters to extended experimental data. For predictions of beta-decay half-lives a model of the beta-strength function, based on Finite Fermi Systems Theory, was used. Calculations for an extended list of short-lived nuclei were performed. It was shown that the accuracy of beta-decay half-lives increases for shorter values of $T_{1/2}$. This permit to apply the derived data to r-process calculations and predict the

IOP Conf. Series: Journal of Physics: Conf. Series **940** (2017) 012053 doi:10.1088/1742-6596/940/1/012053

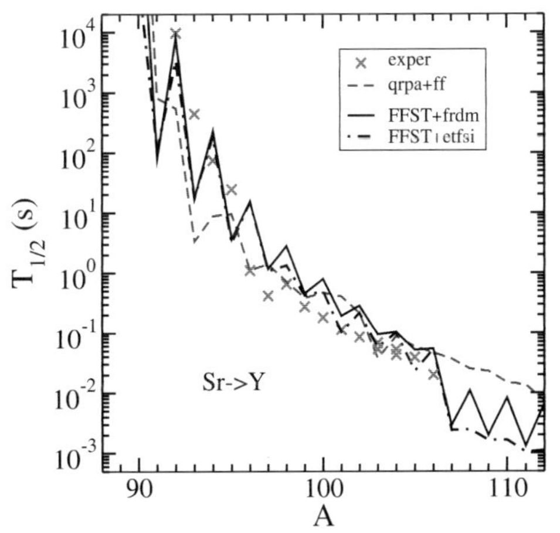

Figure 1. $T_{1/2}^{ffst}$ for neutron-rich Sr isotopes, utilizing FRDM [18] (line) and ETFSI [19] (dash-dotted line) mass predictions. Experimental results (crosses), and predictions from the FRDM + QRPA (dashed black line) [9], are shown.

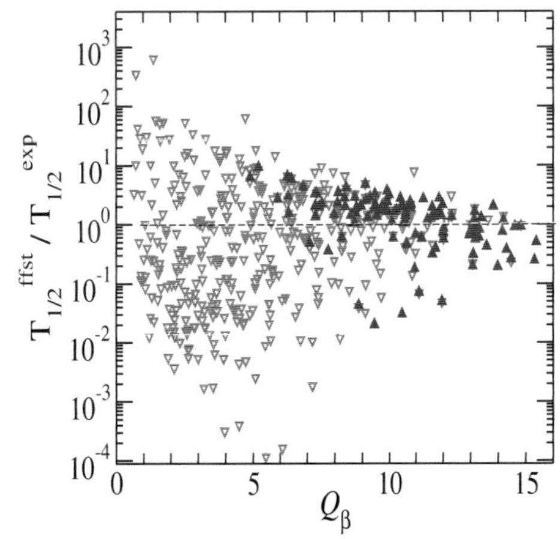

Figure 2. Ratios of predicted ($T_{1/2}^{ffst}$) and experimental ($T_{1/2}^{exp}$) values of the beta-decay half-lives for the nuclei with Z from 36 to 82. The ratio $T_{1/2}^{ffst}/T_{1/2}^{exp}$ for new experimental data [17] also shown as triangles up.

abundances of heavy nuclei in high neutron density environments. For major part of nuclei heavier rubidium the half-lives predictions of heaviest neutron-rich isotopes in times in average less, than the values predicted by other models. The self-consistent predictions of beta-decay half-lives and probabilities of delayed processes [16] can be applied for r-process modelling in different astrophysical scenarios.

This work was supported in part by Russian Fond of Basic Research 13-02-12106-ofi_m, 14-22-03040-of_m, and Swiss national scientific Fond, project IZ73Z0_152485 SCOPES.

References

[1] Gaponov Yu V and Lyutostanskii Yu S 1972 *JETP Lett.* **15** 120
[2] Gaponov Yu V and Lyutostanskii Yu S 1981 *Sov. J. Part. Nucl.* **12** 528
[3] Takahashi K, Yamada M and Kondoh T 1973 *At. Data Nucl. Data Tables* **12** 1014
[4] Lutostansky Yu S and Tikhonov V N 2014 *Bull. Rus. Acad. Sci. Phys.* **78** 373
[5] Migdal A B 1967 *Theory of Finite Fermi Systems* (New York: Interscience)
[6] Borzov I N 2006 *Nucl. Phys.* A **777** 645
[7] Borzov I N, Fayans S A, Kromer E and Zawischa D 1996 *Z. Phys.* A **355** 117
[8] Möller P, Nix J R and Kratz K-L 1997 *At. Data Nucl. Data Tables* **66** 131
[9] Moller P, Pfeifer B and Kratz K-L 2003 *Phys. Rev.* C **67** 055802
[10] Eichler M, Arcones A, Kelic A, Korobkin O, Langanke K, Marketin T, Martinez-Pinedo G, Panov I, Rauscher T, Rosswog S, Winteler C, Zinner N and Thielemann F-K 2015 *The Astrophysical Journal* **808** 30
[11] Aleksankin V G, Lyutostanskii Yu S and Panov I V 1981 *Sov. J. Nucl. Phys.* **34** 804
[12] Panov I V, Lutostanskii Yu S and Thielemann F-K 2015 *Bull. Of RAS, phys.* **79** 437
[13] Ejiri H, Ikeda K and Fujita J 1968 Phys. Rev. B **176** 1277
[14] Arita K, Ejiri H and Fujita J 1976 *Supplement of Prog. Theor. Phys.* **60** 47
[15] Panov I V, Korneev I Yu and Thielemann F-K 2008 *Astronomy Letters* **34** 189
[16] Panov I V, Lutostanskii Yu S and Thielemann F-K 2013 *Phys. At. Nucl.* **76** 88
[17] Lorusso G, Nishimura S, Xu Z Y et al. 2015 *Phys. Rev. Lett.* **114** 192501
[18] Möller P, Nix J R, Myers W D, and Swiatecki W J 1995 *At. Data Nucl. Data Tables* **59** 185
[19] Aboussir Y, Pearson J M, Dutta A K, Tondeur F 1995 *At. Data Nucl. Data Tables* **61** 127

IOP Conf. Series: Journal of Physics: Conf. Series **940** (2017) 012054 doi:10.1088/1742-6596/940/1/012054

Neutrino nucleosynthesis in core-collapse Supernova explosions

A. Sieverding[1], L. Huther[1], G. Martínez-Pinedo[1,2], K. Langanke[2], A. Heger[3]

[1] Institut für Kernphysik (Theoriezentrum), Technische Universität Darmstadt, Schlossgartenstraße 2, 64289 Darmstadt, Germany

[2] Gesellschaft für Schwerionenforschung Darmstadt, Planckstr. 1, D-64259 Darmstadt, Germany

[3] Monash Centre for Astrophysics, School of Physics and Astronomy, Monash University, Victoria 3800, Australia

E-mail: asiever@theorie.ikp.physik.tu-darmstadt.de

Abstract. The neutrino-induced nucleosynthesis (ν process) in supernova explosions of massive stars of solar metallicity with initial main sequence masses between 15 and 40 $M_\odot$ has been studied. A new extensive set of neutrino-nucleus cross-sections for all the nuclei included in the reaction network is used and the average neutrino energies are reduced to agree with modern supernova simulations. Despite these changes the ν process is found to contribute still significantly to the production of the nuclei ^{7}Li, ^{11}B, ^{19}F, ^{138}La and ^{180}Ta, even though the total yields for those nuclei are reduced. Furthermore we study in detail contributions of the ν process to the production of radioactive isotopes ^{26}Al, ^{22}Na and confirm the production of ^{92}Nb and ^{98}Tc.

1. Introduction

Core-Collapse-Supernova explosions are the most energetic events known to astronomers. The conditions created by the passage of the supernova shockwave through the chemically enriched layers of massive stars allow for important nucleosynthesis processes to occur which are crucial for the enrichment of the interstellar medium and chemical evolution of the universe. A major part of the energy released by the core-collapse leaves the stellar core as neutrinos of all flavors. Those neutrinos can interact with the shock-heated material and affect the nucleosynthesis. In recent years multi-dimensional supernova simulations have given new insights concerning the neutrino radiation. Neutrinos with energies of the order of 10 MeV can lead to nuclear excitations beyond the particle separation threshold, affecting the chemical composition and density of free nucleons. Furthermore, electron neutrinos and electron antineutrinos can be captured by nuclei, leading to the inverse process of β-decay.

Following the approach of previous studies we choose a highly parametrized supernova model as described in reference [1] which allows a exploration of the parameter space of the neutrino properties and the explosion. The key ingredient of the explosive nucleosynthesis is the peak temperature of the shock heated material. This is estimated by an analytic expression that has been shown to agree reasonably well with hydrodynamical simulation. The decrease of temperature and density are modeled to be exponential on a density dependent timescale.

Table 1. Production factors relative to solar abundances from reference [8], normalized to ^{16}O production. Shown are the results obtained without neutrinos, with our choice of neutrino temperatures ("Low energies", $T_{\nu_e} = 2.8$ MeV, $T_{\bar{\nu}_e} = T_{\bar{\nu}_{\mu,\tau}} = T_{\nu_{\mu,\tau}} = 4.0$ MeV), and with the choice of ref. [5] ("High energies", $T_{\nu_e} = T_{\bar{\nu}_e} = 4.0$ MeV, $T_{\nu_{\mu,\tau}} = T_{\bar{\nu}_{\mu,\tau}} = 6.0$ MeV) for 25 $M_\odot$ progenitor model.

Star	Nucleus	no ν	Low energies	High energies
25 $M_\odot$	^{7}Li	0.0005	0.11	0.55
	^{11}B	0.003	0.80	2.61
	^{15}N	0.08	0.10	0.13
	^{19}F	0.06	0.24	0.43
	^{138}La	0.03	0.63	1.14
	^{180}Ta	0.14	1.80	2.81

Neutrinos are assumed to have Fermi-Dirac spectra with chemical potential $\mu_\nu = 0$, characterized by a neutrino temperature T_ν related to the average energy by $\langle E_\nu \rangle \approx 3.14\ T_\nu$. Neutrino luminosities are modeled with and exponential decrease on the timescale of 3 s. A total energy of 10^{53} erg is distributed equally among the 6 neutrino flavors, leading to number fluxes according to the average energy of the prescribed to the corresponding neutrino type.
Our set of neutrino-nucleus cross-sections has been calculated for both, charged- and neutral-current reactions, based on Random Phase Approximation (RPA) [2]. Spallation products in the case of states above particle separation threshold are computed based on statistical models [3].

We use solar metallicity supernova progenitors calculated by A. Heger et al. [4], with main sequence masses ranging between 15 and 40 $M_\odot$. The stellar models have been evolved up to core-collapse, providing the initial conditions and composition for the further evolution parametrized as stated above.

2. Canonical ν process nuclei

The ν process is known to have significant effects on the production of ^{7}Li, ^{11}B, ^{19}F, ^{138}La and ^{180}Ta. Previous studies of neutrino nucleosynthesis have assumed average neutrino energies of $\langle E_{\nu_e,\bar{\nu}_e} \rangle = 12.6$ MeV, $\langle E_{\nu_x} \rangle = 18.8$ MeV, where ν_x corresponds to ν_μ, $\bar{\nu}_\mu$, ν_τ and $\bar{\nu}_\tau$ [5]. Using instead significantly reduced values of $\langle E_{\nu_e} \rangle = 8.8$ MeV, $\langle E_{\bar{\nu}_e,\nu_x} \rangle = 12.6$ MeV as suggested by detailed simulations e.g. in references [6, 7] we still see a significant increase in the production of ^{7}Li, ^{11}B, ^{19}F, ^{138}La and ^{180}Ta due to neutrino-induced reactions (see Table 1).
We observe, that the the 15 $M_\odot$ progenitor model overproduces ^{19}F even without neutrinos mainly via the reaction chain $^{18}\mathrm{O}(p,\alpha)^{15}\mathrm{N}(\alpha,\gamma)^{19}\mathrm{F}$ operating on ^{18}O at the lower edge of the He-shell. Due to the stellar structure this process does not work for the more massive progenitor models and here the ν process can make a major contribution.

A direct comparison between the present results and the results of reference [5] tends to be misleading, since differences appear due to the different treatment of hydrodynamic variables. Work to disentangle changes due to different neutrino interactions from changes due to the hydrodynamics is in progress.

3. Impact on radioactive nuclei

Observations of γ-rays allow direct access to the production of ^{26}Al in the galaxy. Supernova explosions are assumed to give a significant contribution to the overall abundance of ^{26}Al [9].

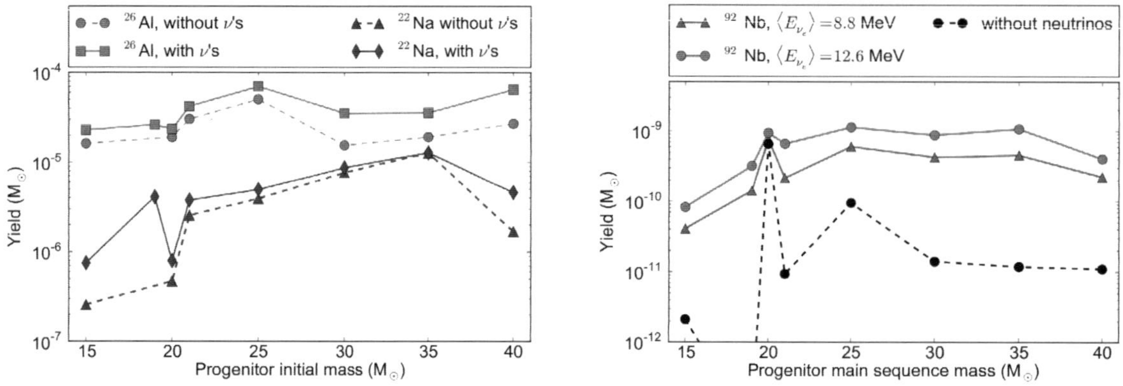

(a) ^{26}Al yield for a set of progenitor stars.

(b) ^{92}Nb yield for a set of progenitor stars.

Figure 1. Shown are the results of calculations without neutrinos (dashed lines) and including neutrinos with different spectra. Electron neutrinos and anti-neutrinos are assumed have a distribution with $T_{\nu_e} = T_{\bar{\nu}_e} = 4$ MeV.

Previous studies have already reported an enhancement of the production of ^{26}Al by 30% to 50% [10] due to neutrinos. This effect is largely due to neutral-current spallation reactions that increase the abundance of free protons, enhancing capture reactions on ^{25}Mg in the Oxygen-Neon shells. In our calculations we even find an increase in the production of ^{26}Al of up to a factor of 2 for the 30 and 35 M$_\odot$ progenitor models. Especially in those cases the charged-current reaction ^{26}Mg(ν_e,e$^-$)^{26}Al is found to be the dominating channel for the increased production.

The decay of ^{22}Na it is relevant for the description of supernova lightcurves [11]. In our calculations the production of ^{22}Na is increased up to a factor of 3 , mainly due to increased proton captures on ^{22}Ne and ^{22}Ne(ν_e,e$^-$)^{22}Na.
In total we find also for ^{22}Na an increase of the relative importance of charged-current channels, as discussed in more detail in reference [12].
We can also confirm significant contributions of the ν process to the production of^{92}Nb and ^{98}Tc due to electron neutrino absorption on the corresponding isobars, as also discussed in [13].
Furthermore, we find a significant enhancement of the production of the short-lived radioisotope ^{36}Cl. The characteristic γ-rays from the decay of ^{44}Ti and ^{60}Fe are also used as tracers for active nucleosynthesis sites [9, 14]. The effect of neutrino interactions on the yields of ^{44}Ti and ^{60}Fe have been found to be at most 2% in the case of ^{44}Ti and even less for ^{60}Fe.

4. References

[1] S. E. Woosley, D. H. Hartmann, R. D. Hoffman, W. C. Haxton, *The nu-process*, ApJ **356** 272.

[2] E. Kolbe, K. Langanke, G. Martínez-Pinedo, P. Vogel, *Neutrino nucleus reactions and nuclear structure*, Journal of Physics G Nuclear Physics **29** 2569.

[3] L. Huther, *Nucleosynthesis in neutrino-driven winds*, Ph.D. thesis, TU Darmstadt, Darmstadt, Germany (2013).

[4] T. Rauscher, A. Heger, R. D. Hoffman, S. E. Woosley, *Nucleosynthesis in Massive Stars with Improved Nuclear and Stellar Physics*, ApJ **576** 323.

[5] A. Heger, E. Kolbe, W. C. Haxton, K. Langanke, G. Martínez-Pinedo, S. E. Woosley, *Neutrino nucleosynthesis*, Physics Letters B **606** 258.

[6] T. Fischer, S. C. Whitehouse, A. Mezzacappa, F.-K. Thielemann, M. Liebendörfer, *The neutrino signal from protoneutron star accretion and black hole formation*, A&A **499** 1, [0809.5129].

[7] L. Hüdepohl, B. Müller, H.-T. Janka, A. Marek, G. G. Raffelt, *Neutrino Signal of Electron-Capture Supernovae from Core Collapse to Cooling*, Physical Review Letters **104** (25) 251101, [0912.0260].

[8] K. Lodders, *Solar System Abundances and Condensation Temperatures of the Elements*, ApJ **591** 1220.

[9] M. Limongi, A. Chieffi, *The Nucleosynthesis of* ^{26}Al *and* ^{60}Fe *in Solar Metallicity Stars Extending in Mass from 11 to 120* M_{solar}*: The Hydrostatic and Explosive Contributions*, ApJ **647** 483.

[10] F. X. Timmes, S. E. Woosley, D. H. Hartmann, R. D. Hoffman, T. A. Weaver, F. Matteucci, *26Al and 60Fe from Supernova Explosions*, ApJ **449** 204.

[11] F.-K. Thielemann, M.-A. Hashimoto, K. Nomoto, *Explosive nucleosynthesis in SN 1987A. II - Composition, radioactivities, and the neutron star mass*, ApJ **349** 222.

[12] A. Sieverding, L. Huther, K. Langanke, G. Martínez-Pinedo, A. Heger, *Neutrino Nucleosynthesis of radioactive nuclei in supernovae*, ArXiv e-prints [1505.01082].

[13] M.-K. Cheoun, *et al.*, *Neutrino induced reactions for* ν*-process nucleosynthesis of* ^{92}Nb *and* ^{98}Tc, Phys. Rev. C **85** (6) 065807.

[14] B. W. Grefenstette, *et al.*, *Asymmetries in core-collapse supernovae from maps of radioactive* ^{44}Ti *in Cassiopeia A*, Nature **506** 339.

IOP Conf. Series: Journal of Physics: Conf. Series **940** (2017) 012055 doi:10.1088/1742-6596/940/1/012055

Alpha-Induced Production Cross Sections of $^{77,\,79}$Kr and ^{77}Br

Zuzana Slavkovská[1], Stefan Fiebiger[1], Ulrich Giesen[2], Tanja Heftrich[1], René Reifarth[1], Stefan Schmidt[1], Benedikt Thomas[1], Mario Weigand[1]

[1]Goethe University Frankfurt, Germany

[2]Physikalisch-Technische Bundesanstalt Braunschweig, Germany

Abstract. The production cross sections of $^{77,\,79}$Kr and ^{77}Br following the α-irradiation of natural selenium were determined between the α-energies of 11 MeV and 15 MeV using the activation technique. The irradiation of natural selenium targets with He^{2+} ions extracted from a cyclotron was conducted at Physikalisch-Technische Bundesanstalt in Braunschweig in Germany. The spectroscopic analysis of the reaction products was performed using a HPGe detector. As the α-beam was stopped inside the targets, the thick target yields were determined. The corresponding energy-dependent cross sections were calculated from the difference of the thick target yields at various beam energies. The determined values were compared to theoretical predictions based on the TALYS code.

1. Motivation

Measurements of reaction cross sections help to constrain models predicting stellar reaction rates and can therefore improve our understanding of the stellar nucleosynthesis. The α-induced production cross sections of the isotopes $^{77,\,79}$Kr and ^{77}Br (Figure 1) help to understand the chemical evolution of the universe.

Figure 1. Excerpt of the charts of nuclides showing the production of ^{77}Kr via the ^{74}Se(α,n) reaction, ^{79}Kr via the ^{76}Se(α,n) reaction and ^{77}Br via the ^{74}Se(α,p) reaction.

IOP Conf. Series: Journal of Physics: Conf. Series **940** (2017) 012055 doi:10.1088/1742-6596/940/1/012055

2. Selenium Target Production

The selenium targets for the activation at the Physikalisch-Technische Bundesnstalt (PTB) in Braunschweig, Germany [1], were produced via the irradiation of a stable metallic ^{82}Se sample by a He-beam. As the planned α-beam were entirely stopped in the target material, a thick target layer was produced by melting selenium onto an aluminium backing. In total, nine backings with a diameter of 35 mm and a thickness of 1.5 mm and a gap in the middle with a diameter of 10 mm and a height of 0.5 mm were produced (Figure 2).

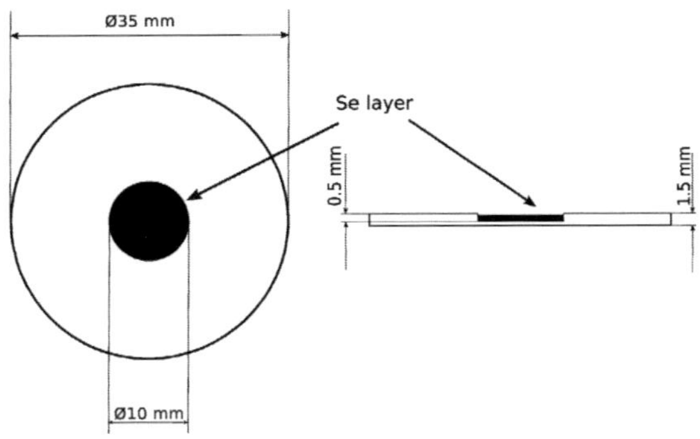

Figure 2. Geometry of the selenium backings used for the irradiation at PTB in Braunschweig.

Selenium powder was placed in the gap and heated in an oven to its melting point of 221°C. Because of the high surface tension of selenium, the powder did not melt smoothly, but built droplets. These were spread mechanically using a spattle after reducing the oven temperature at approximately 100°C. More selenium powder was put in the gap as a part of it stick on the spattle. The heating procedure was repeated to form a smooth layer of selenium, which is essential for an activation experiment.

3. Experiment at Physikalisch-Technische Bundesanstalt

The prepared selenium targets were activated in February 2014 at the PTB in Braunschweig, Germany, and irradiated at the energy-variable PTB cyclotron [2]. Doubly-charged He ions were extracted from the cyclotron to irradiate the targets in the activation chamber designed as a Faraday cup [2]. A wobbled α-beam was used to increase the size of the illumination spot on the targets. A water cooling system was used to reduce the target temperature. In total, nine samples were irradiated at different α-energies between 11 MeV and 15 MeV.

Krypton stays trapped in selenium as long as the temperature stays below 50°C [3], so the temperature during the irradiation had to stay below this temperature. For this reason, four activations were performed to test the thermal stability of selenium. The currents of $I = 4\ \mu$A and $I = 1\ \mu$A led to selenium losses, hence a current of $I = 500$ nA was found to be appropriate for the irradiations. Five activations with activation times between 0.5 h and 6.8 h were performed. The spectroscopic analysis of the reaction products ^{77,79}Kr and ^{77}Br was conducted using a High Purity Germanium (HPGe) detector at PTB in Braunschweig.

4. Results

As the α-beam was stopped inside the targets, the thick target yields were determined. The corresponding energy-dependent cross sections at the α-energies of 12 MeV and 14 MeV were calculated from the difference of the thick target yields at various energies:

$$\sigma(E_1, E_2) \propto Y_{\text{thick}}(E_2) - Y_{\text{thick}}(E_1). \tag{1}$$

The resulting preliminary α-induced production cross section of ^{77}Kr at an α-energy of 12 MeV is 160 mb and at an α-energy of 14 MeV 400 mb. The α-induced production cross section of ^{77}Br at an α-energy of 12 MeV is 62 mb and at an α-energy of 14 MeV 120 mb. The α-induced production cross section of ^{79}Br at an α-energy of 12 MeV is 250 mb and at an α-energy of 14 MeV 400 mb .

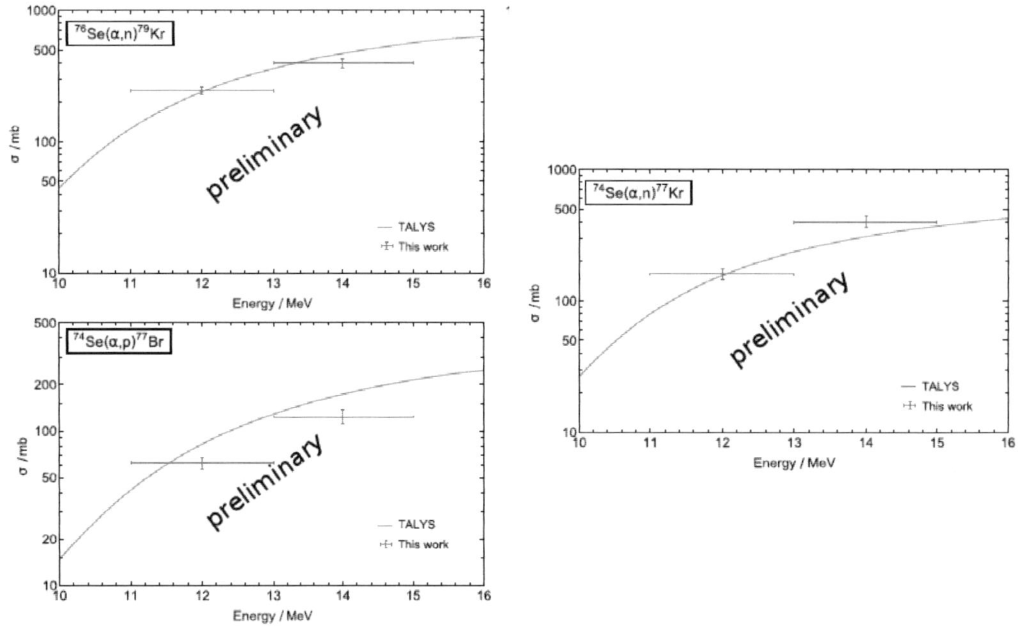

Figure 3. The α-induced production cross sections of $^{77,\,79}$Kr and ^{77}Br at 12 MeV and 14 MeV compared to the theoretical predictions of TALYS-1.6. The energy uncertainty of ±1 MeV originates from the subtraction of the thick target yields.

Finally, the resulting values of the α-induced production cross sections were compared to theoretical predictions based on the TALYS-1.6 code [4]. The theoretical predictions were found to be in a good agreement with the measured cross sections (Figure 3).

5. Acknowledgement

This project was supported by the Europian Research Council under the Europian Union's Seventh Framework Programme (FP/2007-2013)/ERC Grant Agreement n. 615126, HIC for FAIR and NAVI.

References

[1] Brede H, Cosack M, Dietze G, Gumpert H, Jahr R, Kutscha M, Schlegel-Bickmann D and Schölermann H 1980 *Nuclear Instruments and Methods* **169** 349–358

[2] Sauerwein A, Becker H W, Dombrowski H, Elvers M, Endres J, Giesen U, Hasper J, Hennig A, Netterdon L, Rauscher T, Rogalla D, Zell K O and Zilges A 2011 *Phys. Rev. C* **84** 045808

[3] Suzuki K, Blessing G, Qaim S and Stöcklin G 1982 *The International Journal of Applied Radiation and Isotopes* **33** 1445 – 1448 ISSN 0020-708X

[4] Koning A J, Hilaire S and Duijvestijn M C 2005 *International Conference on Nuclear Data for Science and Technology* (*American Institute of Physics Conference Series* vol 769) ed Haight R C, Chadwick M B, Kawano T and Talou P pp 1154–1159

IOP Conf. Series: Journal of Physics: Conf. Series **940** (2017) 012056 doi:10.1088/1742-6596/940/1/012056

Test of a SiPM-scintillator-based muon detector at the Gran Sasso National Laboratory

I Stark[1], V Könye[1], R Antolini[2], F Arneodo[3], A Boeltzig[4], A Candela[2], A Di Giovanni[3], Zs Fülöp[5] and M Junker[2,4]

[1] Eötvös Loránd University (ELTE), H-1117 Budapest, Hungary
[2] Gran Sasso National Laboratory of INFN (LNGS), Assergi(AQ), Italy
[3] New York University in Abu Dhabi, Abu Dhabi, UAE
[4] Gran Sasso Science Institute (GSSI), L'Aquila, Italy
[5] Institute for Nuclear Research (Atomki), H-4001 Debrecen, POB.51., Hungary

E-mail: konyeviktor@caesar.elte.hu

Abstract. The aim of the present work was to determine the characteristics of a tracking scintillation detector by measuring cosmic muons above- and underground at the Gran Sasso National Laboratory (LNGS). The detector has been designed for demonstrational purposes and had never been involved in scientific research before. It is a compact, layered detector using plastic scintillator bars and silicon photomultipliers (SiPM). In the present setup no energy information is used. In the underground laboratory we had to deal with a very low rate of muons (6 orders of magnitude less than on the surface). For this reason we optimized the trigger configuration. During the evaluation process we designed and optimized a maximum-likelihood line fitting method which is able to deal with the high dark count rate and the non optimized detection efficiency of the detector. We also carried out simulations to calculate the actual flux and intensity from the measured ones. Finally, we compared our results with the literature and they were found to be in agreement.

1. Introduction

The primary role of underground laboratories such as LNGS is to provide a low background environment suitable for the low cross-section experiments conducted inside. Precise detectors capable of detecting the background (such as cosmic muons) can serve as a veto-system in many of the performed experiments, and therefore are under constant development.

The detector we worked with can be used for muon track reconstruction. As it wasn't involved in scientific research before, a lot of work was done to understand its properties, and to prepare it for data collection. From the measured data and the found properties the angular intensity aboveground and the flux above- and underground was determined.

2. Setup

In the detector the cosmic ray detection is based on scintillation. The used scintillator bars are made from extruded polystyrene (STYRON 663W) doped with PPO and POPOP made by Fermilab [1]. The polystyrene bars are extruded in the form of $64 \times 4 \times 1$ cm^3 bricks with a hole at the longest symmetry axis. In this hole a wavelength shifter fiber is inserted which is used to collect the light at one end of the scintillator. This is important for the photodetection,

IOP Conf. Series: Journal of Physics: Conf. Series **940** (2017) 012056 doi:10.1088/1742-6596/940/1/012056

which in this case is more sensitive to the shifted wavelength. For photon detection silicon photomultipliers (SiPMs) were used. The used SiPMs made by AdvanSiD (ASD-SiPM1S-M) have an array of 10×10 Geiger-mode avalanche photodiodes (GM-APD) with a total size of $1\,\mathrm{mm}^2$. These diodes are capable of detecting few photons, but only in counting mode (without energy information).

The detector (figure 1) is built in a way to have a naturally built-in Cartesian coordinate system. There are 10 layers of scintillator bars on top of each other. Every layer is composed of two orthogonal half-layers one below the other (figure 2). Each half-layer consists of 16 parallel bars, so the whole detector has 320 bars connected to 320 SiPMs. Each bar is 1 cm high so a layer is 2 cm. The distance between the layers is 10.9 cm.

A signal coming from a SiPM is preamplified and the analogue sum of a half-layer is collected on the controller board present on the top of the detector. The triggering mechanism uses this summed value and gives a trigger when the selected half-layers have signals above a predefined threshold. When the trigger occurs, the digitized signals of each SiPM are driven to LED matrices present on two panels. This way the passing particles become visible in real-time. For data analysis the digitized signals can be acquired trough a PC. Seeing the bars which the muon has crossed gives the possibility to reconstruct the three dimensional track of the detected particle.

Figure 1. The muon telescope.

Figure 2. Illustration of the orthogonal half-layers

3. Data analysis

3.1. Performance of the detector

The noise rate for each bar was evaluated by collecting about 13 million events underground. Assuming that practically all of the signals underground are noise events, the rate can be calculated by counting how many times a given SiPM was giving a signal.

For evaluating the efficiency of each bar about 100 000 muons were measured aboveground. After fitting the lines of muons and selecting the events caused by muons, the efficiencies were calculated by counting how many times a given SiPM was giving a signal out of the times a muon passed the relevant bar. In the line fitting method the efficiencies of the bars are used, therefore prior efficiency values were estimated from vertical tracks.

The efficiency is strongly dependent on the zenith angle of the muon tracks, because the more inclined the angle of the track is, the longer the muon travels through the bars. The angular dependence of the efficiency of each bar was measured using the method described before, but restricting it to muons that come from certain solid angles.

IOP Conf. Series: Journal of Physics: Conf. Series **940** (2017) 012056 doi:10.1088/1742-6596/940/1/012056

3.2. Line fitting and sorting algorithm

In the present work the trajectories of muons are assumed to be straight lines. In our case the widely used method of least squares cannot be applied, because of the high noise rates and the fact, that the information about the track is not a set of points in space but bars.

For this reason a maximum likelihood method was used: for numerous lines where the muon could have passed through the detector the probability of measuring the given dataset is calculated. The line for which this probability is the highest will be the fitted line. The probability can be determined from the efficiency and noise rate of each bar.

Underground the ratio of measured muons to random coincidences is 1:300 000. Therefore it is crucial to differentiate the two. An event was claimed to be caused by a muon, if there are at least 7 signals on the fitted line and no more than 9 elsewhere.

4. Simulations

Two simulations were carried out. In the first one a 100 million random coincidences were simulated based on the measured noise rates. After fitting and selecting them, none of them was claimed to be a muon event. This means, as there were 33 million events collected underground, that the used selection criteria is strong enough to use in further data analysis.

In the second simulation measured datasets caused by muons coming from different directions were simulated, based on the noise rates and the angular dependent efficiencies. For each incoming solid angle range it was calculated how many times it was measured to come from different solid angle ranges. This can be represented as a matrix called the efficiency matrix (figure 3). The figure shows that most of the simulated muons were measured in the solid angle where they came from (purple line). Although, a few were measured in other directions too. Note, that the purple colour refers to only 18%, so about 80% of the simulated muons weren't measured, because they weren't triggered.

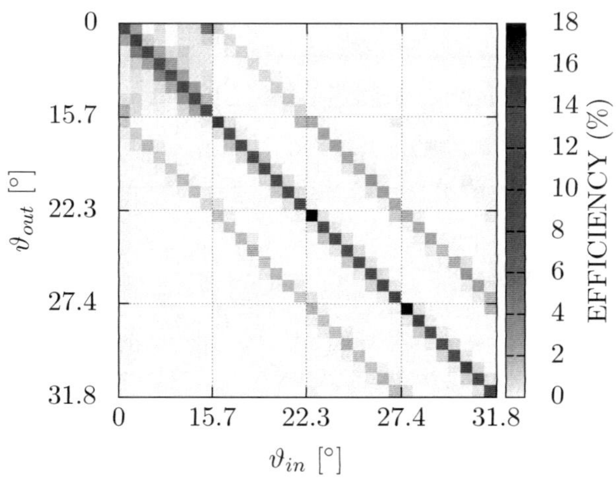

Figure 3. A part of the efficiency matrix, which shows in what percentage were simulated muons coming from given solid angles (columns), measured in different solid angles (rows). Within the dotted lines the solid angles are in the same zenith angle range. The 8 pixels in them refer to 8 different azimuthal angle range. The figure shows 4×4 different zenith angle ranges, subdivided into 8×8 azimuthal angle ranges each.

5. Results

With the previously described efficiency matrix ($\mathcal{E}$), having the measured data in each solid angle ($\mathbf{N}_m$) the corrected data ($\mathbf{N}$) can be calculated by inverting the efficiency matrix:

$$\mathcal{E} \cdot \mathbf{N} = \mathbf{N}_m \quad \Rightarrow \quad \mathcal{E}^{-1} \cdot \mathbf{N}_m = \mathbf{N} \tag{1}$$

The corrected data can be used to calculate the zenith angle dependence of the angular intensity. These calculated intensity values can be seen compared to the values calculated from the measured rates (figure 4).

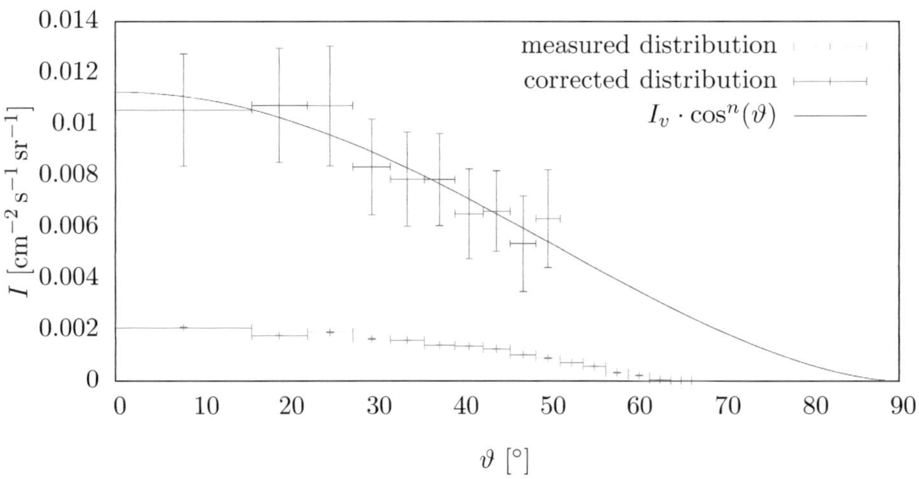

Figure 4. The zenith angle dependence of the angular intensity of muons at the surface. The fitted vertical angular intensity is $I_v = (1.1 \pm 0.2)\,\mathrm{cm^{-2}\,s^{-1}\,sr^{-1}}$ and the fitted exponent is $n = 1.7 \pm 0.2$.

By integrating the fitted angular intensity the aboveground flux of the muons was calculated: $F_a = (1.9 \pm 0.3) \times 10^{-2}\,\mathrm{cm^{-2}\,s^{-1}}$. The flux at the Gran Sasso Laboratory can be estimated from the count rates above- and underground. Table 1 shows the measured muons.

Table 1. Collected muon data under- and aboveground

	registered events	muons	T [s]	measured count rate [s^{-1}]
underground	32 848 892	95	3 610 429	2.631×10^{-5}
aboveground	169 397	80 792	6158	1.312×10^{1}

The underground muon reduction was estimated by dividing the count rates: $(2.0 \pm 0.2) \times 10^{-6}$. This assumes that the muon angular and energy distribution is the same in both locations. Multiplying this with the aboveground flux, we get the underground flux: $F_u = (3.8 \pm 0.7) \times 10^{-8}\,\mathrm{cm^{-2}\,s^{-1}}$. This is compatible within uncertainty with the official data of the Gran Sasso Laboratory: $F = (3.41 \pm 0.01) \times 10^{-8}\,\mathrm{cm^{-2}\,s^{-1}}$ [2].

Acknowledgments
Support of the experimental work by the people of LNGS is gratefully acknowledged. I S and V K want to express their gratitude to the Erasmus Programme for giving them the opportunity to work at Gran Sasso for three months.

References

[1] Pla-Dalmau A, Bross A D and Rykalin V V 2003 Extruding plastic scintillator at Fermilab
[2] Bellini G *et al.* (Borexino) 2012 *JCAP* **1205** 015 (*Preprint* 1202.6403)

IOP Conf. Series: Journal of Physics: Conf. Series **940** (2017) 012057 doi:10.1088/1742-6596/940/1/012057

Simple potential model for interaction of dark particles with massive bodies

Nurgali Takibayev

al-Farabi Kazakh National University, Almaty 050040, Kazakhstan

E-mail: `takibayev@gmail.com`

Abstract. A simple model for interaction of dark particles with matter based on resonance behavior in a three-body system is proposed. The model describes resonant amplification of effective interaction between two massive bodies at large distances between them. The phenomenon is explained by catalytic action of dark particles rescattering at a system of two heavy bodies which are understood here as the big stellar objects. Resonant amplification of the effective interaction between the two heavy bodies imitates the increase in their mass while their true gravitational mass remains unchanged. Such increased interaction leads to more pronounced gravitational lensing of bypassing light. It is shown that effective interaction between the heavy bodies is changed at larger distances and can transform into repulsive action.

1. Introduction

It is well known, that the dark matter mystery emerged at the analysis of astronomical data when considerable discrepancy between orbital velocities of stellar objects at periphery of large galaxies compared to velocities calculated for usual gravitational fields has been found. This phenomenon was explained by the dark matter action that resulted in increased mass of such stellar objects. Discovered effect of gravitational lensing supported such assumption [1, 2].

In this presentation a model for resonance amplification of interaction between two stellar objects is proposed. The amplification is stimulated by additional interaction of the massive objects with dark matter particles (hereinafter dark particles). Our consideration is conducted in the framework of non-relativistic quantum theory of scattering in the three-body system, more exactly, for describing the interactions of two heavy bodies in presence of a dark particle of a very small mass. The solutions distinctly demonstrate that the effective interaction of these massive objects has the resonance behavior with the changing of distance between them.

So, at relatively small distances the catalytic amplification is negligibly small and the effective interaction here coincides with the direct (gravitational) interaction between the heavy bodies. At larger distances, the catalytic action of the dark particles increases and the effective interaction between the heavy bodies also increases. At even larger distances, the catalytic action becomes smaller and then negative; so instead of attraction between the heavy bodies, it may cause their repulsion. The last to a certain extent would imitate the effects of dark energy.

We assume that the dark particles do not interact with each other; the massive bodies interact not only via gravity, but also through exchange of dark particles. For simplicity, we would not consider the three-body type forces.

IOP Conf. Series: Journal of Physics: Conf. Series **940** (2017) 012057 doi:10.1088/1742-6596/940/1/012057

2. Three-body quantum-mechanical scattering problem

Solutions for scattering amplitudes in the three-body problem are based on solutions for scattering amplitudes in two-body subsystems [3, 4]. The i-pair potential of interaction between the dark particle and a heavy body could be taken in the separable form: $V_i = |\nu_i > \lambda_i < \nu_i|$, where $< \nu_i|k >$is the form-factor of the potential, λ_i is the coupling constant. Then the two-body t_i-matrix equals: $t_i = |\nu_i > \eta_i < \nu_i|$, where $\eta_i^{-1} = \lambda_i^{-1} - < \nu_i|G_0|\nu_i >$, $G_0 = (E_0 - H_0)^{-1}$ is the free Green's function, $E_0 = k_0^2/2m$.

For illustration, we consider the simple potential $< \nu_i|k >= Nk/(\beta^2+k^2)$, where constants N and β have been taken to obtain the virtual level with the extremely small energy $E_v = -\kappa_v^2/2m$, when $\eta_i^{-1}(E_0) = 0$ for $k_0 = -i\kappa_v$. We use the units $\hbar = 1$ and $c = 1$.

Another simplification is related to the Born-Oppenheimer approximation which leads to the simple solutions of the three-body problem [4]. The task therefore is reduced to determination of interactions in the three body system: the dark particle interacts with each of the heavy bodies and the two heavy ones interact gravitationally. In the first case the equations for component of T-matrix are

$$T_{ij} = T_i\delta_{ij} + |\nu_i > \eta_i M_{ij}\eta_j < \nu_j| \; , \tag{1}$$

$$M_{ij}(\vec{r},\vec{r}') = J_{ij}(\vec{r})\delta(\vec{r}+\vec{r}') + \sum_{l=2,3} J_{il}(\vec{r})\eta_l M_{lj}(-\vec{r},\vec{r}') \; , \tag{2}$$

$$J_{ij}(\vec{r}) = \int d\vec{k}\exp(i\vec{r}\vec{k})\frac{\nu_i(\vec{k})\nu_j(\vec{k})}{E_0 - E_k + i\gamma} \; ; j \neq i \; . \tag{3}$$

The solutions can be written in the form: $M_{ij}(\vec{r},\vec{r}') = M^+(\vec{r})\delta(\vec{r}+\vec{r}') + M^-(\vec{r})\delta(\vec{r}-\vec{r}')$, where

$$M^+_{i\neq j}(\vec{r}) = D^{-1}_{ii}J_{ij}(\vec{r}) \; , \quad M^-_{ii}(\vec{r}) = D^{-1}_{ii}J_{ij}(\vec{r})\eta_j J_{ji}(-\vec{r}) \; , \tag{4}$$

$$D = I - J_{ij}(\vec{r})\eta_j J_{ji}(-\vec{r})\eta_i \; . \tag{5}$$

There are two important variables: k_0 is the wavenumber of the incident dark particle and $d = 2r$ is the distance between the two heavy bodies.

It is very important that if the peculiarities of two-body amplitude of dark particle with heavy body are situated close to the physical area of scattering.

Below we denote with the index dh the solutions where only the dark matter particle interactions are taken into account. The indices i, j are omitted for simplicity.

Now we have to include in the consideration the potential V_1, which acts directly between the two heavy bodies. Using the Lippmann-Schwinger equation for two potentials, the wavefunction of the system can be written in the form:

$$|\Psi >= |\Psi_{dh} > +G_{dh}V_1|\Psi >= |\Psi_{dh} > +G_0(I + T_{dh}G_0)V_1|\Psi >= |\Psi_{dh} > +G_0V_{ef}|\Psi > \; , \tag{6}$$

where $\Psi(\vec{r})$ is the total wave function of the three-body system. Note, the total effective potential corresponds to the distorted potential V_1:

$$V_{ef} = V_1 + |\nu_{dh} > \eta_{dh}[1 + (M^+_{dh} + M^-_{dh})\eta_{dh}] < \nu_{dh}|)G_0]V_1 \; . \tag{7}$$

This effective potential as well as the amplitude $M^{\pm}_{dh}(r)$ would depend resonantly on both the distance between the heavy particles and the energy of the dark particle [5]. The enhancement factor could be written as $\xi =< V_{ef} > / < V_1 >$, where $< V >$ means taking the average values with consideration of the density of dark matter particles. Assuming that the density of number of dark matter particles is comparable with density of massive particles, one can write:

$$\xi = 1 + (1 - \eta_{dh}/\lambda_{dh})[1 + (M^+_{dh} + M^-_{dh})\eta_{dh}] \; . \tag{8}$$

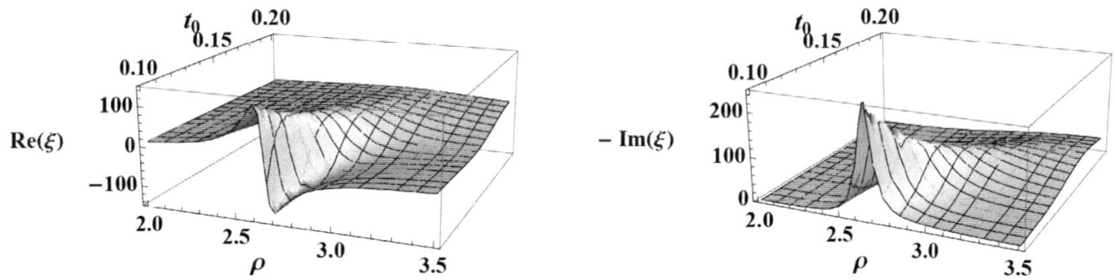

Figure 1. Real and imaginary parts of $\xi(t_0, \rho)$; $\lambda_{dh} = -0.95$; the resonance at $\rho \approx 2.65$.

Let us assess the characteristic distances. For example in our model, resonance takes place at the following parameters: $\lambda_{dh} = -0.95$, $\rho = r \cdot \beta = 2.65$. In the case of the distance between the massive bodies is about $r \approx 2.65 \cdot 10^{22} cm$, then $k_0 \approx 10^{-23} cm^{-1}$, $t_0 \cdot = k_0/\beta$.

Taking into account the resonant behavior of the effective interaction at such distances, one can get from (8) the following multiplication of the interaction potential for the factor: $\xi \approx 100$ at $\rho = 2.65$.

This important outcome demonstrates that the interactions with the dark particles can substantially change the character of forces acting between the heavy bodies (Fig.1).

The orbital velocity of the peripheral body becomes higher than that at normal gravity owing to the enhancement factor. Moreover, this flux would glue up these two heavy bodies and, with respect to other particles and fields (gamma quanta, for instance) such a system would appear as a single object. The system would have the effective mass much higher than its own mass.

Such mechanism can contribute to gravitational lensing of electromagnetic radiation. Really, we can include in the equation (6) the additional interactions with gammas and obtain the additional enhancement of these interactions.

It is remarkable that the enhancement factors can be negative at certain values in the case of two-body attractive potentials. For example, in the case of $\lambda_{dh} = -0.95$ we have $\xi \approx -80 - i \cdot 50$ at $\rho = 2.85$. Then the resulting potential between two heavy bodies looks like a repulsive force.

Conclusion

It is important to note that the catalytic impact of dark particles on the interaction between the two heavy bodies becomes very strong only at large distances between these bodies. This effect has to be much stronger in the case the energy of two-body (dark particle and heavy body) virtual state of the subsystem is situated closer to the physical scattering region ($\lambda_{dh} = -1 + \delta$, when $\delta \to 0_+$). It is remarkable that if the distance between two heavy bodies is increasing, then the attractive forces can transform to the repulsive ones. This changing can lead to the formation of structures, for example, star clusters. If we assume the existence of several types of dark particles, they can produce many different structures such as galaxies, galaxy clusters, etc.

In reality, the interactions between dark particles and a heavy body can be more complex and the catalytic action may happen to be far from simple.

References

[1] Bertone G., Merritt D., 2005 *Modern Physics Letters* B **20** 1021
[2] Franson J.D., 2014 *New Journal of Physics* **16** 065008
[3] Faddeev L.D., 1965 *Mathematical Aspects of the Three Body Problem in Quantum Scattering Theory*, Davey, New York
[4] Takibayev N., 2008 *Physics of Atomic Nuclear* **71** 460
[5] Takibayev N., 2014 *Advanced Studies in Theoretical Physics* **8** 875 (arXiv:1401.2738v3[quant-ph])

IOP Conf. Series: Journal of Physics: Conf. Series **940** (2017) 012058 doi:10.1088/1742-6596/940/1/012058

Excited nuclei, resonances and reactions in neutron star crusts

N Takibayev[1], D Nasirova[1], K Katō[2] and V Kurmangaliyeva[1]

[1] IETP, al-Farabi Kazakh National University, Almaty 050040, Kazakhstan
[2] Nuclear Data Centre, Hokkaido University, Sapporo, Japan

E-mail: takibayev@gmail.com

Abstract. The short review of research results concerning the study of reactions and processes that occur in the neutron star crusts is given. The peculiarities of electron capture reactions by a nucleus in overdense crystalline structures have been demonstrated for various nuclei, in particular some even-even nuclei at electron capture reactions give daughter nuclei in excited states. Excited nuclei due to nonlinear interactions lead to a high-order harmonic generation. High energy gammas interact with charged particles, give a neutrino radiation and also knock out nucleons from neighbour nuclei.

It is also shown that interactions of neutrons with two and more nuclei in an overdence lattice give a large number of new resonance states. These resonances result in a formation of specific local oscillations in the corresponding layers of the lattice. The periodic enhancement of these processes in the dependence on the elemental composition of the primary neutron star matter is considered.

1. Introduction

We consider processes, reactions and few-body effects that occur in the crystalline structures of neutron star crusts in the region of density $10^6 g/cm^3$ - $10^{13} g/cm^3$. In this area the substance is the overdense crystalline lattice formed with bare nuclei placed in the degenerate electron Fermi liquid. It is the unique environment for the development of interrelated reactions, nonlinear phenomena and new resonance states.

The main sources of information about the neutron stars are the characteristics of radiation from the surface and the near surroundings; the thin atmosphere and land surface, strong magnetic field, accreting outside material from the nebula or the nearest star, and so on.

It is noticeable, that describing the processes in the neutron star crusts one can use the quantum theory for the quasiparticles supplemented with methods provided by physics of nonlinear phenomena, and the quantum theory of few-body systems, taking into account resonance states. In this short paper, we explain two our recent studies.

2. Processes and reactions in the overdense structures of neutron star envelopes

Considering a high density region in the envelopes of neutron stars, one should pay attention to the fact that the electrons displaced off the atomic orbitals form the degenerate electron Fermi liquid. In such a place, instead of undergoing the β-decay, absorption of electrons by nuclei with the emission of neutrinos will take place [1]. In this area, the atomic nuclei still retain their individual properties, as the distance between them are much greater than the size of the nuclei themselves.

Based on the different variants of the primary chemical composition of neutron stars, we investigate the chains of electron capture reactions and the dependence on the depth in the envelopes. The reactions have different characteristics for even-even, even-odd, odd-even and odd-odd initial nuclei. For example, the reactions starting from the $^{56}_{26}Fe^{(0+)}$ nuclei are expressed as the following chain:

$$^{56}_{26}Fe + e^- \rightarrow ^{56}_{25} Mn^{(3+)} + \nu \; (Q = 3.7MeV) \; ; \; ^{56}_{25}Mn + e^- \rightarrow ^{56}_{24} Cr^{(0+)} + \nu \; (Q = 1.63MeV)$$
$$^{56}_{24}Cr + e^- \rightarrow ^{56}_{23} V^{(1+)} + \nu \; (Q = 9.2MeV) \; ; \; ^{56}_{23}V + e^- \rightarrow ^{56}_{22} Ti^{(0+)} + \nu \; (Q = 7.14MeV)....$$

In this chain of reactions, the threshold energy Q for each of the odd proton number of the parent nuclei is less than the threshold energy of the previous reaction. It means that the second reaction for $^{56}_{25}Mn^{(3+)}$ is open and could be realized immediately after the first, because the electron energy $E > Q$.

However, taking into account the quantum numbers of the parent and daughter nuclei in the second reaction, i.e. for $^{56}_{25}Mn^{(3+)}$ and $^{56}_{24}Cr^{(0+)}$, one can see that direct transitions between the nuclei in ground states would be difficult (the second order forbidden). But the transition in the excited state is open: $^{56}_{25}Mn^{(3+)} + e^- \rightarrow ^{56}_{24} Cr^{*(2+)} + \nu \; (Q = 1.01MeV)$. The formation of an excited state of $^{56}_{24}Cr^{*(2+)}$ becomes therefore the most probable. This excited state is long-lived, since excited energy is less than the Fermi-electron energy. Moreover, the wavelength of the gamma is much greater than the distance between the nuclei. Therefore, the emission of the gamma by the excited nucleus is improbable.

In this case the transfer of energies to the environment represents a very complex process, which involves other factors such as the existence of neighboring excited nuclei, tunneling effects and emergence of quasi-particles, as well as the generation of Fermi electrons.

Excited nuclei in a high number density will contribute to the high-order harmonic generation of gamma rays. In turn, the high-energy gammas could cause reactions with the ejection of neutrons from neighboring nuclei and produce neutrino-antineutrino pairs [2].

The avalanche process - induced discharge of coherent radiation or excess of excitation of nuclei followed by the release of the mass of free neutrons occurs until reaching the critical density values of the excited nuclei. One can expect that some of these powerful emissions can cause "glitches" in the emission of pulsars.

For odd isobars, the situation is simpler: each subsequent reaction has an energy threshold above the energy threshold of the previous reaction, so the new nuclei remain in the ground state in the chain of electron capture reactions.

3. Neutron resonances in overdense lattice

It is known that a neutron-nucleus scattering at low energies has the resonance behavior and a number of resonances increases rapidly in keV-energy region [3]. Breit-Wigner's two-body resonances described by $t(E) = -(2\pi\rho)^{-1}\Gamma_2/(E - E_{R,2} + i\Gamma_2/2)$, where $E = k^2/2m$ is the neutron energy, $\pi\rho = 2mk/4\pi$, $E_{R,2} = k_2^2/2m$ and Γ_2 are the energy and the width of resonance, respectively. Here the index 2 denotes two-body values and we take $\hbar = 1$.

Free neutrons collide with nuclei fixed at nodes of the lattice, and as a result a neutron resonates with nuclei and makes a re-scattering few-body resonance state. For example, we discuss a simple case of isolated neutron resonances and the subsystem of two heavy nuclei. The new resonance states could be described as [4]

$$E_{R,3} = E_{R,2} \pm \Gamma_2 \cos(k_3 \cdot r)/(2 \cdot k_{R,2} \cdot r) \; , \; \Gamma_3 = \Gamma_2 \cdot (1 \pm \sin(k_3 \cdot r)/(k_{R,2} \cdot r) \; , \qquad (1)$$

where the index 3 corresponds to the three-body quantities such as $E_{R,3} = k_3^2/2m$. Note that all resonances are situated at the second Riemann sheet of the complex energy.

IOP Conf. Series: Journal of Physics: Conf. Series **940** (2017) 012058 doi:10.1088/1742-6596/940/1/012058

It is very important that the characteristics of the three-body resonances depend not only on the resonance energy and width of the neutron-nucleus two-body subsystem but also on the distance between two heavy nuclei.

Then, one can introduce the effective interaction between two nuclei in the form $\xi_n \cdot V^{ef}$, where ξ_n is a number density of free neutrons in the local layer of the crystalline structure. Following the well-known Hellmann-Feynman relationship $dE_n^{ef}/d\xi_n =< \chi_n|V^{ef}|\chi_n > / < \chi_n|\chi_n >$, where the wave function of the neutron χ_n can be taken such as a particle in a potential well, we can write the expression of the effective energy as $E_n^{ef} \approx \xi_n < \chi_n|V^{ef}|\chi_n > / < \chi_n|\chi_n >$ for a small value of ξ_n in an adiabatic approximation. Note that in this approximation $E_n^{ef} = E_{R,2} - i\Gamma_2/2 \pm J(d/2)$, where $J(r) = \Gamma_2 exp(ikr)/(2rk_{R,2})$ contains the main dependence on the distance between the nuclei. Here, d is the lattice parameter.

Enhancement of this specific interaction leads to changing of balance of forces in the corresponding layers of the overdense crystal, and causes local oscillations in these layers.

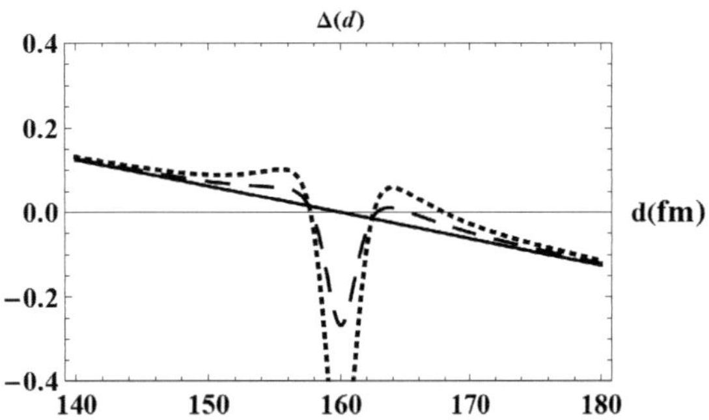

Figure 1. The sum of pressures in relative terms: $\Delta = (P_G + P_C + P_{res}^{ef})/P_C$. The balance of gravitational, Coulomb and Neutron resonance forces gives $\Delta(d_0) = 0$. The solid line corresponds the neutron density $\xi_n = 0$, i.e. $P_{res}^{ef} = 0$, the dashed line - $\xi_n = 6 \cdot 10^{-5} fm^{-3}$ and the dotted line - $\xi_n = 1.44 \cdot 10^{-4} fm^{-3}$

Pressure can be determined from the general expression [1]: $P = n^2 \partial(E_{tot}/n)/\partial n$, where n^{-1} is the volume per one baryon, E_{tot} is the total energy density of the respective forces. Note that P_{res}^{ef} is almost zero everywhere, but increases significantly only near the $d \approx d_{res}$. The net action of the forces leads to a new equilibrium value of $d_0 \neq d_{res}$, which may not have a monotonic behavior and there may be local density oscillations [5].

Figure 1 shows the deviation of the equilibrium values of d_0 in the absence and the presence cases of neutron resonances in the local layers of the overdense crystal. The resulting local oscillations allow us to predict that the radiation accompanying the few-body neutron resonances will have an oscillatory behavior. This may explain the micro-pulse emission of the periodic pulses observed in some pulsars (for example, PSR 1133+16) [6, 7, 8].

Conclusion

We here presented the two interesting phenomena which are expected to occur in neutron star envelopes. When we take into account the nonlinear and few-body interactions, they give various phenomena furthermore. It concerns the neutrino flow pulsations, star-quakes and so on.

References

[1] Shapiro S.L., Teukolsky S.A., 1983 *Blach Holes, White Dwarfs, and Neutron Stars: the Physics of Compact Objects*, Wiley, New York
[2] Takibayev N., Kato K. et. al. 2013 *Few-Body Systems* **54** doi 10.1007/s00601-013-0598-0
[3] Mughabghab S.F., 2006 *Atlas of Neutron Resonances*, Elsevier BV, Amsterdam
[4] Takibayev N., 2010 *EPJ Web of Conferences* **3** 05028
[5] Takibayev N., 2011 *Few-Body Systems* **50** 311
[6] Smirnova T.V., Tul'bashev S.A., Boriakoff V. 1994 *Astronomy and Astrophysics* **286** 807-814
[7] Kramer M., Johnston S., W. van Straten 2002 arXiv:astro-ph/0203126v18 Mar 2002
[8] Raithel C.A., Shannon R.M. et. al. 2015 arXiv: 1503.04490v1 [astro-ph.SR] 15 Mar 2015

IOP Conf. Series: Journal of Physics: Conf. Series **940** (2017) 012059 doi:10.1088/1742-6596/940/1/012059

Study of the ^{2}H(p,γ)^{3}He reaction in the BBN energy range at LUNA

Davide Trezzi for the LUNA collaboration

Università degli Studi di Milano & INFN, Milan, Italy

E-mail: davide.trezzi@mi.infn.it

Abstract. Using Big Bang Nucleosynthesis with the recent cosmological parameters obtained by the Planck collaboration, a primordial deuterium abundance value D/H = (2.65 ± 0.07) × 10^{-5} is obtained. This one is a little bit in tension with astronomical observations on metal-poor damped Lyman alpha systems where D/H = (2.53 ± 0.04) × 10^5. In order to reduce the BBN calculation uncertainty, a measurement of the ^{2}H(p,γ)^{3}He cross section in the energy range 10-300 keV with a 3% accuracy is thus desirable. Thanks to the low background of the underground Gran Sasso Laboratories, and to the experience accumulated in more than twenty years of scientific activity, LUNA (Laboratory for Underground Nuclear Astrophysics) planned to measure the ^{2}H(p,γ)^{3}He fusion cross section at the BBN energy range in 2015-2016. A feasibility test of the measurement has been recently performed at LUNA. In this paper, the results obtained will be shown. Possible cosmological outcomes from the future LUNA data will be also discussed.

1. Introduction

About three minutes and half after the Big Bang during the so called "Big Bang Nucleosynthesis" (BBN) era, thanks to the cosmic expansion, the Universe pass through a phase in which nuclear forces were able to synthesize light elements like Hydrogen, Helium, Lithium and Beryllium. From the theoretical point of view, the Universe expansion is described by the ΛCDM (Lambda Cold Dark Matter) model whose parameters are constrained by cosmological measurements. Using the experimental cross section of the reaction involved, it's thus possible to calculate the abundances of the primordial isotopes produced during the BBN. This values can be compared with the astronomical observations obtained in ancient astrophysical objects. Using the recent cosmological parameter values obtained by the PLANCK [1], a good agreement between BBN calculations and astronomical observations has been obtained except for Lithium (the so called "Lithium problem"). In the case of primordial deuterium, an accurate abundance determination has been obtained both from the astronomical and nuclear-astrophysical point of view. D/H = (2.53 ± 0.04) × 10^5 value has been recently obtained by Cooke et al. [2] in metal-poor damped Lyman alpha systems. This one is a little bit in tension with BBN calculations D/H = (2.65 ± 0.07) × 10^{-5} [3]. In order to clarify the actual situation, a reduction of the uncertainty is thus mandatory. From the BBN side, the main contribution to the error budget comes from the nuclear sector, i.e. the measurement of the cross sections involved in the deuterium production and destruction. It can be produced by neutron capture on proton and destructed with the ^{2}H(p,γ)^{3}He, ^{2}H(^{2}H,p)^{3}H, ^{2}H(^{2}H,n)^{3}He nuclear reactions. Up to now, the main contribution to the D/H error budget is due to the uncertainty on the ^{2}H(p,γ)^{3}He cross section [3]. This cross section has been measured in the past by many authors but the energy range relevant for the BBN (10 keV < E < 300 keV) is up to now covered only by a limited subset [4, 5] of them providing an overall 6-10% error on the S-factor [3]. As suggested by many authors [1, 3, 6], a new precise measurement of the ^{2}H(p,γ)^{3}He cross section in the BBN energy of interest is thus desirable. Thanks to the experience accumulated in about two decades of scientific activity and to the low natural background provided by the Laboratori Nazionali del Gran Sasso (LNGS), a new direct measurement of the ^{2}H(p,γ)^{3}He cross section has been planned at LUNA (Laboratory for Underground Nuclear Astrophysics) at the BBN energy range between the end of the 2015 and the beginning of 2016. A feasibility test has been performed at LUNA in October 2014. In the next sections a description of the apparatus used and the results obtained will be shown.

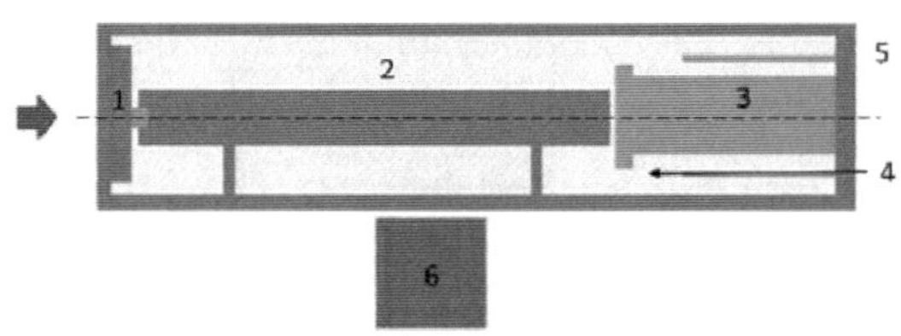

Figure 1. Experimental setup: The proton beam enters in the gas target filled of deuterium through the collimator (1), passes through the steel pipe (2) and reaches the calorimeter (3). The deuterium gas inlet is below the calorimeter (4). The pressure is kept constant by means of a feedback system controlled by pressure gauge (5). The γ-rays produced by the ^{2}H(p,γ)^{3}He reaction are detected by a HPGe detector (6).

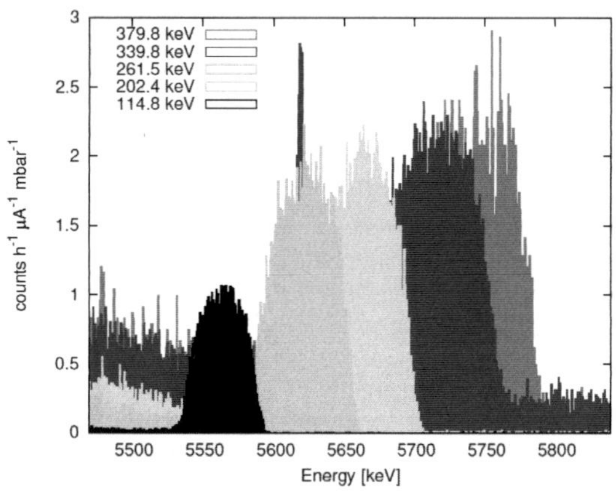

Figure 2. The ^{2}H(p,γ)^{3}He counting rate taken at LUNA during the 2014 feasibility test. The ^{2}H(p,γ)^{3}He peak is broad due to the Doppler effect of the emitted gammas and the recoil of the ^{3}He nuclei. Its width depends on the setup geometry. The narrow peak at 6.1 MeV for the 340 keV run is due to the ^{19}F(p,$\alpha\gamma$)^{16}O reaction on fluorine contaminant.

2. Experimental Setup

The ^{2}H(p,γ)^{3}He reaction has been studied impinging a proton beam provided by the LUNA 400 kV accelerator on a windowless gas target filled with 0.3 mbar deuterium. The beam exhibits a current of about 150 μA and an energy varying between 50 and 400 keV with low spread and long term stability. The current has been measured at the 5% level using a calibrated constant temperature gradient calorimeter. A scratch of the setup is shown in figure 1. The gamma rays produced during the reaction have been collected by a High Purity Germanium Detector (HPGe) with a 137% relative efficiency. No shielding has been provided during the test. As a matter of fact, the natural background, due to cosmic rays and environmental radioactivity, is reduced by the LNGS rock shield. The reduction is strongest for the high energy gammas produced by cosmic rays. Gammas that came from the environmental radioactivity have a cut off at 2.6 MeV, well below the Region of Interest (RoI) for the ^{2}H(p,γ)^{3}He reaction and thus negligible for uor purposes. The RoI can be calculated from kinematic and it is within 5.5,MeV and 5.8 MeV. During the test we could provide only an upper limit < 2 counts/day for the natural background rate in the ^{2}H(p,γ)^{3}He. This energy region is really close to the 6.1 MeV gamma peak coming from the ^{19}F(p,$\alpha\gamma$)^{16}O reaction on the Fluorine contaminant present on the beam stop i.e. the calorimeter endcap. This reaction has a strong resonance at 340 keV beam energy.

IOP Conf. Series: Journal of Physics: Conf. Series **940** (2017) 012059 doi:10.1088/1742-6596/940/1/012059

An on-resonance run has been taken during the test in order to estimate the maximum beam-induced background. This has been found to be 2.9×10^3 counts/C. Direct capture on Carbon has been also found, but the gamma rays produced are at energy lower than the ^{2}H(p,γ)^{3}He RoI. Another possible source of background is the ^{2}H(^{2}H,n)^{3}He between deuterons elastically scattered by the proton beam and deuteron at rest in the gas target. These fast neutrons produce high energy gammas through (n,n'γ) reaction with the surrounding materials. In order to reduce the proton-deuteron scattering a 3 cm in diameter steal pipe has been mounted along the beam line. However the gamma spectrum acquired during the ^{2}H(p,γ)^{3}He test don't show any neutron induced background neither produced in the gas target nor with implanted deuterium on the beam collimator and calorimeter endcap.

3. Data Analysis and Results

The ^{2}H(^{2}H,n)^{3}He gamma ray spectra has been acquired for proton energies: 379.8, 339.8, 261.5, 202.4 and 114.8 keV. The noise to signal ratio approaches to zero and the background sources are practically negligible also when the energy is close or at the ^{19}F(p,$\alpha\gamma$)^{16}O resonance. Unfortunately, due to troubles with the LUNA 400 kV accelerator, the measure of the HPGe efficiency at high energy has not been done. Montecarlo simulations in order to estimate this value are under investigation. In this paper no S-factor values are thus provided. The shape of the ^{2}H(^{2}H,n)^{3}He is determined by the kinematic of the process and the angular distribution of the gamma emitted. Starting from the measured peak is thus possible to extract not only the total cross section but also the differential one. During the test done at LUNA, it was validated the procedure in order to do this. However, due to the low statistics acquired no differential cross section has been provided in this paper.

4. Conclusions

The ^{2}H(p,γ)^{3}He reaction has been studied at LUNA during a feasibility test. Possible beam induced backgrounds as well as unknown systematics have been investigated. In 2015 and 2016 this reaction will be measured at LUNA with high accuracy. New experimental techniques in order to reduce systematics to the 3% level are under investigation. This reduction will be achieved also using two different experimental setups and detectors. In a first phase, an extended windowless gas target with a HPGe detector will be used in order to obtain the differential cross section of the ^{2}H(p,γ)^{3}He reaction in the BBN energy range. In a second phase, a 4π BGO detector will be used in order to investigate the lowest energies achieved by the LUNA 400 kV accelerator.

Future cosmological scenarios will depend on the degree of precision in the primordial deuterium abundance estimation. A 20% increase in the ^{2}H(p,γ)^{3}He cross section is requested in order to have an agreement between observed and calculated primordial deuterium abundance when the ^{4}He abundance is well reproduced by the BBN theory. In contrary, it's necessary to modify the Universe expansion rate by means of an increase of cosmological model degree of freedoms like, for example, the number of neutrino families N_{eff}.

References

[1] P. A. R. Ade et al. (Planck Collaboration), ArXiv e-prints, February 2015.
[2] R. Cooke at al., Astrophys. J. **781** (2014) 31
[3] E. Di Valentino et al., Phys. Rev. D **90** (2014) 023543
[4] L. Ma et al., Phys. Rev. C **55**, (1997) 588
[5] G.M. Bailey et al., Can J. Phys. **48** (1970) 3059
[6] A. Coc, F. Hammache and J. Kiener, Eur. Phys. J. A, **51** (2015) 34

IOP Conf. Series: Journal of Physics: Conf. Series **940** (2017) 012060 doi:10.1088/1742-6596/940/1/012060

Development of a detector in order to investigate (n,γ)-cross sections by ToF method with a very short flight path

C Wolf[1], J Glorius[1], R Reifarth[1], M Weigand[1]
[1]Goethe University Frankfurt, Frankfurt, Germany

E-mail: clemens.c.wolf@gmail.com

Abstract. The determination of neutron capture cross sections of some radioactive isotopes like ^{85}Kr is very important to improve the knowledge about the s process. Based on its own radioactive decay these isotopes can only be used in small samples inside a TOF facility, which is why the neutron flux of these facilities has to be very high. Unfortunately the neutron flux of the FRANZ setup at Goethe University Frankfurt, which will offer the highest neutron flux in astrophysical energy regions (keV region) [1], is still to low to investigate isotopes like ^{85}Kr. Therefore a new setup called NAUTILUS is under development, which will reduce the flight path from 80 cm to a few centimeter to enhance the angular coverage of the sample and therefore increase the neutron flux by a factor of nearly 100. This implies a higher intensity of the γ-flash energy inside the detector and the neutron induced background. Hence the geometry, the scintillator material and the moderator were optimized by GEANT3 simulations.

1. Motivation

The FRANZ facility will generate the highest neutron flux in the astrophysical energy regions world wide [2]. Unfortunately this neutron flux will be still too low to investigate neutron capture cross sections of some interesting isotopes (like ^{85}Kr) [3]. Thus, the idea is to increase the neutron flux by reducing the flight path from 80 cm to a few centimeter and in doing so enhance the angular coverage of the sample [4].
Before this new setup can be implement two main challenges have to be managed, which result from the fact that the neutrons will be produced inside the detector.Consequently the γ-flash can interfere with the promt γ's from the sample. Also no neutron collimator can be used. This results in a high neutron induced background, which has to fade away as much as possible until the next neutron pulse after 4 μs. This work concentrates on optimizing the geometry, the szintilator and moderator by GEANT3 γ- and neutron simulations [5].

2. Simulations

As the detector has to detect all γ's of the (n,γ)-reaction to identify the isotopes on which the capture occurred a 4π detector is the best choice [5]. This detector is segmented in 162 moduls, which effects a good distribution of the γ-flash energy [6]. In the simulation the inner radius, the thickness of the crystals and the flightpath of the neutrons were varied.Additionally the effect of different scintillator materials (BaF_2, $LaBr_3$, $LaCl_3$) was tested. The simulations were split into two parts: γ-simulations for efficiency and neutron simulations for background studies.

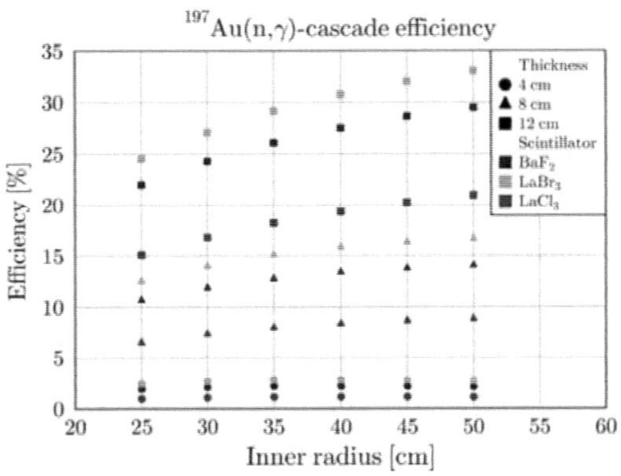

Figure 1. Cascade efficiency for different scintillators and geometries.

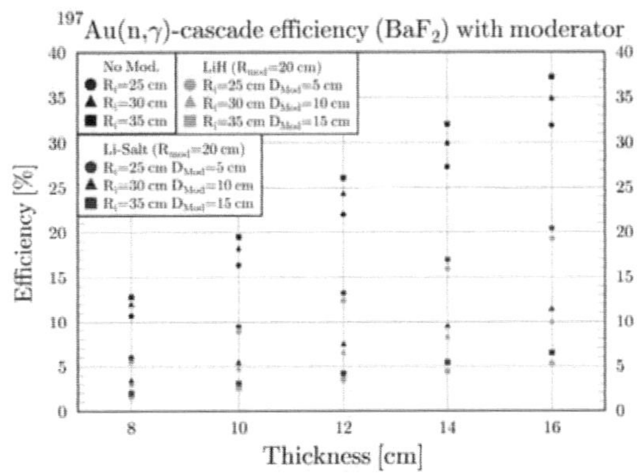

Figure 2. Cascade efficiency for a BaF_2 scintillator with different moderators.

2.1. Gamma Simulations

In figure 1 all results of simulations of γ-efficiency are shown. For these simulations $5 \cdot 10^6$ ^{197}Au(n,γ)-cascades were started in the middle of the detector. The efficiency increases with thicker crystals and bigger inner radius, which is a effect of constant wide gaps (0.1 cm TefAl layer) between the crystals. Figure 1 also demonstrates that $LaBr_3$ is the best choice regarding the efficiency followed by BaF_2. Single γ simulations show that these two materials have nearly the same efficiency up to 2 MeV, but for higher energies the higher Z of Br explans the better property of $LaBr_3$. For a high γ-efficiency the detector should have a big inner radius and the crystals should be as thick as possible.

2.2. Neutron Simulations

At the beginning of the neutron simulations the flight path was investigated. This flightpath has to be as short as possible to have a high neutron flux at the sample but also the time between the γ-flash and (n,γ)-reaction has to be as long as possible. The best compromise was achieved with a flightpath of 8 cm ($4{\cdot}10^7 \frac{\mathrm{n}}{\mathrm{s{\cdot}cm^2}}$ with 16 ns flight time). At this point 10^7 neutrons with energy- and momentum distribution correspondig to a neutron spectrum resulting from the interaction of a 1920 keV proton beam with ^{7}Li were started for each simulation [7]. Figure 3 shows taht the scintillator material strongly affects the interaction time of the neutron background with the setup. This figure also shows that the backgrounds corresponding events can be as late as the next neutron pulse (green box). Therefore the background events for 15 pulses were accumalted for the background studies (after 4,8,12... μs).
As illustrated in figure 4 the background increases dramatically with thicker crystals. In contrast, bigger inner radius has no influence (except for $LaCl_3$). The comparison of the materials show that the background of BaF_2 depends strongly on the crystal thickness. As the background shouldn't be higher than 15 events per 10^5 neutrons the crystals have to be thinner than 8 cm which effects a too low γ-efficiency. Hence an neutron absorber has to be used.

2.3. Simulations with absorber

An absorber in form of a hollow sphere was placed into the detector in order to decrease the neutron background [4]. In these simulations two different moderator materials were tested: enriched Lithium hydride (^{6}LiH), which is the best but also a hazardous absorber, and Li-salt ($C_12H_2O_4(^6Li)_2$) which is used at nTOF setup [8].
The first simulations with both moderators revealed that BaF_2, which was the preferred

scintillator because of its fast light component, provides good results hence these simulations were only done with BaF_2. Figure 4 shows that the absorber has a big influence on the background and with it much thicker crystals can be used (12 cm with an 10 cm Li-Salt or 5 cm LiH absorber). Unfortunately the absorber decreases the cascade efficiency as shown in figure 2. Nevertheless the use of an absorber is recommended because the crystals can be choose much thicker. Expectedly LiH is the best choice because much thinner material can be used.

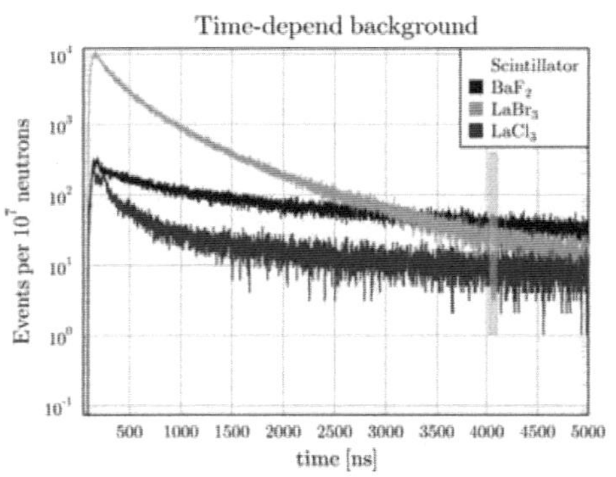

Figure 3. Time depend background for the different scintillators. Green tagged box shows the time interval of the next neutron pulse.

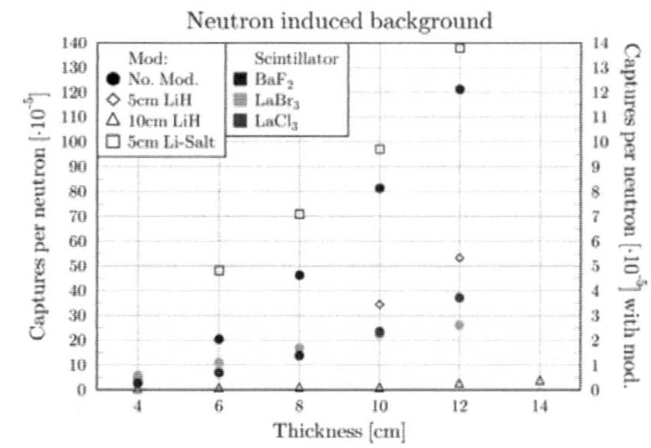

Figure 4. Neutron induced background for different scintillators with and without moderator.

3. Conclusion

The simulations showed that a ToF-facility with an ultra-short flight path is feasible. The best choice for the detector is a 4π BaF_2 detector with a flight path of 8 cm and a 5 cm thick ^{6}LiH absorber. The inner radius has to be as big as possible and the thickness of the crystals should be 12 cm. In this case the neutron flux at the sample will be $4{\cdot}10^7\frac{n}{s{\cdot}cm^2}$ with 16 ns flight time, the cascades efficiency of ^{197}Au(n,γ)-cascades will be around 15 % and less then 0.8 background events per neutron pulse is expected.

References

[1] Reifarth R, Chau L, Heil M, Käppeler F, Meusel O, Plag R, Ratzinger U, Schempp A and Volk K 2009 *Publications of the Astronomical Society of Australia* **26** 255–258
[2] Reifarth R, Lederer C and Käppeler F 2014 *Journal of Physics G Nuclear Physics* **41** 053101
[3] Couture A and Reifarth R 2007 *Atomic Data and Nuclear Data Tables* **93** 807
[4] Reifarth R, Haight R C, Heil M, Käppeler F and Vieira D J 2004 *Nucl. Instr. Meth. A* **524** 215
[5] Heil M, Reifarth R, Fowler M M, Haight R C, Käppeler F, Rundberg R S, Seabury E H, Ullmann J L, Wilhelmy J B, Wisshak K and Voss F 2001 *Nucl. Instr. Meth. A* **459** 229
[6] Reifarth R, Bredeweg T A, Alpizar-Vicente A, Browne J C, Esch E I, Greife U, Haight R C, Hatarik R, Kronenberg A, O'Donnell J M, Rundberg R S, Ullmann J L *et al.* 2004 *Nucl. Instr. Meth. A* **531** 530
[7] Reifarth R, Heil M, Käppeler F and Plag R 2009 *Nucl. Inst. Meth. A* **608** 139
[8] Guerrero C, Abbondanno U, Aerts G, Alvarez H, Alvarez-Velarde F, Andriamonje S, Andrzejewski J, Assimakopoulos P, Audouin L, Badurek G *et al.* 2009 *Nucl. Instr. Meth. A* **608** 424

Acknowledgments

This project was supported by the European Research Council under the European Union's Seventh Framework Programme (FP/2007-2013)/ERC Grant Agreement n. 615126, HIC for FAIR and NAVI.